Die Grundlehren der mathematischen Wissenschaften

in Einzeldarstellungen
mit besonderer Berücksichtigung
der Anwendungsgebiete

Band 163

Josef Stoer · Christoph Witzgall

Convexity and Optimization in Finite Dimensions I

Springer-Verlag New York · Heidelberg · Berlin 1970

Prof. Dr. Josef Stoer
Universität Würzburg

Dr. Christoph Witzgall
Boeing Scientific Research Laboratories, Seattle

Printed in Germany. Title No. 5146

Foreword

Dantzig's development of linear programming into one of the most applicable optimization techniques has spread interest in the algebra of linear inequalities, the geometry of polyhedra, the topology of convex sets, and the analysis of convex functions. It is the goal of this volume to provide a synopsis of these topics, and thereby the theoretical background for the arithmetic of convex optimization to be treated in a subsequent volume.

The exposition of each chapter is essentially independent, and attempts to reflect a specific style of mathematical reasoning.

The emphasis lies on linear and convex duality theory, as initiated by Gale, Kuhn and Tucker, Fenchel, and v. Neumann, because it represents the theoretical development whose impact on modern optimization techniques has been the most pronounced. Chapters 5 and 6 are devoted to two characteristic aspects of duality theory: conjugate functions or polarity on the one hand, and saddle points on the other. The Farkas lemma on linear inequalities and its generalizations, Motzkin's description of polyhedra, Minkowski's supporting plane theorem are indispensable elementary tools which are contained in chapters 1, 2 and 3, respectively. The treatment of extremal properties of polyhedra as well as of general convex sets is based on the far reaching work of Klee. Chapter 2 terminates with a description of Gale diagrams, a recently developed successful technique for exploring polyhedral structures.

The first two chapters require only an elementary knowledge of linear algebra and analytic geometry. Some familiarity with basic topological concepts and the theory of real functions, however, will be needed by readers of the remaining chapters.

It would have been impossible to complete this volume without the continuous encouragement and generous sponsorship from which the authors were privileged to benefit, and for which they express their deep appreciation and gratitude.

The effort has been partly sponsored by the Princeton-I.B.M. Mathematics research project, the Deutsche Forschungsgemeinschaft and the U.S. Army Research Office Durham (DA-31-124-ARO-D257). The National Bureau of Standards in Washington, D.C., the Mathematische Institut der Technischen Hochschule München, and the Boeing

Scientific Research Laboratories in Seattle, have supported the project in every respect.

The authors wish to thank Professors F. L. Bauer and A. W. Tucker for their support and encouragement. Parts of the manuscript have been read by A. J. Goldman, R. T. Rockafellar, D. M. Bulman, J. Levy, and P. R. Meyers, who contributed valuable criticisms and suggestions. In particular, we are indebted to J. Zowe, who undertook the arduous task of reading the proofs. We shall never forget our indefatigable typists, too numerous to mention by name.

Finally we wish to thank the Springer-Verlag for its extraordinary patience and accommodation of all our wishes.

Würzburg, Seattle
January 1970

J. Stoer, C. Witzgall

Contents

CHAPTER 3

Convex Sets

CHAPTER 4

Convex Functions

CHAPTER 5

Duality Theorems

CHAPTER 6

Saddle Point Theorems

Introduction

A system of linear equations is given by

$$\text{(I.1)} \qquad \begin{array}{c} a_{11}x_1 + \cdots + a_{1n}x_n = b_1 \\ \cdots \qquad \cdots \\ a_{m1}x_1 + \cdots + a_{mn}x_n = b_m. \end{array}$$

Here the coefficients a_{ik}, $i=1,\ldots,m$, $k=1,\ldots,n$, and b_i, $i=1,\ldots,m$, denote elements of a *field*

$$R.$$

Elements x_k, $k=1,\ldots,n$, form a solution of (I.1) if they satisfy every one of the equations.

Similarly, a system of linear inequalities may be given by

$$\text{(I.2)} \qquad \begin{array}{c} a_{11}x_1 + \cdots + a_{1n}x_n \geqslant b_1 \\ \cdots \qquad \cdots \\ a_{m1}x_1 + \cdots + a_{mn}x_n \geqslant b_m. \end{array}$$

Of course, there are other kinds of systems of linear inequalities, since the other order symbols

$$\leqslant, \; >, \; <$$

also may be used for formulating linear inequalities. In any case, however, the coefficients a_{ik} must be chosen from an *ordered field* R.

The reader who is not familiar with the notion of general ordered fields may consider R to be the ordered field of real numbers. In fact, every ordered field that satisfies the Axiom of Archimedes, which states that

(I.3) *each element of R is surpassed by an integer*

is automatically isomorphic in an order preserving way to a subfield of the field of real numbers. Every field contains a smallest subfield. In case of ordered fields, this smallest subfield is isomorphic to the field of rational numbers, and will be identified with the latter. This identification is unique since the field of rationals admits only the trivial automorphism. Zero is the largest element which is smaller than every positive element.

There are ordered fields which are not commutative (see for instance Pickert [*1*]). These fields are complicated and not very important. While commutativity is not essential for most of the developments of this tract, failure to require commutativity causes some notational inconvenience and occasionally hampers the geometric interpretation of arithmetic statements. For this reason, *commutativity is assumed throughout this book.*

Whenever possible we shall use matrix notation, with which the reader is assumed to be familiar. Thus we write

$$AX=B$$

instead of (I.1), thereby putting

$$\text{(I.4)} \qquad A:=\begin{vmatrix} a_{11} \cdots a_{1n} \\ \cdots \quad \cdots \\ a_{m1} \cdots a_{mn} \end{vmatrix}, \qquad B:=\begin{vmatrix} b_1 \\ \vdots \\ b_m \end{vmatrix}, \qquad X:=\begin{vmatrix} x_1 \\ \vdots \\ x_n \end{vmatrix}.$$

Arrays of this kind are well known as *matrices*. Matrices that consist of one row or one column only are frequently called *row vectors*, *column vectors*, or just *vectors*. The terms *row* and *column* are also used to denote row and column vectors respectively. Two matrices are of the same

$$\text{(I.5)} \qquad \textit{size} \ m\times n,$$

if both are of m rows and n columns, that is, if they are both $m\times n$-matrices. The *transpose*

$$\text{(I.6)} \qquad A^T:=\begin{vmatrix} a_{11} \cdots a_{m1} \\ \cdots \quad \cdots \\ a_{1n} \cdots a_{mn} \end{vmatrix}$$

of the $m\times n$-matrix A in (I.4) is an $n\times m$-matrix. The transpose of a column is a row, and vice versa.

We shall use the order symbols

$$\geqslant, \ \leqslant, \ >, \ <$$

between matrices of the same size. The symbols are understood to hold componentwise. Thus we write

$$AX\geqslant B$$

instead of (I.2).

The term "matrix" was introduced by Sylvester [*1*], connoting rectangular arrays of numbers. This interpretation still prevails. However, there exist competing concepts, in particular, the one advanced by Bourbaki [*1*]. According to this concept, a vector is any *indexed set* or *family* $\{a_i\}_{i\in I}$, and a matrix is a doubly indexed set $\{a_{ij}\}_{i\in I, j\in J}$. Note that in this concept there is no inherent "order" of rows and columns. Also it makes no sense to distinguish between "row" and "column" vectors.

In many situations this second matrix concept offers definite advantanges. If in this tract preference is given to the first concept, it is for three reasons: The first concept is still more common. The "indexing by position" is the most important practical way of writing matrices explicitly. Finally, an ordering of rows and columns will be required to describe lexicographic techniques in connection with the Simplex Method.

Both matrix concepts are closely related, and every statement based on one concept is easily reformulated in terms of the other.

We mention a third concept, which is described for instance, by MacDuffee [*1*]. Here matrices are regarded as the elements of an abstract "matrix algebra". This concept is not suited for our purposes since we shall employ matrix operations that are not operations of the algebra.

The theory of matrices has been enhanced recently by the notion of *matroids*, introduced by Whitney [*2*], and extensively studied by Tutte [*1*], [*2*], and others. Here the combinatorial behavior of matrices is abstracted.

Now we introduce several supplementary conventions and notations as for instance,

$$\{A\} := \{a_{ij} | a_{ij} \text{ elements of } A\}.$$

The use of matrix partitions

$$A = (B, C)$$

leads to considering

(I.7) *empty matrices*

A with $\{A\} = \emptyset$. For the sake of an easy extension of matrix operations, we shall introduce one empty matrix of each size

$$\begin{aligned} &0\times 0, \\ &0\times n, \quad n = 1, 2, \dots, \\ &m\times 0, \quad m = 1, 2, \dots. \end{aligned}$$

Multiplication of the empty $0\times m$-matrix with any $m\times n$-matrix is defined to yield the empty $0\times n$-matrix. The product of the empty $m\times 0$-matrix with the empty $0\times n$-matrix, however, is defined to be a nonempty matrix, namely the zero matrix of size $m\times n$. With these definitions, the rules

$$(B, C)\left|\begin{matrix} X \\ Y \end{matrix}\right| = BX + CY, \qquad Z(B, C) = (ZB, ZC)$$

hold even if B and X, of C and Y, are empty matrices.

An *identity matrix* is a square matrix whose entries are zero except along the main diagonal where the entries are one. Following a common practice, we denote any identity matrix by I:

$$\text{(I.8)} \qquad I := \begin{vmatrix} 1 & & 0 \\ & \ddots & \\ 0 & & 1 \end{vmatrix}.$$

All the columns I are *unit columns*, that is, columns whose only nonzero component has the value one. The size of the identity matrix I usually follows from the context. Analogously, we indicate by

$$\text{(I.9)} \qquad A = 0$$

that A is a zero matrix of the same size as A. In other words, (I.9) means that all elements of A vanish. By writing

$$\text{(I.10)} \qquad A \neq 0$$

we negate (I.9), thereby expressing that A contains at least one nonzero element. According to this convention, (I.10) can only hold if A is not empty. Therefore

(I.11) $\quad A = 0 \quad$ *holds for every empty matrix.*

We shall use this convention extensively.

Further, we shall need a somewhat uncommon notion of submatrices. Consider an $m \times n$-matrix A with columns $A_1, \ldots, A_n$:

$$A = [A_1, \ldots, A_n].$$

Let

$$J := [j_1, \ldots, j_d], \qquad 0 \leqslant d \leqslant n$$

be a row vector of pairwise different column indices j_i, $1 \leqslant j_i \leqslant n$. Then we define the $m \times d$-matrix

$$A_J := [A_{j_1}, \ldots, A_{j_d}].$$

If $d = 0$, then J is empty, and A_J is considered to be the empty $m \times 0$-matrix. If

$$K := \begin{vmatrix} k_1 \\ \vdots \\ k_d \end{vmatrix}, \qquad 0 \leqslant d \leqslant m$$

is a column vector of pairwise different row indices of A, we define analogously

$$_K A = \begin{vmatrix} _{k_1}A \\ \vdots \\ _{k_d}A \end{vmatrix}$$

where $_iA$ denotes the ith row of A. Obviously $(_KA)_J = {}_K(A_J)$. Hence we may drop the parenthesis and write

$$_KA_J,$$

denoting the

(I.12) $\qquad$ *K, J-submatrix*

of A.

All index vectors subscripted to the left of a matrix are column vectors, those subscripted to the right, row vectors. Therefore we shall not always specify the type of the index vectors. We may, for instance, use the same letter S to denote both S and S^T. With these notational conventions we have the following rules:

(I.13) $$(_KA)^T = (A^T)_K,$$

(I.14) $$_{(_KS)}A = {}_K(_SA),$$

(I.15) $$_SI_S = I,$$

(I.16) $$_K(A+B) = {}_KA + {}_KB,$$

(I.17) $$_K(AB) = (_KA)B.$$

(I.18) *If the union* $\{K\} \cup \{J\}$ *covers all row indices of the matrices* A *and* B*, then* $_KA = {}_KB$ *and* $_JA = {}_JB$ *imply* $A = B$.

(I.19) *If* K *and* J *partition the set of column indices of* A*, and the set of row indices of* B*, respectively, then*

$$AB = (A_K)(_KB) + (A_J)(_JB).$$

Following the above conventions, we may write $_kB$ for the kth component b_k of a vector B. Analogously, $_iA_k$ denotes the element a_{ik} of a matrix A. A_1 denotes the first column of a matrix A. However, we shall use subscripts also for merely distinguishing between several vectors, as for instance $A_0, X_1, \ldots$. It will follow from the context in which way the symbols should be interpreted.

Further notational conventions concern algebraic operations between two subsets $\mathcal{M}$ and $\mathcal{N}$ of the space R^n of n-vectors. By $\mathcal{M}+\mathcal{N}$ we denote the set of all possible sums of vectors in $\mathcal{M}$ and vectors in $\mathcal{N}$, i.e.

$$\mathcal{M}+\mathcal{N} := \{X+Y \mid X \in \mathcal{M},\ Y \in \mathcal{N}\}.$$

Similarly

$$\mathcal{M}-\mathcal{N} := \{X-Y \mid X \in \mathcal{M},\ Y \in \mathcal{N}\}.$$

Instead of $\mathcal{M}+\{X\}$, we write

$$\mathcal{M}+X,$$

and if $\lambda \in R$, then

$$\lambda M := \{\lambda X \mid X \in M\}.$$

The notation $M - N$ is frequently used in the literature for denoting the set-theoretical difference of two sets. In order to avoid misunderstanding, we shall denote the set theoretical difference of M and N by $M \sim N$, i.e.

$$M \sim N := \{X \in M \mid X \notin N\}.$$

The set-theoretical inclusion signs $\supset$ and $\subset$ will be used only for strict inclusion and will imply inequality of the sets concerned:

$$M \subset N \quad \text{implies} \quad M \neq N.$$

If equality of the sets M and N is permitted, then we use the notation

$$M \subseteq N.$$

CHAPTER 1

Inequality Systems

This chapter examines finite systems of linear inequalities and equations over a commutative ordered field as for instance the field of real numbers. It uses the oldest and most straightforward approach. There is no concern for geometrical interpretations or arithmetical efficiency as in other chapters. The systems are viewed as sets of relations, either true or false, and their logical structure and interdependence is investigated.

This approach is most clearly represented in Kuhn's paper: "Solvability and Consistency for Linear Equations and Inequalities" [*2*]. As a proof-theoretical tool, Kuhn employs a generalization of the classical elimination procedure for systems of linear equations. The concept of this generalization is due to Fourier [*1*]. The resulting theorem, which we call therefore the Kuhn-Fourier theorem, although it is also contained in the fundamental results of Motzkin [*2*], governs the solvability of general systems of linear relations, and serves us as the foundation of the theory of linear inequalities. Its advantages over other equivalent theorems,—besides its straightforward analogy to a well-known theorem on linear equations—, lies in the automatic fashion in which other results of the theory are obtainable from it. In other words, it establishes a technique. In the subsequent development of this chapter, "complementary slackness" will be the central notion. For information about systems of inequalities with an infinite number of variables, the reader is referred to the comprehensive study of systems of linear inequalities by Fan [*4*].

1.1. Linear Combinations of Inequalities

Consider a system of linear relations

$$(1.1.1) \qquad (\mathscr{S}): \begin{cases} AX > A_0, \\ BX \geqslant B_0, \\ CX = C_0. \end{cases}$$

A relation

$$d_1 x_1 + \cdots + d_n x_n \ \rho \ d_0,$$

where ρ stands for one of the symbols $>, \geqslant, =$, is called a

(1.1.2) *consequence relation* or, briefly, a *consequence*

of the system $(\mathscr{S})$, if it is true for all X which solve $(\mathscr{S})$. If $(\mathscr{S})$ is not solvable, then every relation is a consequence of $(\mathscr{S})$. It will be one of our central problems to characterize the set of all consequence relations of a given system. Therefore we start out examining the most important formation rule, namely the *linear combination* of given relations. Not every linear combination yields a consequence relation. We shall specify a class of linear combinations which do yield consequence relations, and call these combinations "legal".

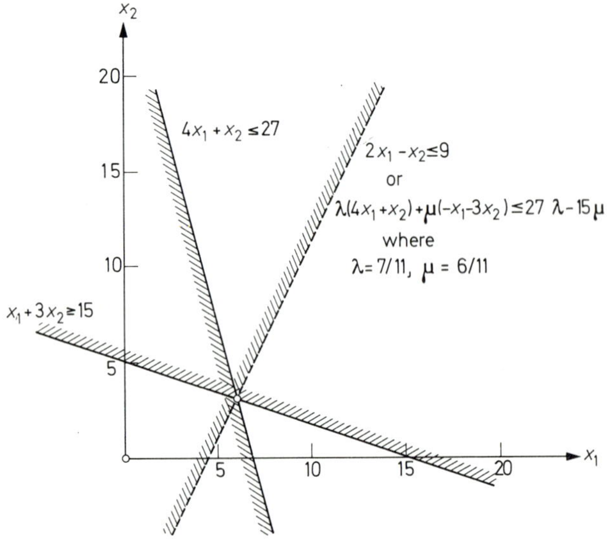

Fig. 1. Legal linear combination of inequalities

Consider for instance, the two relations

(1.1.3)
$$\begin{aligned} b_1 x_1 + \cdots + b_n x_n &\geqslant b_0, \\ \bar{b}_1 x_1 + \cdots + \bar{b}_n x_n &\geqslant \bar{b}_0, \end{aligned}$$

and their linear combination (Fig. 1)

(1.1.4) $$(\lambda b_1 + \mu \bar{b}_1) x_1 + \cdots + (\lambda b_n + \mu \bar{b}_n) x_n \geqslant \lambda b_0 + \mu \bar{b}_0.$$

given the system of linear relations

$$(\mathcal{L}): \begin{cases} AX > A_0 \\ BX \geqq B_0 \\ CX = C_0 \end{cases}$$

the relation

$$DX = d_1 x_1 + \ldots + d_m x_m \ \rho \ d_0$$

is a legal linear combination of the relations in $(\mathcal{L})$ if

$$(D, d_0) = U(A, A_0) + V(B, B_0) + W(C, C_0)$$

$$U \geqq 0, \ V \geqq 0$$

and

$$\rho = \begin{cases} > \text{ if } U \neq 0 \\ \geqq \text{ if } U = 0 \text{ but } V \neq 0 \\ = \text{ if } U = 0 \text{ and } V = 0 \end{cases}$$

If the coefficients λ and μ are nonnegative, then (1.1.4) is clearly a consequence of the relations (1.1.3). If λ or μ is negative, then (1.1.4) is in general not a consequence relation.

Next we consider the relations

$$a_1 x_1 + \cdots + a_n x_n > a_0,$$
$$b_1 x_1 + \cdots + b_n x_n \geqslant b_0.$$

The combination

$$(\lambda a_1 + \mu b_1) x_1 + \cdots + (\lambda a_n + \mu b_n) x_n > \lambda a_0 + \mu b_0 \tag{1.1.5}$$

will be a consequence if $\lambda > 0$ and $\mu \geqslant 0$. For $\lambda = 0$ and $\mu > 0$ the relation (1.1.5) may or may not be a consequence, depending on the constituents of the combination.

These and similar considerations lead to the following definition: The relation

$$DX = d_1 x_1 + \cdots + d_n x_n \quad \rho \quad d_0$$

is a

(1.1.6) *legal linear combination*

of the relations in $(\mathscr{S})$ if

$$(D, d_0) = U(A, A_0) + V(B, B_0) + W(C, C_0), \qquad U \geqslant 0, \qquad V \geqslant 0,$$

and (recall convention (I.10) and its consequence (I.11))

$$\rho = \begin{cases} > & \text{if} \quad U \neq 0, \\ \geqslant & \text{if} \quad U = 0 \quad \text{but} \quad V \neq 0, \\ = & \text{if} \quad U = 0 \quad \text{and} \quad V = 0. \end{cases}$$

Clearly, every legal linear combination of relations in $(\mathscr{S})$ is a consequence of $(\mathscr{S})$.

Observe also that the formation of legal linear combinations is a transitive process. More precisely, given a family of relations

$$D_i X \rho_i D_{i0}, \qquad i = 1, \ldots, k,$$

which are legal linear combinations of relations in $(\mathscr{S})$, then each legal linear combination of relations $D_i X \rho_i D_{i0}$ is again a legal linear combination of relations in $(\mathscr{S})$. In what follows, this transitivity property will play an important role.

A fundamental theorem about the solvability of a linear equation system $DX = D_0$ states (compare Bourbaki [*1*]): $DX = D_0$ is solvable if and only if every linear dependence of the left hand sides of the equations holds also for the right hand sides.

This theorem may be extended to systems of general linear relations. To this end, we have to extend the concept of linear dependence. The system $(\mathscr{S})$, defined by (1.1.1), possesses a

(1.1.7) *legal linear dependence* (U, V, W)

if this vector does not vanish and if it yields as legal linear combination a relation

(1.1.8) $$0x_1 + \cdots + 0x_n \;\; \rho \;\; d_0$$

with zero coefficients on the left hand side[1].

Relations of the type (1.1.8) are either always true or always false. This is relevant for the formulation of the following theorem, which is the promised extension of the solvability theorem for linear equations, (Fourier [*1*], Kuhn [*2*], Motzkin [*2*].

(1.1.9) **Theorem of Kuhn-Fourier.** *A system $(\mathscr{S})$ of linear relations is solvable if and only if each legal linear dependence within the system $(\mathscr{S})$ leads to a zero relation* (1.1.8) *which is always true. In particular, the system $(\mathscr{S})$ is solvable if it admits no legal linear dependence.*

Specialized versions of theorem (1.1.9) have been generalized to infinite dimensions for instance by Braunschweiger [*1*], and to matrix variables by Bellman and Fan [*1*].

A proof will be given in section 1.3. We shall finish this section by pointing out a connection between theorem (1.1.9) and a typical separation theorem. Consider, for example, the systems

$$(\mathscr{S}_1):\quad AX \geqslant A_0, \qquad (\mathscr{S}_2):\quad BX \geqslant B_0.$$

(1.1.10) **Theorem.** *If both systems $(\mathscr{S}_1)$ and $(\mathscr{S}_2)$ are solvable, but have no solutions common to both, then they are separated by a plane $\{X \mid DX = d\}$ such that $DX > d$ for all solutions of $(\mathscr{S}_1)$, and $DX < d$ for all solutions of $(\mathscr{S}_2)$.*

Proof. Since there exist no solutions common to both $(\mathscr{S}_1)$ and $(\mathscr{S}_2)$, the combined system

$$AX \geqslant A_0, \qquad BX \geqslant B_0$$

is not solvable. According to the Kuhn-Fourier theorem, there exist therefore U and V, at least one of them nonzero, such that

$$UA + VB = 0, \qquad U \geqslant 0, \qquad V \geqslant 0, \qquad UA_0 + VB_0 > 0.$$

[1] This concept differs from the concept of "positive linear dependence" introduced by Davis [*3*].

UA does not vanish. Indeed, this would imply $VB=0$, and since both $(\mathscr{S}_1)$ and $(\mathscr{S}_2)$ are solvable, it would follow that $UA_0 \leqslant 0$ and $VB_0 \leqslant 0$, contradicting the last of the above relations. Hence $\{X \mid DX=d\}$ with $D:=UA=-VB$ and $d:=\frac{1}{2}(UA_0-VB_0)$ is a plane with the properties required by the theorem. □

1.2. Fourier Elimination

We shall now describe Fourier's algorithm for solving systems of linear relations

$$(\mathscr{S}): \begin{cases} AX > A_0, \\ BX \geqslant B_0, \\ CX = C_0. \end{cases}$$

This algorithm will generalize the well known elimination procedure for systems of linear equations.

For a moment let us dwell upon the general principle of elimination procedures: A system $(\mathscr{S})$ of relations for the unknowns $(x_1, \ldots, x_n)^T$ is replaced by a new system $(\mathscr{S}')$ in which one of the unknowns, say x_n, does not occur anymore: x_n has been "eliminated". The systems $(\mathscr{S})$ and $(\mathscr{S}')$ are linked together by the requirement that $(x_1, \ldots, x_{n-1})^T$ solve $(\mathscr{S}')$ if and only if there exists an x_n such that $(x_1, \ldots, x_{n-1}, x_n)^T$ solves $(\mathscr{S})$. Then every solution of $(\mathscr{S})$ can be found by successive eliminations.

In order to obtain clues concerning the structure of a suitable system $(\mathscr{S}')$ which meets the above requirements, suppose that $(\mathscr{S})$ contains two inequalities

$$a_{i1}x_1 + \cdots + a_{in}x_n > a_{i0},$$
$$a_{j1}x_1 + \cdots + a_{jn}x_n > a_{j0},$$

with $a_{in} > 0$ and $a_{jn} < 0$. Then we may write as well:

$$\frac{a_{i0}}{a_{in}} - \frac{a_{i1}}{a_{in}}x_1 - \cdots - \frac{a_{i,n-1}}{a_{in}}x_{n-1} < x_n,$$

$$x_n < \frac{a_{j0}}{a_{jn}} - \frac{a_{j1}}{a_{jn}}x_1 - \cdots - \frac{a_{j,n-1}}{a_{jn}}x_{n-1}.$$

Thus one relation bounds x_n below, the other above. Both may be combined to yield a relation which involves only $x_1, \ldots, x_{n-1}$:

$$\left(\frac{a_{i1}}{a_{in}} - \frac{a_{j1}}{a_{jn}}\right)x_1 + \cdots + \left(\frac{a_{i,n-1}}{a_{in}} - \frac{a_{j,n-1}}{a_{jn}}\right)x_{n-1} > \left(\frac{a_{i0}}{a_{in}} - \frac{a_{j0}}{a_{jn}}\right).$$

This relation is necessary for $(x_1, \ldots, x_{n-1})^T$ to yield by extension a solution $(x_1, \ldots, x_n)^T$ of $(\mathscr{S})$. We collect all relations of this type to form the following system:

(1.2.1) $(\mathscr{S}')$:

$$\left(\frac{a_{i1}}{a_{in}} - \frac{a_{j1}}{a_{jn}}\right) x_1 + \cdots + \left(\frac{a_{i,n-1}}{a_{in}} - \frac{a_{j,n-1}}{a_{jn}}\right) x_{n-1} > \left(\frac{a_{i0}}{a_{in}} - \frac{a_{j0}}{a_{jn}}\right)$$

for all pairs i,j with $a_{in} > 0$, $a_{jn} < 0$,

$$\left(\frac{a_{i1}}{a_{in}} - \frac{b_{j1}}{b_{jn}}\right) x_1 + \cdots + \left(\frac{a_{i,n-1}}{a_{in}} - \frac{b_{j,n-1}}{b_{jn}}\right) x_{n-1} > \left(\frac{a_{i0}}{a_{in}} - \frac{b_{j0}}{b_{jn}}\right)$$

for all pairs i,j with $a_{in} > 0$, $b_{jn} < 0$,

$$\left(\frac{a_{i1}}{a_{in}} - \frac{c_{j1}}{c_{jn}}\right) x_1 + \cdots + \left(\frac{a_{i,n-1}}{a_{in}} - \frac{c_{j,n-1}}{c_{jn}}\right) x_{n-1} > \left(\frac{a_{i0}}{a_{in}} - \frac{c_{j0}}{c_{jn}}\right)$$

for all pairs i,j with $a_{in} > 0$, $c_{jn} \neq 0$,

$$\left(\frac{b_{i1}}{b_{in}} - \frac{a_{j1}}{a_{jn}}\right) x_1 + \cdots + \left(\frac{b_{i,n-1}}{b_{in}} - \frac{a_{j,n-1}}{a_{jn}}\right) x_{n-1} > \left(\frac{b_{i0}}{b_{in}} - \frac{a_{j0}}{a_{jn}}\right)$$

for all pairs i,j with $b_{in} > 0$, $a_{jn} < 0$,

$$\left(\frac{b_{i1}}{b_{in}} - \frac{b_{j1}}{b_{jn}}\right) x_1 + \cdots + \left(\frac{b_{i,n-1}}{b_{in}} - \frac{b_{j,n-1}}{b_{jn}}\right) x_{n-1} \geqslant \left(\frac{b_{i0}}{b_{in}} - \frac{b_{j0}}{b_{jn}}\right)$$

for all pairs i,j with $b_{in} > 0$, $b_{jn} < 0$,

$$\left(\frac{b_{i1}}{b_{in}} - \frac{c_{j1}}{c_{jn}}\right) x_1 + \cdots + \left(\frac{b_{i,n-1}}{b_{in}} - \frac{c_{j,n-1}}{c_{jn}}\right) x_{n-1} \geqslant \left(\frac{b_{i0}}{b_{in}} - \frac{c_{j0}}{c_{jn}}\right)$$

for all pairs i,j with $b_{in} > 0$, $c_{jn} \neq 0$,

$$\left(\frac{c_{i1}}{c_{in}} - \frac{a_{j1}}{a_{jn}}\right) x_1 + \cdots + \left(\frac{c_{i,n-1}}{c_{in}} - \frac{a_{j,n-1}}{a_{jn}}\right) x_{n-1} > \left(\frac{c_{i0}}{c_{in}} - \frac{a_{j0}}{a_{jn}}\right)$$

for all pairs i,j with $c_{in} \neq 0$, $a_{jn} < 0$,

$$\left(\frac{c_{i1}}{c_{in}} - \frac{b_{j1}}{b_{jn}}\right) x_1 + \cdots + \left(\frac{c_{i,n-1}}{c_{in}} - \frac{b_{j,n-1}}{b_{jn}}\right) x_{n-1} \geqslant \left(\frac{c_{i0}}{c_{in}} - \frac{b_{j0}}{b_{jn}}\right)$$

for all pairs i,j with $c_{in} \neq 0$, $b_{jn} < 0$,

$$\left(\frac{c_{i1}}{c_{in}} - \frac{c_{j1}}{c_{jn}}\right) x_1 + \cdots + \left(\frac{c_{i,n-1}}{c_{in}} - \frac{c_{j,n-1}}{c_{jn}}\right) x_{n-1} = \left(\frac{c_{i0}}{c_{in}} - \frac{c_{j0}}{c_{jn}}\right)$$

for all pairs i,j with $c_{in} \neq 0$, $c_{jn} \neq 0$,

$$\begin{aligned}
a_{k1}x_1 + \cdots + a_{k,n-1}x_{n-1} &> a_{k0} &&\text{for all } k \text{ with } && a_{kn}=0,\\
b_{k1}x_1 + \cdots + b_{k,n-1}x_{n-1} &\geq b_{k0} &&\text{for all } k \text{ with } && b_{kn}=0,\\
c_{k1}x_1 + \cdots + c_{k,n-1}x_{n-1} &= c_{k0} &&\text{for all } k \text{ with } && c_{kn}=0.
\end{aligned}$$

Some of these equations are superfluous. It is clearly sufficient to distinguish one equation

$$c_{s1}x_1 + \cdots + c_{sn}x_n = c_{s0}$$

with $c_{sn} \neq 0$, and to compare it with the other equations. However, we have chosen the above form of the system for the sake of uniformity.

We proceed to show that the system $(\mathscr{S}')$ is suited for an elimination procedure. To this end, it remains to be established that for each solution $(x_1, \ldots, x_{n-1})^T$ of $(\mathscr{S}')$ there exists an x_n such that $(x_1, \ldots, x_{n-1}, x_n)^T$ solves $(\mathscr{S})$.

Let then $(x_1, \ldots, x_{n-1})^T$ be any solution of $(\mathscr{S}')$. As we have seen above,

$$\frac{a_{i0}}{a_{in}} - \frac{a_{i1}}{a_{in}}x_1 - \cdots - \frac{a_{i,n-1}}{a_{in}}x_{n-1} < x_n$$

must hold for every i with $a_{in} > 0$. Consequently $\alpha' < x_n$ must hold, where

(1.2.2)

$$\alpha' := \max\left\{\frac{a_{i0}}{a_{in}} - \frac{a_{i1}}{a_{in}}x_1 - \cdots - \frac{a_{i,n-1}}{a_{in}}x_{n-1} \;\middle|\; \text{for all } i \text{ with } a_{in} > 0\right\}.$$

Similarly we define

$$\alpha'' := \min\left\{\frac{a_{j0}}{a_{jn}} - \frac{a_{j1}}{a_{jn}}x_1 - \cdots - \frac{a_{j,n-1}}{a_{jn}}x_{n-1} \;\middle|\; \text{for all } j \text{ with } a_{jn} < 0\right\},$$

$$\beta' := \max\left\{\frac{b_{i0}}{b_{in}} - \frac{b_{i1}}{b_{in}}x_1 - \cdots - \frac{b_{i,n-1}}{b_{in}}x_{n-1} \;\middle|\; \text{for all } i \text{ with } b_{in} > 0\right\},$$

$$\beta'' := \min\left\{\frac{b_{j0}}{b_{jn}} - \frac{b_{j1}}{b_{jn}}x_1 - \cdots - \frac{b_{j,n-1}}{b_{jn}}x_{n-1} \;\middle|\; \text{for all } j \text{ with } b_{jn} < 0\right\},$$

$$\gamma' := \max\left\{\frac{c_{i0}}{c_{in}} - \frac{c_{i1}}{c_{in}}x_1 - \cdots - \frac{c_{i,n-1}}{c_{in}}x_{n-1} \;\middle|\; \text{for all } i \text{ with } c_{in} \neq 0\right\},$$

$$\gamma'' := \min\left\{\frac{c_{j0}}{c_{jn}} - \frac{c_{j1}}{c_{jn}}x_1 - \cdots - \frac{c_{j,n-1}}{c_{jn}}x_{n-1} \;\middle|\; \text{for all } j \text{ with } c_{jn} \neq 0\right\}.$$

The minimum of the empty set is defined to be $+\infty$, the maximum, $-\infty$.

Now we distinguish two cases:

Case I: there is a $c_{in} \neq 0$. *Case II:* there is no $c_{in} \neq 0$.

In case I we find the relations

$$\begin{array}{lll} \alpha' < x_n, & \beta' \leqslant x_n, & \gamma' = x_n, \\ x_n < \alpha'', & x_n \leqslant \beta'', & \gamma'' = x_n, \end{array} \tag{1.2.3}$$

to be necessary for x_n in order to yield an extension of the given solution $(x_1, \ldots, x_{n-1})^T$ of $(\mathscr{S}')$ to a solution of $(\mathscr{S})$. It follows from the definition of $\alpha', \ldots, \gamma''$ that the above relations are also sufficient.

If the system (1.2.3) is solvable at all, then its solution must be $x_n = \gamma'$. Hence the following relations are necessary and sufficient for the solvability of (1.2.3):

$$\alpha' < \gamma', \quad \beta' \leqslant \gamma', \quad \gamma'' < \alpha'', \quad \gamma'' \leqslant \beta'', \quad \gamma' = \gamma''.$$

But these relations follows immediately from $(\mathscr{S}')$. For instance, we have

$$\gamma' - \gamma'' = \frac{c_{i0}}{c_{in}} - \frac{c_{j0}}{c_{jn}} - \left(\frac{c_{i1}}{c_{in}} - \frac{c_{j1}}{c_{jn}}\right) x_1 - \cdots - \left(\frac{c_{i,n-1}}{c_{in}} - \frac{c_{j,n-1}}{c_{jn}}\right) x_{n-1} = 0$$

for the maximizing i and the minimizing j. Similarly

$$\gamma' - \alpha' = \frac{c_{i0}}{c_{in}} - \frac{a_{k0}}{a_{kn}} - \left(\frac{c_{i1}}{c_{in}} - \frac{a_{k1}}{a_{kn}}\right) x_1 - \cdots - \left(\frac{c_{i,n-1}}{c_{in}} - \frac{a_{k,n-1}}{a_{kn}}\right) x_{n-1} > 0,$$

where the subscript k characterizes the maximum element of (1.2.2). Hence $x_n = \gamma'$ is the solution of (1.2.3).

In case II we arrive at a smaller set of necessary and sufficient relations for x_n:

$$\begin{array}{ll} \alpha' < x_n, & \beta' \leqslant x_n, \\ x_n < \alpha'', & x_n \leqslant \beta''. \end{array} \tag{1.2.4}$$

From $(\mathscr{S}')$ we conclude again:

$$\alpha' < \alpha'', \quad \alpha' < \beta'', \quad \beta' < \alpha'', \quad \beta' \leqslant \beta''.$$

Therefore the following subcases cover all possibilities:

II a) $\alpha' \geqslant \beta'$ and $\beta'' \geqslant \alpha''$, II b) $\alpha' \geqslant \beta'$ and $\alpha'' > \beta''$,

II c) $\beta' > \alpha'$ and $\beta'' \geqslant \alpha''$, II d) $\beta' > \alpha'$ and $\alpha'' > \beta''$.

The solution interval of (1.2.4) is

in case II a: $\alpha' < x_n < \alpha''$,

in case II b: $\alpha' < x_n \leqslant \beta''$,

in case II c: $\beta' \leqslant x_n < \alpha''$,

in case II d: $\beta' \leqslant x_n \leqslant \beta''$.

Thus it is shown that

(1.2.5) *each solution* $(x_1, \ldots, x_{n-1})^T$ *of* $(\mathscr{S}')$ *can be extended to yield a solution* $(x_1, \ldots, x_{n-1}, x_n)^T$ *of* $(\mathscr{S})$.

In other words, the transition from $(\mathscr{S})$ to $(\mathscr{S}')$ is one step of an elimination procedure. We call it

(1.2.6) *Fourier elimination.*

It allows the determination of every solution of a given system of linear relations. The solution set of $(\mathscr{S}')$ can be interpreted as the orthogonal projection of the solution set of $(\mathscr{S})$ into the coordinate plane $\{X \mid x_n = 0\}$.

The significance of Fourier elimination is more theoretical than practical. However, efforts are directed towards making the procedure more effective (Abadie [*1*]).

1.3. Proof of the Kuhn-Fourier Theorem

In this section it will be convenient to express Fourier elimination in a slightly different manner: The system $(\mathscr{S}')$ (1.2.1) will be replaced by the system $(\mathscr{S}^1)$ in which the unknowns $x_1, \ldots, x_{n-1}$ have the same coefficients as in $(\mathscr{S}')$ but x_n is written down with zero coefficients.

$(\mathscr{S}^1)$:

$$\left(\frac{a_{i1}}{a_{in}} - \frac{a_{j1}}{a_{jn}}\right) x_1 + \cdots + \left(\frac{a_{i,n-1}}{a_{in}} - \frac{a_{j,n-1}}{a_{jn}}\right) x_{n-1} + 0 x_n > \left(\frac{a_{i0}}{a_{in}} - \frac{a_{j0}}{a_{jn}}\right)$$

for all pairs i, j with $a_{in} > 0$, $a_{jn} < 0$,

$$\left(\frac{a_{i1}}{a_{in}} - \frac{b_{j1}}{b_{jn}}\right) x_1 + \cdots + \left(\frac{a_{i,n-1}}{a_{in}} - \frac{b_{j,n-1}}{b_{jn}}\right) x_{n-1} + 0 x_n > \left(\frac{a_{i0}}{a_{in}} - \frac{b_{j0}}{b_{jn}}\right)$$

for all pairs i, j with $a_{in} > 0$, $b_{jn} < 0$,

$$\left(\frac{a_{i1}}{a_{in}} - \frac{c_{j1}}{c_{jn}}\right) x_1 + \cdots + \left(\frac{a_{i,n-1}}{a_{in}} - \frac{c_{j,n-1}}{c_{jn}}\right) x_{n-1} + 0 x_n > \left(\frac{a_{i0}}{a_{in}} - \frac{c_{j0}}{c_{jn}}\right)$$

for all pairs i, j with $a_{in} > 0$, $c_{jn} \neq 0$,

$$\left(\frac{b_{i1}}{b_{in}} - \frac{a_{j1}}{a_{jn}}\right) x_1 + \cdots + \left(\frac{b_{i,n-1}}{b_{in}} - \frac{a_{j,n-1}}{a_{jn}}\right) x_{n-1} + 0 x_n > \left(\frac{b_{i0}}{b_{in}} - \frac{a_{j0}}{a_{jn}}\right)$$

for all pairs i, j with $b_{in} > 0$, $a_{jn} < 0$,

$$\left(\frac{b_{i1}}{b_{in}} - \frac{b_{j1}}{b_{jn}}\right) x_1 + \cdots + \left(\frac{b_{i,n-1}}{b_{in}} - \frac{b_{j,n-1}}{b_{jn}}\right) x_{n-1} + 0 x_n \geqslant \left(\frac{b_{i0}}{b_{in}} - \frac{b_{j0}}{b_{jn}}\right)$$

for all pairs i, j with $b_{in}>0$, $b_{jn}<0$,

$$\left(\frac{b_{i1}}{b_{in}}-\frac{c_{j1}}{c_{jn}}\right)x_1+\cdots+\left(\frac{b_{i,n-1}}{b_{in}}-\frac{c_{j,n-1}}{c_{jn}}\right)x_{n-1}+0x_n \geqslant \left(\frac{b_{i0}}{b_{in}}-\frac{c_{j0}}{c_{jn}}\right)$$

for all pairs i, j with $b_{in}>0$, $c_{jn}\neq 0$,

$$\left(\frac{c_{i1}}{c_{in}}-\frac{a_{j1}}{a_{jn}}\right)x_1+\cdots+\left(\frac{c_{i,n-1}}{c_{in}}-\frac{a_{j,n-1}}{a_{jn}}\right)x_{n-1}+0x_n > \left(\frac{c_{i0}}{c_{in}}-\frac{a_{j0}}{a_{jn}}\right)$$

for all pairs i, j with $c_{in}\neq 0$, $a_{jn}<0$,

$$\left(\frac{c_{i1}}{c_{in}}-\frac{b_{j1}}{b_{jn}}\right)x_1+\cdots+\left(\frac{c_{i,n-1}}{c_{in}}-\frac{b_{j,n-1}}{b_{jn}}\right)x_{n-1}+0x_n \geqslant \left(\frac{c_{i0}}{c_{in}}-\frac{b_{j0}}{b_{jn}}\right)$$

for all pairs i, j with $c_{in}\neq 0$, $b_{jn}<0$,

$$\left(\frac{c_{i1}}{c_{in}}-\frac{c_{j1}}{c_{jn}}\right)x_1+\cdots+\left(\frac{c_{i,n-1}}{c_{in}}-\frac{c_{j,n-1}}{c_{jn}}\right)x_{n-1}+0x_n = \left(\frac{c_{i0}}{c_{in}}-\frac{c_{j0}}{c_{jn}}\right)$$

for all pairs i, j with $c_{in}\neq 0$, $c_{jn}\neq 0$,

$$\begin{aligned}
&a_{k1}x_1+\cdots+a_{k,n-1}x_{n-1}+0x_n>a_{k0} &&\text{for all } k \text{ with } && a_{kn}=0,\\
&b_{k1}x_1+\cdots+b_{k,n-1}x_{n-1}+0x_n\geqslant b_{k0} &&\text{for all } k \text{ with } && b_{kn}=0,\\
&c_{k1}x_1+\cdots+c_{k,n-1}x_{n-1}+0x_n=c_{k0} &&\text{for all } k \text{ with } && c_{kn}=0.
\end{aligned}$$

If $(x_1,\dots,x_{n-1})^T$ solves $(\mathscr{S}')$, then $(x_1,\dots,x_{n-1},x_n)^T$ solves $(\mathscr{S}^1)$ for arbitrary x_n. Consequently, the interpretation of the solutions of $(\mathscr{S}')$ and $(\mathscr{S}^1)$ is a slightly different one: the solutions of $(\mathscr{S}')$ are to be extended to yield solutions of $(\mathscr{S})$, whereas a solution of $(\mathscr{S}^1)$ requires specification of its last component. But as before, $(\mathscr{S})$ is solvable if and only if $(\mathscr{S}^1)$ is solvable.

Since the formation of legal linear combinations is a transitive process, every relation of $(\mathscr{S}^1)$ is a legal linear combination (1.1.6) of relations of $(\mathscr{S})$. For instance,

$$\left(\frac{a_{i1}}{a_{in}}-\frac{b_{j1}}{b_{jn}}\right)x_1+\cdots+\left(\frac{a_{i,n-1}}{a_{in}}-\frac{b_{j,n-1}}{b_{jn}}\right)x_{n-1}+0x_n > \left(\frac{a_{i0}}{a_{in}}-\frac{b_{j0}}{b_{jn}}\right),$$

where $a_{in}>0$, $b_{jn}<0$, is a combination of the two relations

$$\begin{aligned}
a_{i1}x_1+\cdots+a_{in}x_n&>a_{i0},\\
b_{j1}x_1+\cdots+b_{jn}x_n&\geqslant b_{j0},
\end{aligned}$$

with the positive weights $\dfrac{1}{a_{in}}$, $\dfrac{-1}{b_{jn}}$. Successive elimination yields a sequence of systems

$$(\mathscr{S}),(\mathscr{S}^1),(\mathscr{S}^2),\dots,(\mathscr{S}^n).$$

Again each relation of $(\mathscr{S}^n)$ is a legal linear combination of relations of $(\mathscr{S})$, and again $(\mathscr{S})$ is solvable if and only if $(\mathscr{S}^n)$ is solvable.

As to the system $(\mathscr{S}^n)$, it evidently no longer contains unknowns with nonzero coefficients. Hence it must be of the form

$$(\mathscr{S}^n): \quad \begin{array}{ll} 0x_1 + \cdots + 0x_n > a_{i0}^{(n)}, & i=1,\ldots,p^{(n)}, \\ 0x_1 + \cdots + 0x_n \geqslant b_{i0}^{(n)}, & i=1,\ldots,q^{(n)}, \\ 0x_1 + \cdots + 0x_n = c_{i0}^{(n)}, & i=1,\ldots,r^{(n)}. \end{array}$$

Now, we have the tools for proving the Kuhn-Fourier theorem, of which we repeat the nontrivial part:

The system $(\mathscr{S})$ is solvable if every legal linear dependence leads to a zero-relation (1.1.8) *which is always true.*

If all legal linear dependences of $(\mathscr{S})$ are zero-relations which are always true, then this holds in particular for the zero-relations of which $(\mathscr{S}^n)$ consists. Hence $(\mathscr{S}^n)$ is solvable, and so is $(\mathscr{S})$. □

1.4. Consequence Relations. The Farkas Lemma

In this section, we shall characterize the set of consequence relations of a given system

$$(1.4.1) \qquad (\mathscr{S}): \begin{cases} AX > A_0, \\ BX \geqslant B_0, \\ CX = C_0. \end{cases}$$

In section (1.1) we have introduced legal linear combinations (1.1.6) as the most important way of deriving consequence relations. We shall now list two other formation rules. We call one of them the

(1.4.2) *weakening*

of a relation. It consists in either (i) replacing ">" by "⩾", or "=" by "⩾", or (ii) diminishing the right hand side and applying the law of transitivity:

$$a_{i1}x_1 + \cdots + a_{in}x_n \;\; \rho_1 \;\; a_{i0} \qquad \text{and} \qquad a_{i0} \;\; \rho_2 \;\; \bar{a}_{i0}$$

implies

$$a_{i1}x_1 + \cdots + a_{in}x_n \;\; \rho_3 \;\; \bar{a}_{i0}$$

where

$$\rho_3 := \begin{cases} ">" \begin{cases} \text{if} & \rho_1 = ">" \quad \text{and} \quad \rho_2 = "\geqslant", \\ \text{or} & \rho_1 = "\geqslant" \quad \text{and} \quad \rho_2 = ">", \end{cases} \\ "\geqslant" \;\; \text{if} \quad \rho_1 = "\geqslant" \quad \text{and} \quad \rho_2 = "\geqslant". \end{cases}$$

Finally, we may derive the equation $BX = b_0$ by

(1.4.3) *coupling*

the two inequalities $BX \geqslant b_0$ and $-BX \geqslant -b_0$.

The three formation rules yield all consequence relations. Indeed, we have

(1.4.4) **Theorem.** *If the system $(\mathscr{S})$ of linear relations is nonvoid and solvable, then every consequence inequality of $(\mathscr{S})$ can be obtained by legal linear combination and subsequent weakening. Every consequence equation can then be derived by coupling two consequence inequalities.*

Proof. Let

$$DX = d_1 x_1 + \cdots + d_n x_n > d_0$$

be a consequence relation of $(\mathscr{S})$. Then the system

$$\begin{aligned} AX &> A_0, \\ BX &\geqslant B_0, \\ CX &= C_0, \\ -DX &\geqslant -d_0, \end{aligned}$$

has no solution. Hence by virtue of the Kuhn-Fourier theorem there exists a legal linear dependence which leads to a contradictory zero-relation. We denote the coefficient vector of this linear dependence by

$$(U, V, W, z).$$

Since it is a legal linear combination we must have $z \geqslant 0$. But z does not vanish; for otherwise the contradictory relation would be a consequence of $(\mathscr{S})$, which was supposed to be solvable. We may therefore assume that $z = 1$, and that the contradictory relation reads as follows:

$$0X \;\; \rho \;\; UA_0 + VB_0 + WC_0 - d_0. \tag{1.4.5}$$

Here U, V, W are such that

$$UA + VB + WC - D = 0. \tag{1.4.6}$$

Since there is at least one inequality, namely the last one, which actually enters the combination, ρ cannot stand for equality. Therefore

$$\rho = \begin{cases} \text{“}>\text{”} & \text{if} \quad U \neq 0, \\ \text{“}\geqslant\text{”} & \text{otherwise.} \end{cases}$$

In order to yield a contradictory relation, the right hand side of (1.4.5) has to satisfy

$$UA_0+VB_0+WC_0-d_0 \;\; \bar{\rho} \;\; 0, \tag{1.4.7}$$

where

$$\bar{\rho} = \begin{cases} \text{"}\geqslant\text{"} & \text{if} \quad U \neq 0, \\ \text{"}>\text{"} & \text{otherwise}. \end{cases}$$

Now it follows from (1.4.6) that

$$DX \;\; \rho_0 \;\; UA_0+VB_0+WC_0,$$

with

$$\rho_0 := \begin{cases} \text{"}>\text{"} & \text{if} \quad U \neq 0, \\ \text{"}\geqslant\text{"} & \text{if} \quad U=0 \quad \text{and} \quad V \neq 0, \\ \text{"}=\text{"} & \text{if} \quad U=0 \quad \text{and} \quad V=0, \end{cases}$$

is a legal linear combination of relations of $(\mathscr{S})$. It can be weakened to yield $DX > d_0$. Indeed, (1.4.7) implies

$$DX \;\; \rho_0 \;\; UA_0+VB_0+WC_0 \;\; \bar{\rho} \;\; d_0,$$

and at least one of the two relation symbols ρ_0 and $\bar{\rho}$ stands for "$>$".

In the same manner, one obtains every consequence relation of the type

$$DX = d_1x_1 + \cdots + d_nx_n \geqslant d_0$$

by legal linear combination and subsequent weakening.

Finally, if the equation

$$DX = d_1x_1 + \cdots + d_nx_n = d_0$$

is a consequence relation of $(\mathscr{S})$, then so are the two relations

$$DX \geqslant d_0, \qquad -DX \geqslant -d_0$$

to which the above equation can be weakened. By coupling these inequalities we regain the original equation. This completes the proof of the theorem. ▯

Note that for systems of infinitely many linear relations the three formation rules of legal linear combination, weaking, and coupling do not yield all consequence relations. For instance, $x \geqslant 0$ is a consequence relation of the infinite system

$$x > -\frac{1}{n}, \qquad n=1,2,\ldots,$$

but the relation cannot be derived by the above rules of inference.

The well-known lemma of Farkas [*1*] is a specialization of theorem (1.4.4) to the case of a homogeneous system of inequalities.

the system is homogeneous if the right hand side is zero

(1.4.8) **Lemma of Farkas.** *Each homogeneous consequence relation* $DX \geqslant 0$ *of a homogeneous system*

$$(\mathscr{H}): \quad AX \geqslant 0$$

is a linear combination of inequalities of $(\mathscr{H})$ *with nonnegative coefficients.*

Proof. According to theorem (1.4.4), the relation $DX \geqslant 0$ may be obtained by legal linear combination of inequalities of $(\mathscr{H})$ and subsequent weakening. The latter must be ruled out since it destroys homogeneity. Since $(\mathscr{H})$ contains no equations, all coefficients of a legal linear combination are nonnegative. This proves the theorem. □

1.5. Irreducibly Inconsistent Systems

A system $(\mathscr{S})$ of linear relations is called

(1.5.1) *irreducibly inconsistent*

if it is nonsolvable, but each of its proper subsystem is solvable. A proper subsystem arises from $(\mathscr{S})$ by deletion of at least one, but possibly more, of its linear relations. If the same relation occurs more than once, then deleting may just reduce the multiplicity with which this relation occurs; nevertheless, the resulting system is considered a proper subsystem. Irreducibly inconsistent systems have been studied by Carver [*1*], Motzkin [*2*], and Fan [*4*]. They play a role in the theory of approximations (Rice [*1*]).

According to the Kuhn-Fourier theorem, a linearly independent system, that is, a system without any linear dependencies of its left hand sides is always solvable, since there are also no legal linear dependencies. Hence a nonsolvable system all of whose proper subsystems are linearly independent is irreducibly inconsistent. The converse of this statement forms the nontrivial part of the following theorem, which is essentially due to Carver.

(1.5.2) **Theorem.** *A nonsolvable system*

$$(\mathscr{S}): \begin{cases} AX > A_0, \\ BY \geqslant B_0, \\ CZ = C_0, \end{cases}$$

is irreducibly inconsistent if and only if the left hand sides of all proper subsystems are linearly independent.

Proof. Assume that $(\mathscr{S})$ is irreducibly inconsistent. We then have to show that all proper subsystems of $(\mathscr{S})$ are linearly independent. Since $(\mathscr{S})$ is nonsolvable there exists, according to the Kuhn-Fourier theorem, a legal linear dependence

(1.5.3) $(U,V,W)\neq 0,\quad$ with $\quad UA+VB+WC=0,\quad U,V\geqq 0.$

such that

$$UA_0+VB_0+WC_0\begin{cases}\geqq 0 & \text{if} \quad U\neq 0,\\ >0 & \text{if} \quad U=0 \quad \text{and} \quad V\neq 0,\\ \neq 0 & \text{if} \quad U=0 \quad \text{and} \quad V=0.\end{cases}$$

Without restriction of generality we may assume:

(1.5.4)
$$UA_0+VB_0+WC_0\begin{cases}\geqq 0 & \text{if} \quad U\neq 0,\\ >0 & \text{if} \quad U=0.\end{cases}$$

We observe that

(1.5.5) *none of the components of* (1.5.3) *vanishes,*

because this would imply the existence of a nonsolvable proper subsystem. Note that $U=0$ need not contradict (1.5.5) since U may be empty. The same holds for V and W.

Assume now some proper subsystem to be linearly dependent. This amounts to assuming the existence of a linear dependence

(1.5.6) $$(R,S,T)\neq 0$$

of the entire system $(\mathscr{S})$, at least one component of which vanishes. This linear dependence will be used to modify (1.5.3) so as to yield a legal linear dependence with at least one zero component, still leading to a contradictory zero relation. This then contradicts (1.5.5).

We proceed to show that we may assume without restriction of generality that (1.5.6) satisfies

(1.5.7) *U is linearly independent of R, provided U and R are not both empty.*

(1.5.8) *Either $R=S=0$, or $\{R\}\cup\{S\}$ contains a positive element.*

(1.5.9) $$RA_0+SB_0+TC_0\leqq 0.$$

Indeed, if $U=\theta R>0$ for some θ, then we replace the vector (1.5.6) by

$$(0, V-\theta S, W-\theta T).$$

This vector does not vanish, because $\theta R>0$ implies $R>0$, and the zero component of (1.5.6) must therefore occur in S or T; by (1.5.5), on the other hand, $\{V\}\cup\{W\}$ contains no zero element. We have thus found a linear dependence of a proper subsystem, and it satisfies (1.5.7). Condition (1.5.8) can always be achieved by a multiplication of (1.5.6) by -1 if necessary. Remains condition (1.5.9). It holds automatically if all elements of $\{R\} \cup \{S\}$ are nonnegative. Indeed, (1.5.6) is in this case a legal linear dependence of a proper subsystem. Since the latter is solvable, (1.5.6) must yield a zero relation that is always true. In other words,

$$RA_0 + SB_0 + TC_0 \begin{cases} <0 & \text{if} \quad R \neq 0, \\ \leqslant 0 & \text{if} \quad R=0 \quad \text{and} \quad S \neq 0, \\ =0 & \text{if} \quad R=0 \quad \text{and} \quad S=0. \end{cases}$$

If $\{R\} \cup \{S\}$ contains a negative element, then (1.5.9) can be achieved by multiplying (1.5.6) by -1, if necessary, without destroying (1.5.8).

We now define

(1.5.10) $$(\bar{U}, \bar{V}, \bar{W}) := (U, V, W) - \theta(R, S, T),$$

where, —and this definition is justified by (1.5.8)—:

$$\theta := \min \left\{ \frac{u_i}{r_i}, \frac{v_i}{s_i} \,\middle|\, r_i > 0, \; s_i > 0 \right\}.$$

Then (1.5.10) is a legal linear dependence of ($\mathscr{S}$) with at least one zero component. It remains to be shown that it leads to a contradictory zero relation. In other words, we have to verify

$$\bar{U} A_0 + \bar{V} B_0 + \bar{W} C_0 \begin{cases} \geqslant 0 & \text{if} \quad \bar{U} \neq 0, \\ >0 & \text{if} \quad \bar{U}=0 \quad \text{and} \quad \bar{V} \neq 0, \\ \neq 0 & \text{if} \quad \bar{U}=0 \quad \text{and} \quad \bar{V}=0, \end{cases}$$

or the slightly stronger statement,

(1.5.11) $$\bar{U} A_0 + \bar{V} B_0 + \bar{W} C_0 \begin{cases} \geqslant 0 & \text{if} \quad \bar{U} \neq 0, \\ >0 & \text{if} \quad \bar{U}=0. \end{cases}$$

Since $\bar{U} A_0 + \bar{V} B_0 + \bar{W} C_0 = (U A_0 + V B_0 + W C_0) - \theta(RA_0 + SB_0 + TC_0)$, relation (1.5.11) follows from (1.5.4) and (1.5.9) immediately, except if $\bar{U}=0$ and $U \neq 0$. But this has been ruled out by (1.5.7). □

(1.5.12) **Corollary to Theorem (1.5.2).** *An irreducibly inconsistent system contains at most $n+1$ relations, where n is the number of variables in the system.*

This follows from the well-known fact that more than n relations of n variables are always linearly dependent (compare section 2.2). □

1.6. Transposition Theorems

In this section we shall deal with systems of homogeneous linear relations

$$(\mathscr{H}): \begin{cases} AX > 0, \\ BX \geqslant 0, \\ CX = 0, \end{cases}$$

only. For such systems, the Kuhn-Fourier theorem (1.1.9) takes a very simple form:

(1.6.1) *A system $(\mathscr{H})$ of homogeneous linear relations is solvable if and only if there is no legal linear dependence (U, V, W) with $U \neq 0$.*

In particular, $(\mathscr{H})$ is always solvable if A is an empty matrix, that is, if no strict inequalities occur in $(\mathscr{H})$. However, if A is not empty, then the implication[1]

$$\left.\begin{array}{l} UA + VB + WC = 0 \\ U \geqslant 0, \quad V \geqslant 0 \end{array}\right\} \Rightarrow U = 0$$

must hold so that $(\mathscr{H})$ be solvable.

The following two inequality systems, one for X the other for U, where A is nonempty,

(1.6.2)

$AX = 0$	
$X \geqslant 0$	$A^T U \leqslant 0$

are said to form a

dual pair.

The following "transposition theorems" refer to such systems.

(1.6.3) **Transposition Theorem of Gordan** [*1*]. *For $A \neq \emptyset$ the following statements are equivalent:*

(i) $AX = 0$, $X \geqslant 0$ *has a nonzero solution.*
(ii) $A^T U < 0$ *has no solution.*

Proof. According to the corollary (1.6.1) of the Kuhn-Fourier theorem, the system $A^T U < 0$ is solvable if and only if the implication

$$AX = 0, \quad X \geqslant 0 \Rightarrow X = 0$$

holds. For a geometric interpretation of the theorem see section 2.10.

[1] "statement I $\Rightarrow$ statement II" means that statement II follows from statement I.

(1.6.4) **Transposition Theorem of Stiemke** [*1*]. *For $A \neq \emptyset$ the following statements are equivalent:*

(i) $AX=0$, $X>0$ *has no solution,*
(ii) $A^T U \leqslant 0$, $A^T U \neq 0$ *has a solution.*

Proof. According to the corollary (1.6.1) of the Kuhn-Fourier theorem, the system $AX=0$, $X>0$ is solvable if and only if the implication

$$V+A^T U=0, \qquad V \geqslant 0 \;\Rightarrow\; V=0$$

holds. But this is just another way of expressing that the system $A^T U \leqslant 0$ has no solution with $A^T U \neq 0$. ▯

For an interesting application of Stiemke's theorem to a theorem related to the Kirszbraun [*1*] theorem see Minty [*1*].

The transposition theorems of Gordan and Stiemke represent the extremes of a whole spectrum of situations. In order to express these facts in complete generality, we introduce the notion of

(1.6.5) *singular inequalities*

of a system of linear relations. These are inequalities of the type $B^T X \geqslant b$ which are satisfied as equations for all solutions of the system. For example, both inequalities of the system

$$B^T X \geqslant b,$$
$$-B^T X \geqslant -b$$

are singular inequalities. A nonsingular inequality is called (Tucker [*3*]) a

(1.6.6) *slack inequality.*

The following lemma is important in connection with slack inequalities:

(1.6.7) **Lemma.** *Every system*

$$B^T X \geqslant 0,$$
$$C^T X = 0,$$

has a solution $\bar{X}$ which satisfies all slack inequalities $B_i^T X \geqslant 0$ simultaneously as strict inequalities $B_i^T \bar{X} > 0$.

Proof. For every slack inequality $B_i^T X \geqslant 0$ there exists a solution X_i of the system such that $B_i^T X_i > 0$. Then the arithmetic mean $\bar{X}$ of all these solutions X_i is a solution as asserted by the lemma. ▯

The lemma (1.6.7) implies that the system $A^T U < 0$ is nonsolvable only if the system $A^T U \leqslant 0$ has at least one singular inequality. Thus the theorem of Gordan asserts that the system $AX = 0$, $X \geqslant 0$ has slack inequalities if and only if the system $A^T U \leqslant 0$ has singular inequalities.

Now we remark that there is a natural $1-1$ correspondence between the inequalities of the dual systems (1.6.2),

$$x_i \geqslant 0 \leftrightarrow A_i^T U \leqslant 0,$$

where A_i denotes the i-th column of A. It turns out that if $x_i \geqslant 0$ is a slack inequality of its system, then the corresponding inequality $A_i^T U \leqslant 0$ must be singular, and conversely. This phenomenon is called

(1.6.8) *complementary slackness.*

(1.6.9) **Theorem.** *Let $x_i \geqslant 0$ and $A_i^T U \leqslant 0$ be corresponding inequalities of the dual systems* (1.6.2). *Each of these inequalities is slack if and only if its counterpart is singular.*

Proof. $A_i^T U \leqslant 0$ is a slack inequality if and only if the system

$$\begin{aligned} &A_j^T U \leqslant 0 \quad \text{for} \quad j \neq i, \\ &A_i^T U < 0, \end{aligned}$$

is solvable. According to (1.6.1) this is equivalent to

$$AX = 0, \qquad X \geqslant 0 \;\Rightarrow\; x_i = 0,$$

which means that $x_i \geqslant 0$ is a singular inequality. Thus $x_i \geqslant 0$ is singular if and only if $A_i^T U \leqslant 0$ is slack. It follows that $x_i \geqslant 0$ is nonsingular, that is, slack, if and only if $A_i^T U \leqslant 0$ is nonslack, that is, singular. This proves the theorem. □

Theorem (1.6.9) yields the transposition theorems of Gordan and Stiemke by means of lemma (1.6.7).

The dual pair (1.6.2) of systems is a special case of the following general one (see Tucker [3]).

(1.6.10)

$AX + BY = 0$	
$CX + DY \geqslant 0$	$V \geqslant 0$
$X \geqslant 0$	$A^T U + C^T V \leqslant 0$
	$B^T U + D^T V = 0$

It can be left to the reader to verify the law of complementary slackness. Interesting special cases of (1.6.10) are the dual pairs

$CX \geqslant 0$	$V \geqslant 0$
$X \geqslant 0$	$C^T V \leqslant 0$

and, for $K^T = -K$,

$KX \geqslant 0$	$V \geqslant 0$
$X \geqslant 0$	$KV \geqslant 0$

.

A commonly used schematic representation of both systems (1.6.10) simultaneously is

	$X \geqslant 0$	Y	
U	A	B	$=0$
$V \geqslant 0$	C	D	$\geqslant 0$
	$\leqslant 0$	$=0$	

.

Here U and V are considered to be row vectors. The symmetry of the systems (1.6.10) then becomes more apparent.

1.7. The Duality Theorem of Linear Programming

Given a system $(\mathscr{S})$ of linear relations, the problem of finding a solution of $(\mathscr{S})$ that minimizes a given linear function $CX - d$ is called a

(1.7.1) *linear programming problem*, or briefly, a *linear program*.

The solutions of $(\mathscr{S})$ are called

(1.7.2) *feasible*,

the linear function which is to be minimized is called the

(1.7.3) *objective function*.

It assigns a value to each feasible solution. A feasible solution of minimum value is called an

(1.7.4) *optimal solution*.

Minimization can be replaced by maximization, simply by reversing the sign of the objective function.

Linear programs are most frequently studied in the form

(1.7.5)
$$\begin{aligned} &\textit{Minimize} \quad CX-d \quad \textit{subject to}\\ &AX=B,\\ &X\geqslant 0.\end{aligned}$$

This program turns out to be closely related to the following one:

(1.7.6)
$$\begin{aligned} &\textit{Maximize} \quad B^T U-d \quad \textit{subject to}\\ &A^T U\leqslant C^T.\end{aligned}$$

The two programs are said to form a

(1.7.7) *dual pair*

of programs. Notice the analogy to the dual pair of inequality systems (1.6.3). Similarly, duality can be defined for more complex programs, as for instance the pair

(1.7.8)
$$\begin{aligned} &\textit{Minimize} \quad EX+FY-d \quad \textit{subject to}\\ &AX+BY=G,\\ &CX+DY\geqslant H,\\ &X\geqslant 0,\end{aligned}$$

and

(1.7.9)
$$\begin{aligned} &\textit{Maximize} \quad G^T U+H^T V-d \quad \textit{subject to}\\ &V\geqslant 0,\\ &A^T U+C^T V\leqslant E^T,\\ &B^T U+D^T V=F^T,\end{aligned}$$

or its special case

(1.7.10)
$$\begin{aligned} &\textit{Minimize} \quad EX-d \quad \textit{subject to}\\ &CX\geqslant H,\\ &X\geqslant 0,\end{aligned}$$

and

(1.7.11)
$$\begin{aligned} &\textit{Maximize} \quad H^T V-d \quad \textit{subject to}\\ &C^T V\leqslant E^T,\\ &V\geqslant 0.\end{aligned}$$

Let us return to the dual pair of programs (1.7.5) and (1.7.6). For each feasible solution X of the program (1.7.5), and each feasible solution U of the program (1.7.6) we have

(1.7.12)
$$B^T U=U^T B=U^T AX\leqslant CX.$$

A similar relation is readily derived for the dual pairs (1.7.8), (1.7.9) and (1.7.10), (1.7.11). More important, optimal solutions X and U of (1.7.5) and (1.7.6) are characterized by realizing equality in (1.7.12). This is the content of the famous duality theorem of linear programming due to Gale, Kuhn and Tucker [*1*]:

(1.7.13) **Duality Theorem of Linear Programming.** *For dual pairs of linear programs the following statements hold:*

(i) *The value of each feasible solution of the minimization program bounds above all values of feasible solutions of the maximization program, and vice versa.*

(ii) *If both programs have feasible solutions, then both have optimal solutions, and the optimal values of both objective functions are equal.*

(iii) *If one program has an optimal solution, then so has the other (and the optimal values of both objective functions are equal).*

Proof. We prove the theorem for the dual pair (1.7.5), (1.7.6) only. The proof for other pairs of dual programs is analogous and can be left to the reader.

Statement (i) repeats relation (1.7.12). The conclusion part of statement (ii) is equivalent to the solvability of the linear system

$$\begin{aligned} -CX+B^TU&\geqslant 0,\\ AX\qquad &= B,\\ X\qquad &\geqslant 0,\\ -A^TU &\geqslant -C^T. \end{aligned} \tag{1.7.14}$$

If (1.7.14) is not solvable, then according to the Kuhn-Fourier theorem (1.1.9) there exists a contradictory legal linear dependence. The relation $-CX+B^TU\geqslant 0$ must actually enter any such dependence, since without this relation the system (1.7.14) is solvable according to the hypothesis that both programs have feasible solutions. Up to multiples, every contradictory legal linear dependence is therefore of the form $(1, V^T, W^T, Z^T)$ with $-C+V^TA+W^T=0, B^T-Z^TA^T=0, V^TB-Z^TC>0$, $W\geqslant 0$, $Z\geqslant 0$. Hence

$$\begin{gathered} A^TV\leqslant C^T,\\ AZ=B, \quad Z\geqslant 0,\\ B^TV-Z^TC>0, \end{gathered}$$

showing that V and Z are feasible solutions of the two dual programs, respectively. However, for these programs, $B^TV-Z^TC>0$ has been ruled out (1.7.12). Hence the system (1.7.14) possesses no contradictory dependences, and is therefore solvable. This proves (ii).

If program (1.7.5) has an optimal solution X^0, then

$$-CX > -CX^0,$$
$$AX = B,$$
$$X \geqslant 0,$$

is not solvable. Hence there exist U, V such that

$$U^T A + V^T = C, \qquad V \geqslant 0 \quad \text{and} \quad U^T B \geqslant CX^0.$$

Hence U is a feasible solution of the program (1.7.6), and according to (1.7.12) optimal. The analogous proof that the existence of an optimal solution of (1.7.6) entails the existence of an optimal solution of (1.7.5) is left to the reader. □

(1.7.15) **Corollary to Theorem (1.7.13).** *A feasible solution of a linear program is optimal if and only if there exists a feasible solution of the dual program such that both solutions assign the same value to their respective objective functions.*

Let X and U be feasible solutions to the programs (1.7.5) and (1.7.6), respectively. Then

$$CX - B^T U = CX - U^T B = CX - U^T AX = (C - U^T A)X = 0$$

is necessary and sufficient for X and U to be optimal. But the scalar product of two nonnegative vectors vanishes if and only if at least one of each two corresponding components vanishes. Hence every pair of optimal solutions X, U satisfies the following

(1.7.16) **Complementary Slackness Conditions.**

$$x_i > 0 \quad \textit{implies} \quad A_i^T U = c_i,$$
$$A_i^T U < c_i \quad \textit{implies} \quad x_i = 0.$$

It is straightforward to formulate the complementary slackness conditions for more complex programs. For the dual pair (1.7.10), (1.7.11) for instance they read as follows:

(1.7.17)
$$\begin{array}{llllll} {}_iCX > h_i & \textit{implies} & v_i = 0, & v_i > 0 & \textit{implies} & {}_iCX = h_i, \\ x_j > 0 & \textit{implies} & C_j^T V = e_j, & C_j^T V < e_j & \textit{implies} & x_j = 0. \end{array}$$

Contrary to complementary slackness in the case of inequalities (compare (1.6.9) and theorem (1.6.10)), the implications (1.7.16) are not reversable. The following simple program,

$$\begin{aligned} &\textit{Minimize } x_3 \textit{ subject to}\\ &x_1+x_2+x_3=1,\\ &X\geqslant 0, \end{aligned}$$

provides a counterexample. The vector $X=(1,0,0)^T$ is the optimal solution, whereas $U=(0)$ is an optimal solution of the dual program,

$$\begin{aligned} &\textit{Maximize } u \textit{ subject to}\\ &u\leqslant 0,\\ &u\leqslant 0,\\ &u\leqslant 1. \end{aligned}$$

The optimal solutions X and U satisfy both $x_2\geqslant 0$ and $u\leqslant 0$ as equations.

CHAPTER 2

Convex Polyhedra

The first chapter dealt with the logical structure of linear systems. This chapter directs the attention to the geometrical properties of their solution sets. Nevertheless, we shall reestablish some of the results of the first chapter, such as the lemma of Farkas (1.4.8), and the transposition theorem of Gordan (1.6.3). Both are equivalent formulations of what is sometimes called the "key fact" of the theory of linear inequalities. To this class of theorems also belong the theorem of Weyl (2.8.8) and the theorem of Kuhn-Fourier (1.1.9), on which the first chapter was based.

The combinatorial properties of polyhedra are of foremost interest and importance. For a comprehensive treatment we refer the reader to Grünbaum [*6*].

The geometric theory of linear inequalities roots in the work of Minkowski [*2*], [*4*], and has been first developed systematically by Motzkin in his thesis „Beiträge zur Theorie der linearen Ungleichungen" [*2*]. Much of the subsequent developments are due to Tucker and his associates, e.g. Tucker [*3*], [*5*], [*6*] Goldman and Tucker [*1*], [*2*], Goldman [*1*], Gale [*1*], [*4*].

2.1. Means and Averages

A nonvacuous subset S of the vector space R^n forms a subspace if it is closed under the formation of arbitrary finite linear combinations

(2.1.1) $$w_1 X_1 + \cdots + w_s X_s$$

of elements $X_i \in S$. For any nonvacuous $S \subseteq R^n$, the set

$$\mathscr{L} S$$

of all linear combinations of points of S is a subspace. It is called the

(2.1.2) *linear hull*

of S. It will be convenient to define the set consisting of the zero vector 0 exclusively as the linear hull of the empty set:

$$\mathscr{L}\emptyset := \{0\}.$$

By this convention it becomes true that the linear hull of any set $S \in R^n$ is a subspace, more precisely, the smallest subspace containing S. A linear combination (2.1.1) may be expressed as a matrix product XW with

$$X := (X_1, \ldots, X_s) \quad \text{and} \quad W := \begin{bmatrix} w_1 \\ \vdots \\ w_s \end{bmatrix}.$$

The case in which X is the empty $n \times 0$-matrix fits in with the convention (I.11) of the introduction.

Other closure operations employ restricted classes of linear combinations $w_1 X_1 + \cdots + w_s X_s$, namely

(2.1.3) *nonnegative linear combinations* if $w_i \geqslant 0$ for all i,
averages if $\Sigma w_i = 1$,
means if $\Sigma w_i = 1$, and $w_i \geqslant 0$ for all i.

(compare Good [*1*]). The corresponding hulls are the

(2.1.4) *conical* or *positive hull* $\mathscr{C} S$,
spanned manifold or *affine hull* $\mathscr{M} S$,
convex hull $\mathscr{H} S$.

In particular, we define

$$\mathscr{C}\emptyset := \{0\}, \qquad \mathscr{M}\emptyset = \emptyset, \qquad \mathscr{H}\emptyset = \emptyset.$$

S is a

(2.1.5) *convex cone (linear manifold, convex set)*,

if $\mathscr{C} S = S$ ($\mathscr{M} S = S$, $\mathscr{H} S = S$). The empty set is a convex set and a linear manifold, but neither a convex cone nor a linear subspace.

The following notation will be convenient: By

$$(X, Y) := \{Z = \lambda X + (1-\lambda) Y \mid 0 < \lambda < 1\}$$

we shall denote the

(2.1.6) *open segment*

spanned by $X, Y \in R^n$. The reader should be aware that this definition implies

$$(X, X) = \{X\},$$

and differs in this respect from another commonly used definition of open segments, namely $\mathscr{H}\{X, Y\} \sim (\{X\} \cup \{Y\})$.

Now consider any mean $\bar{X} = \sum_{i=1}^{k} w_i X_i$, $\sum_{i=1}^{k} w_i = 1$, $w_1 < 1$. It can be rewritten as $\bar{X} = \lambda X_1 + (1-\lambda) \sum_{i=2}^{k} \bar{w}_i X_i$ where $\lambda := w_1$, $\bar{w}_i := w_i/(1-w_1)$ for $i = 2, \ldots, k$. This shows that every mean can be achieved by successively forming the mean of two points. Thus

(2.1.7) *a set* $S \subseteq R^n$ *is convex if and only if*

$$\{X\} \cup \{Y\} \subseteq S \;\Rightarrow\; (X, Y) \subseteq S.$$

The intersection of a family of convex sets (cones, manifolds, subspaces) is again a convex set (cone, manifold, subspace). Each of the four classes of sets forms a lattice under set inclusion. (See Birkhoff [*1*] for the theory of lattices.)

2.2. Dimensions

A set $S \subseteq R^n$ is called

(2.2.1) *linearly independent*

if it is empty or if there exists no finite subset $\{X_1, \ldots, X_s\}$ of S and no vector $W \neq 0$ such that

$$(X_1, \ldots, X_s) W = 0.$$

Otherwise S is called "linearly dependent". Every subset of a linearly independent set is again linearly independent. Next let L be a linear subspace of the R^n. A set S is a

(2.2.2) *spanning set*

of L if $L = \mathscr{L} S$. If S is a spanning set of L, then so is each set $T \subseteq L$ which contains S.

A linearly independent set S which is contained in a linear subspace L is

maximal

in L if it is not a proper subset of another independent set in L. Analogously, a spanning set S of L is

minimal

if no proper subset of S spans L. Plainly,

(2.2.3) *a spanning set* S *of* L *is minimal if and only if it is linearly independent. A linearly independent subset* S *of* L *is maximal if and only if it spans* L.

Thus each maximal independent subset is a minimal spanning set, and vice versa. Such a set is called a

(2.2.4) *basis*

of a subspace L. The following theorem constitutes a fundamental result of the theory of vector spaces (compare Baer [*1*], Birkhoff and MacLane [*1*], Bourbaki [*1*], Gracub [*1*], Sperner [*2*]).

(2.2.5) **Theorem.** *All bases of a linear subspace* $L \subseteq R^n$ *have the same finite cardinality, called the*

$$\textit{dimension} \ (=\dim L)$$

of L.

This theorem is an immediate consequence of the following

(2.2.6) **Theorem of Steinitz.** *If* S *and* T *are both linearly independent sets in* R^n, *and if* S *contains fewer elements than* T, *then there exists at least one element* $t \in T$ *such that the augmented set* $S \cup \{t\}$ *is also linearly independent*[1].

A constructive proof of the Steinitz theorem can be based on Jordan elimination (for instance Sperner [*2*], Hadley [*1*], Marcus and Minc [*1*]).

We note a few corollaries:

(2.2.7) *A subset* S *of a linear subspace* L *is necessarily linearly dependent if its cardinality exceeds the dimension of* L.

(2.2.8) *If the linear subspace* L_1 *is contained in* L_2, *then* $\dim L_1 \leqslant \dim L_2$.

(2.2.9) *For any two subspaces* L_1 *and* L_2 *of* R^n,

$$\dim(L_1 + L_2) = \dim L_1 + \dim L_2 - \dim(L_1 \cap L_2).$$

We proceed to define dimensions for arbitrary sets S, putting

$$\dim S := \dim \mathscr{M} S. \qquad (2.2.10)$$

The dimension of a linear manifold in turn is defined as follows: if $M = \emptyset$ then $\dim M := -1$. Otherwise let X_0 be any point of a linear manifold M. Then the set

$$L := M - X_0 = \{X - X_0 \mid X \in M\}$$

is the unique linear subspace which runs parallel to M through the origin of R^n. The dimension of M is defined to be the dimension of L. It follows from (2.2.8) that $S \subseteq T$ implies $\dim S \leqslant \dim T$.

---

[1] Another formulation of the Steinitz theorem is that each subset of a linear subspace is a matroid (see Whitney [*2*], Tutte [*1*]).

A convex cone C also possesses minimal subsets S such that $C = \mathscr{C} S$. It is no longer true, however, that the cardinality of such a minimal spanning set depends only on the dimension of C. Nevertheless, the minimum representation of individual points in C can be linked to its dimension. We have

(2.2.11) **Carathéodory's Theorem for Cones.** *Let S be any subset of R^n, and put*

$$d := \dim \mathscr{C} S.$$

Then for each point $X \in \mathscr{C} S$ there exist d points $X_i \in S$ such that $X = \mathscr{C}\{X_1, \ldots, X_d\}$.

Proof. Let $X \in \mathscr{C} S$. Then

$$X = (X_1, \ldots, X_s)\, W,$$

for some $X_i \in S$ and some $W > 0$. If $s \leqslant d$ there is nothing to be proved. If $s > d$, then according to (2.2.7) there exists a $U \neq 0$ such that

$$0 = (X_1, \ldots, X_s)\, U.$$

Without restriction of generality we may assume that U contains a positive component, and we can define a positive number θ by

$$\theta := \min \left\{ \frac{w_i}{u_i} \;\middle|\; u_i > 0 \right\}.$$

Then $V := W - \theta U$ is still nonnegative but contains one zero component. Therefore the linear combination

$$X = (X_1, \ldots, X_s)\, V$$

actually involves only $s-1$ points of S. This process of reduction can be repeated until $s \leqslant d$. □

An analogous theorem holds for convex hulls:

(2.2.12) **Carathéodory's Theorem for Convex Sets.** *Let S be any subset of R^n, and put*

$$d := \dim \mathscr{M} S.$$

Then for each point $X \in \mathscr{H} S$ there exist $d+1$ points $X_i \in S$ such that $X \in \mathscr{H}\{X_1, \ldots, X_{d+1}\}$.

Proof. Consider the set

$$\tilde{S} := \left\{ \begin{pmatrix} X \\ 1 \end{pmatrix} \in R^{n+1} \;\middle|\; X \in S \right\}$$

and verify that the set

$$\begin{pmatrix} X_0 \\ 1 \end{pmatrix} \cup \left\{ \begin{pmatrix} X_0+Y \\ 1 \end{pmatrix} \middle| Y \in T \right\}$$

is a basis of $\mathscr{L}\tilde{S}$ whenever $X_0 \in \mathscr{M}S$ and T is a basis of the linear subspace parallel to $\mathscr{M}S$. Thus $\dim \mathscr{C}\tilde{S} = \dim \mathscr{L}\tilde{S} = d+1$, For each $X \in \mathscr{H}S$, there exist by theorem (2.2.11) $d+1$ points $\begin{pmatrix} X_i \\ 1 \end{pmatrix} \in \tilde{S}$ such $\begin{pmatrix} X \\ 1 \end{pmatrix} \in \mathscr{C}\left\{ \begin{pmatrix} X_i \\ 1 \end{pmatrix}, \ldots, \begin{pmatrix} X_{d+1} \\ 1 \end{pmatrix} \right\}$. It follows that $X \in \mathscr{H}\{X_1, \ldots, X_{d+1}\}$. Furthermore, $X_i \in S$ for $1 \leqslant i \leqslant d+1$ by the definition of $\tilde{S}$. □

The proof of theorem (2.2.12) illustrates a technique which is frequently used for transition from a theorem for convex cones to an analogous theorem for convex sets, because theorems are usually easier to prove for cones. We call this method "homogenization" (see section 2.11).

For generalizations of the theorem of Carathéodory, see Bonnice and Klee [*1*] and Reay [*1*]. The theorem of Hanner and Rådström [*1*] also belongs into this area (see for instance Valentine [*4*]).

2.3. Polyhedra and their Boundaries

The solution set M of a system of linear equations

$$A^T X = B,$$

where A is an $n \times m$-matrix and $B \in R^m$, may be interpreted as a subset of the vector space R^n. It is easily verified that M is a linear manifold. If $A_i \neq 0$, then the solutions sets of a single linear equation

$$A_i^T X = b_i$$

and of a single linear inequality

$$A_i^T X \leqslant b_i$$

are a

(2.3.1) *hyperplane* or briefly, *plane*, and a *halfspace*,

respectively. The solution set P of a finite system of linear inequalities

$$A^T X \leqslant B$$

is a convex, possibly vacuous, set, and is called a

(2.3.2) *convex polyhedron* or briefly, *polyhedron*.

We call any finite system of linear inequalities $A^T X \leqslant B$ which has *P* as its solution set a

(2.3.3) *representation*

of *P*. For the remainder of this section, we shall consider a fixed representation for each polyhedron *P*. If $A_i \neq 0$ we call the plane described by the equation

$$A_i^T X = b_i$$

a (Fig. 2)

(2.3.4) *boundary plane.*

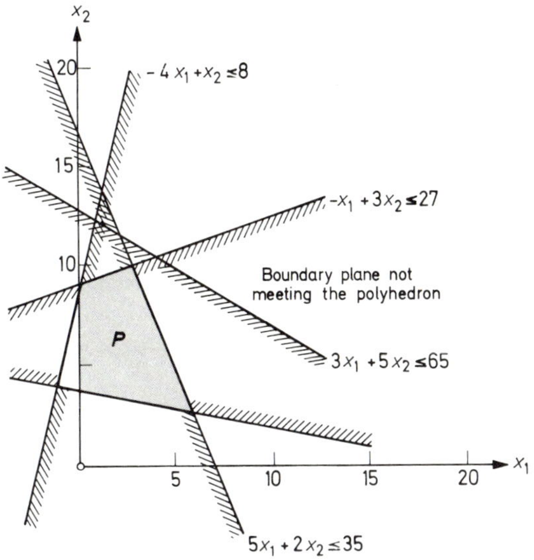

Fig. 2. Boundary planes

Boundary planes need not have any points in common with the polyhedron *P*. On the other hand, a boundary plane may contain the entire polyhedron. We call a boundary plane with the latter property

(2.3.5) *singular.*

Analogously we shall speak of *singular inequalities* or *constraints.* Nonsingular inequalities are also called *slack inequalities* (compare (1.6.5) and (1.6.6)).

The intersection of a polyhedron P with some boundary planes is a

(2.3.6) *face*

of the polyhedron. We always consider the empty set a face. A face consisting of exactly one point is a

(2.3.7) *vertex.*

A face of a polyhedron is again a polyhedron. If $\bar{F}$ is a face of the face F of a polyhedron P, then $\bar{F}$ is also a face of P. A polyhedron P is always a face of itself since it is the intersection of P with R^n, the latter being the intersection of the empty set of boundary planes. We call a face which is different from P a

(2.3.8) *proper face.*

Note that different inequality systems describing the same polyhedron may give rise to different sets of boundary planes. In other words, the notion of boundary planes applies to inequality systems rather than to polyhedra. The definition of faces (2.3.6) is based on a particular set of boundary planes. That is, this definition depends on the representation of a polyhedron by an inequality system. However, we shall see in the next section that this dependence is not an actual one.

We distinguish between

(2.3.9) *boundary points* and *inner points*

of a polyhedron depending on whether they belong to a proper face or not. Inner points satisfy every nonsingular inequality as a strict inequality.

Inner points need not be interior points of a polyhedron $P \subseteq R^n$ with respect to the euclidean topology of R^n. They are interior points with respect to the relative topology induced in $\mathscr{M}P$ by the euclidean topology (compare chapter 3). Inner points are also called "algebraically interior points" (Köthe [*1*]).

(2.3.10) **Lemma.** *Every nonvacuous polyhedron* $P=\{X \mid A^T X \leqslant B\}$ *has inner points.*

Proof. For every nonsingular constraint $A_i^T X \leqslant b_i$ there exists a point $X_i \in P$ with $A_i^T X_i < b_i$. The arithmetic mean $\bar{X}$ of all these points is an inner point of P (compare lemma (1.6.7)). If all boundary planes are singular then every point of P is an inner point. □

It is possible to move an inner point a little bit in every direction which is compatible with the singular constraints without leaving the polyhedron. This is the content of the following

(2.3.11) **Lemma.** *If U is an inner point of P, and Y an arbitrary point of the intersection of all singular boundary planes of P, then there exists an $\varepsilon > 0$ such that*

$$U + \varepsilon(Y - U) \quad \text{and} \quad U - \varepsilon(Y - U)$$

are both in P.

Proof. If $A_i^T(Y-U)=0$ for all inequalities, then there is nothing to be shown. Otherwise choose

$$0 < \varepsilon \leqslant \min\left\{\frac{b_i - A_i^T U}{|A_i^T(Y-U)|} \;\middle|\; A_i^T(Y-U) \neq 0\right\}. \quad \square$$

The following lemma justifies the term "boundary points":

(2.3.12) **Lemma.** *If U is an inner point of P, and Y a point which lies on the intersection of all singular boundary planes and which does not belong to P, then the open segment (U, Y) contains a boundary point X of P.*

Proof. $X := U + \varepsilon(Y-U)$, where

$$\varepsilon := \min\left\{\frac{b_i - A_i^T U}{A_i^T(Y-U)} \;\middle|\; A_i^T(Y-U) > 0\right\},$$

is a boundary point. $\square$

Finally we have the

(2.3.13) **Accessibility Lemma for Polyhedra.** *If U is an inner point, and X an arbitrary point of P, then all points on the open segment (U, X) are inner points of P.*

This lemma is a trivial for polyhedra. An analogous result will be proved for general convex sets in section 3.2.

It will be convenient to have a short notation for the set of all inner points of a polyhedron. Therefore we shall frequently write

$$P^I \tag{2.3.14}$$

for the set of all inner points (2.3.9) of a polyhedron P. Note that $P^I = \emptyset$ if and only if $P = \emptyset$ by lemma (2.3.10). P^I is convex.

2.4. Extreme and Exposed Sets

In this section we shall recognize the faces of a polyhedron as its "extreme subsets". This will prove that, in fact, faces do not depend on the representation by an inequality system, since extreme subsets of a convex set K will be defined in a manner that depends only on K itself. Other important subsets of a convex set are its "exposed subsets." We shall see that for polyhedra the notions "extreme subset" and "exposed subset" coincide.

Let K be any convex set. A convex subset W of K is called an

(2.4.1) *extreme subset*

of K if none of its points are included in an open segment spanned by two points of K which are not both in W. In other words, W is extreme if $W=\emptyset$, or if W is convex and the following implication holds

$$\left.\begin{array}{l} X, Y \in K \\ (X, Y) \cap W \neq \emptyset \end{array}\right\} \Rightarrow X, Y \in W. \tag{2.4.2}$$

An extreme subset consisting of exactly one point is an

(2.4.3) *extreme point.*

A convex set K is always an extreme subset of itself. Any extreme subset W of K is the intersection of K with the manifold $\mathscr{M} W$:

$$W = \mathscr{M} W \cap K \quad \text{(Fig. 3).} \tag{2.4.4}$$

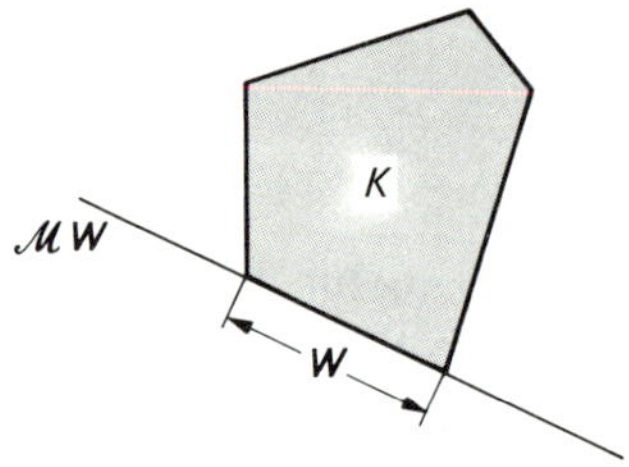

Fig. 3. $W = \mathscr{M} W \cap K$

Proof. Any $X \in \mathscr{M} W \cap K$ is an average of a finite number of points $Y_i \in W$:

$$X = w_1 Y_1 + \cdots + w_k Y_k, \quad \sum_i w_i = 1.$$

We want to show that X lies in W. This is trivial if all w_i are nonnegative. Otherwise put

$$\alpha := -\sum_{i: w_i < 0} w_i > 0, \quad Z_1 := \frac{1}{1+\alpha} \sum_{i: w_i \geqslant 0} w_i Y_i, \quad Z_2 := \frac{-1}{\alpha} \sum_{i: w_i < 0} w_i Y_i .$$

We then have

$$X \in K, \qquad Z_2 \in K, \qquad \text{and} \qquad \frac{X + \alpha Z_2}{1+\alpha} = Z_1 \in (X, Z_2) \cap W.$$

Since W is extreme, this implies $X \in W$ according to (2.4.2). Thus $\mathcal{M}W \cap K \subseteq W$. The converse is trivial. □

Now let the system $A^T X \leqslant B$ represent the polyhedron P. In the following discussion, the term "boundary plane " will be used with respect to this particular system.

It is easy to see that

(2.4.5) *every face of a polyhedron is extreme.*

Consider a face F of the polyhedron P. It is an intersection of P with some boundary planes, say

$$E_i = \{X \mid A_i^T X = b\}, \qquad i = 1, \ldots, k.$$

Then for every pair $X, Y \in P$ with $(X, Y) \cap F \neq \emptyset$ there exist coefficients $\lambda, \mu > 0$ with $\lambda + \mu = 1$ such that

$$b_i = \lambda A_i^T X + \mu A_i^T Y$$

holds for every $i = 1, \ldots, k$. Since $A_i^T X \leqslant b_i$ and $A_i^T Y \leqslant b_i$, this is not possible unless equality is assumed. Thus $X, Y \in P$ and $(X, Y) \cap F \neq \emptyset$ imply $X, Y \in F$. But this implication characterizes extreme subsets.

That every extreme subset of a polyhedron P is a face follows at once from (2.4.4) and the

(2.4.6) **Lemma.** *If W is an extreme subset of the polyhedron $P = \{X \mid A^T X \leqslant B\}$, then the manifold $\mathcal{M}W$ spanned by W is the intersection of all boundary planes $E_i = \{X \mid A_i^T X = b_i\}$ which contain W:*

$$\mathcal{M}W = \bigcap \{E_i \mid E_i \supseteq W\}.$$

Proof. Let

$$I := \begin{bmatrix} i_1 \\ \vdots \\ i_k \end{bmatrix}$$

be a vector consisting of those indices i to which there corresponds a boundary plane containing W. Of course, I may be the empty vector. The proof of the lemma consists of showing that $\mathcal{M}W = \bigcap_{i \in I} E_i$. Obviously, $\mathcal{M}W \subseteq \bigcap_{i \in I} E_i$. In order to verify that $\mathcal{M}W \supseteq \bigcap_{i \in I} E_i$ holds, too, we consider the face

$$F := P \cap \bigcap_{i \in I} E_i \supseteq W.$$

F is the solution set of the system (see (I.12) for notation):

$$A^T X \leqslant B,$$
$$-(A_I)^T X \leqslant -({}_I B).$$

This system gives rise to the same boundary planes E_i as the original system $A^T X \leqslant B$. However, precisely those boundary planes E_i are singular boundary planes of F for which $i \in I$, that is $E_i \supseteq W$ holds. Hence for every $i \notin I$ we may choose a point $Y_i \in W$ with $A_i^T Y_i < b_i$. The arithmetic mean $\bar{Y}$ of these points Y_i is an inner point of F and belongs to W. For $W = P$ this argument was used to prove the existence of inner points (2.3.10).

For any $X \in \bigcap_{i \in I} E_i$ and sufficiently small $\varepsilon > 0$, the points

$$Z_1 := \bar{Y} + \varepsilon(X - \bar{Y}), \qquad Z_2 := \bar{Y} - \varepsilon(X - \bar{Y})$$

lie in F, according to lemma (2.3.11). We then have $\bar{Y} \in (Z_1, Z_2) \cap W$, and consequently $Z_1, Z_2 \in W$, since W is extreme. Hence $X \in \mathcal{M}\{Z_1, Z_2\} \subseteq \mathcal{M} W$.

This proves the lemma and thereby the

(2.4.7) **Theorem.** *The extreme subsets of a polyhedron are its faces. In particular, every extreme point is a vertex and conversely, and is uniquely determined as the intersection of those boundary planes in which it is contained.*

Applied to $W = P$, lemma (2.4.6) yields:

(2.4.8) **Theorem.** *The manifold $\mathcal{M} P$ spanned by a polyhedron P is the intersection of its singular boundary planes.*

This theorem is important for determining the dimension (2.2.10) of P.

The definition (2.3.6) of faces of a polyhedron P was based on a particular representation (2.3.3) of P. Theorem (2.4.7) characterizes faces in terms of the point set P. It establishes therefore the

(2.4.9) **Theorem.** *The number of faces of a polyhedron is finite and independent of the representation of the polyhedron by inequalities.*

A function $f: R^n \to R$ is called a

(2.4.10) *homogeneous linear function* or *linear form*

if $f(\lambda X + \mu Y) = \lambda f(X) + \mu f(Y)$ for all $\lambda, \mu \in R$, $X, Y \in R^n$. A subset W of a convex set K is called an (compare Eggleston [*1*])

(2.4.11) *exposed subset*

of K if there exists a linear form f which assumes precisely on W its maximum over K:

$$W = \{Z \in K \mid f(Z) \geqslant f(X) \quad \text{for all} \quad X \in K\}.$$

For reasons of uniformity, we also consider the empty set exposed.

(2.4.12) **Theorem.** *Every exposed subset W of a polyhedron is a face, and vice versa.*

Proof. Every exposed subset of a convex set K is extreme. Indeed, assume $Z \in W \cap (X, Y)$ where $X, Y \in K$. Then $f(Z) = f(\lambda X + \mu Y) = \lambda f(X) + \mu f(Y)$, with $\lambda, \mu > 0$ and $\lambda + \mu = 1$. Since $f(X), f(Y) \leqslant f(Z)$, this relation can only hold if $f(X) = f(Y) = f(Z)$. Hence $Z \in W \cap (X, Y)$ and $X, Y \in K$ imply $X, Y \in W$. This proves W to be extreme.

There remains only to show that every face is exposed. Consider the face $F := P \cap \bigcap_{i \in I} E_i$, where $E_i := \{X \mid A_i^T X = b_i\}$ are some of the boundary planes. Choose a linear combination $A_0 = \Sigma w_i A_i$, with positive coefficients $w_i > 0$ in such a way that $A_0 \neq 0$. This is possible, except in the trivial case in which all A_i vanish. The linear function $f(X) = A_0^T X$ will have F as its domain of maximality. □

(2.4.13) **Corollary to Theorem (2.4.12).** *If a linear form possesses a maximum value on a polyhedron P, then this maximum is assumed on an entire face of P.*

Using lemma (2.3.11) and notation (2.3.14), the reader verifies readily that

(2.4.14) *if F and G are two different faces of a polyhedron P, then*

$$F^I \cap G^I = \emptyset.$$

Additional properties of extreme and exposed sets of general convex sets can be found in section 3.6.

2.5. Primitive Faces. The Finite Basis Theorem

We call a polyhedron which is *not* the convex hull of its boundary points or, in other words, of its proper faces

(2.5.1) *primitive.*

This concept is due to Klee [*16*][1]. If a polyhedron is not primitive, then by definition it is the convex hull of its proper faces. Each proper face in turn is either primitive or the convex hull of faces smaller than itself. This decomposition process will terminate since the number of faces of a polyhedron is finite (theorem (2.4.9)). Hence

(2.5.2) *every nonvoid polyhedron is the convex hull of its primitive faces.*

Every nonvoid linear manifold is a primitive polyhedron. The same is true of every

(2.5.3) *halfmanifold,*

that is, every nonvoid intersection of a linear manifold M and a half-space H which does not contain M.

(2.5.4) **Theorem** (Klee [*16*]). *A primitive polyhedron is either a nonvoid linear manifold or a halfmanifold.*

Proof. First we show that all boundary points of a primitive polyhedron—if there are any—are contained in a single nonsingular boundary plane. Assume the contrary: There are two boundary planes $E_1 := \{X \mid A_1^T X = b_1\}$, $E_2 := \{X \mid A_2^T X = b_2\}$, and two boundary points V_1, V_2 such that $V_i \in E_i$, $i = 1,2$, but $V_1 \notin E_2$ and $V_2 \notin E_1$. We want to show that under these circumstances every inner point U of P is a mean of boundary points, contradicting our assumption that P is primitive.

Suppose that U is an inner point of P, and choose two positive numbers α_1 and α_2 such that

$$\alpha_1 > \frac{b_1 - A_1^T U}{b_1 - A_1^T V_2}, \qquad \alpha_2 > \frac{b_2 - A_2^T U}{b_2 - A_2^T V_1}.$$

This choice will ensure that the points $Y_1 := U + \alpha_1(V_1 - V_2)$ and $Y_2 := U + \alpha_2(V_2 - V_1)$ do not belong to P (Fig. 4); indeed $A_i^T Y_i > b_i$ is readily verified for $i = 1,2$. The points Y_i satisfy the singular constraints since they are averages of points of P. According to lemma (2.3.12), each of the open segments (U, Y_i) contains a boundary point $\bar{Y}_i$ of P. Since $\bar{Y}_i \in (U, Y_i)$, there exists a $\theta_i > 0$ such that $\bar{Y}_i = U + \theta_i(Y_i - U)$ for $i = 1,2$. Hence $\bar{Y}_1 = U + \theta_1 \alpha_1(V_1 - V_2)$, $\bar{Y}_2 = U + \theta_2 \alpha_2(V_2 - V_1)$, and consequently

$$U = \frac{\theta_2 \alpha_2}{\theta_1 \alpha_1 + \theta_2 \alpha_2} \bar{Y}_1 + \frac{\theta_1 \alpha_1}{\theta_1 \alpha_1 + \theta_2 \alpha_2} \bar{Y}_2 \in (\bar{Y}_1, \bar{Y}_2).$$

[1] Klee uses the term "reducible" for polyhedra which are not primitive.

It remains to be shown that a polyhedron P is a manifold or half-manifold if all its boundary points—if there are any—are contained in a single nonsingular boundary plane.

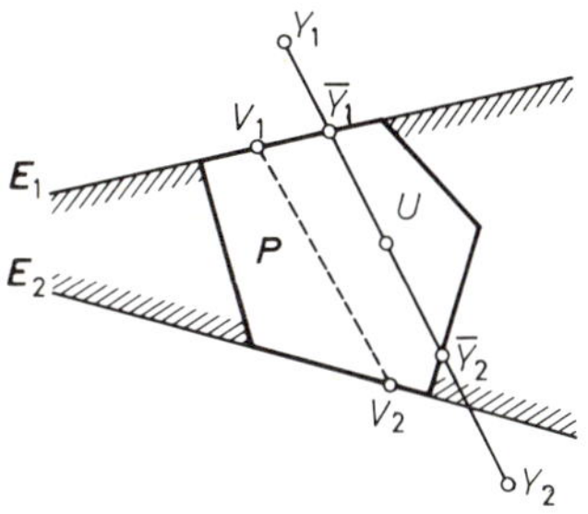

Fig. 4. Construction of Y_i and $\bar{Y}_i$

In the case where there are boundary points, let E be the nonsingular boundary plane which contains all boundary points, and H, where $H \supseteq P$, the corresponding halfspace, and M the intersection of all singular boundary planes of P ($M = R^n$ if there are none). We want to show that

$$P = H \cap M$$

which would exhibit P as a halfmanifold. $P \supseteq H \cap M$ is the nontrivial inclusion. Assume that there is an $X \in H \cap M$ with $X \notin P$. By lemma (2.3.10) there exists an inner point U of P, and by lemma (2.3.12) there exists a boundary point Y of P in the open segment (U, X). Y lies in E since, by hypothesis, E contains all boundary points of P. But $E \cap M$ is a face of $H \cap M$ and therefore extreme (theorem (2.4.7)). According to (2.4.2) we have $U \in E$. But this is not possible since U is an inner point and E a nonsingular boundary plane of P.

In the case where there are no boundary points, the assumption $P \neq M$ leads similarly to a contradiction. ▯

(2.5.5) **Corollary to Theorem (2.5.4).** *Every polyhedron is the convex hull of finitely many manifolds and halfmanifolds.*

The converse is not true: the convex hull, for instance, of a point and a line is not a polyhedron provided the point is not on the line. We proceed to derive from (2.5.5) a generation theorem which will admit the converse. We note

(2.5.6) **Lemma.** *Every linear manifold is the sum of a point and a subspace. Every halfmanifold is the sum of a halfline and a subspace.*

Proof. Trivial for manifolds. Consider a halfmanifold $H \cap M$ with $H := \{X \mid A_0^T X \leqslant b_0\} \not\supseteq M$. By definition (2.5.3), M contains points X, Y such that $A_0^T X \leqslant b_0$ and $A_0^T Y > b_0$. The point

$$V := X + \alpha(Y - X), \qquad \alpha := \frac{b_0 - A_0^T X}{A_0^T Y - A_0^T X}$$

belongs to M and satisfies $A_0^T V = b_0$. Hence $L := E \cap M - V$, where $E := \{X \mid A_0^T X = b_0\}$, is a subspace. We define a halfline $N := \{X = V - \beta(Y - V) \mid \beta \geqslant 0\}$ and assert $H \cap M = N + L$. Assume $X \in H \cap M$. Then there exists $\beta \geqslant 0$ such that $A_0^T (X + \beta(Y - V)) = b_0$. Thus $X + \beta(Y - V) \in E \cap M$, whence

$$X = (X + \beta(Y - V) - V) + (V - \beta(Y - V)) \in L + N.$$

(Note that β is uniquely determined by X. This shows that a halfmanifold is actually a direct sum of a halfline and a subspace: $H \cap M = N \oplus L$.) □

A pair of finite point sets

$$G = \{X_1, \ldots, X_k\}, \qquad H = \{Y_1, \ldots, Y_l\}$$

is called a

(2.5.7) *finite basis*

of a polyhedron P, if $P = \mathscr{H} G + \mathscr{C} H$, that is, if every point of P has a representation

$$u_1 X_1 + \cdots + u_k X_k + v_1 Y_1 + \cdots + v_l Y_l, \qquad \sum_i u_i = 1, \qquad u_i \geqslant 0, \qquad v_i \geqslant 0.$$

We have yet to use the fact that we are in a space of finite dimension. Indeed, the results of section 2.4 and theorem (2.5.4) actually hold for infinitely dimensional spaces,—polyhedra defined by finite inequality systems are however not very interesting in such spaces. Since we are in R^n, one readily obtains the existence of a finite basis for manifolds (for a characterization of such "positive" bases for manifolds see McKinney [*1*] and Reay [*2*], [*3*]). From lemma (2.5.6), one obtains the same for halfmanifolds, and (2.5.5) gives the

(2.5.8) **Finite Basis Theorem.** *Every polyhedron has a finite basis.*

The converse of theorem (2.5.8) is true: every $\mathscr{H} G + \mathscr{C} H$ with finite G, H is a polyhedron. However, this statement lies deeper and expresses the "key fact" of the theory of linear inequalities. Its proof will be presented in section 2.11.

The finite basis theorem has an important consequence for the minimization of linear forms (2.4.10) over a polyhedron,—the linear programming problem:

(2.5.9) **Theorem.** *If a linear function f is bounded above (below) on a nonvoid polyhedron P, then f has a maximum (minimum) on P.*

Proof. Let $G=\{X_1,\ldots,X_k\}$, $H=\{Y_1,\ldots,Y_l\}$ be a basis of P. Then the values which f assumes on P are precisely the numbers of the form

$$\sum_i u_i f(X_i)+\sum_j v_j f(Y_j), \quad \sum_i u_i=1, \quad u_i \geqslant 0, \quad v_j \geqslant 0.$$

This shows that $f(Y_j)\geqslant 0$ must hold if f is bounded below on P. Furthermore, if $f(X_r)$ is the smallest among the values $f(X_i)$, then $f(X_r)=\min\{f(X) \mid X\in P\}$ which was to be shown. □

See section 3.6 for extremal properties of general convex sets in R^n where R is the field of reals. If R is an incomplete field, e.g. the field of rationals, then every convex set for which lemma (2.3.12) holds is the convex hull of its extreme manifolds and halfmanifolds (Klee [25]).

2.6. Subspaces. Orthogonality

The preceding section conceived of polyhedra as convex hulls. In the subsequent sections the central question will be: under which circumstances do convex hulls turn out to be polyhedra? A fundamental result will be Weyl's theorem (2.8.8) that the conical hull of a finite point set is a polyhedral cone. A homogenization process will permit the transition from this result to the analogous one for general polyhedra.

To begin with, let us examine the simpler question as to whether a subspace of the R^n is an intersection of finitely many planes. It is well known that the answer is affirmative, and this fact plays an important role in connection with the concept of an orthogonal complement. Nevertheless it is worthwhile to present a brief treatment of subspaces and their orthogonal complements because it may serve as a guide for the treatment of cones and polarity.

For an arbitrary set $S\in R^n$ we define its

(2.6.1) *orthogonal complement* $S^{\perp}:=\{Y\in R^n \mid Y^T X=0 \quad \text{for all} \quad X\in S\}$.

$S^{\perp}$ is a subspace: $\mathscr{L}(S^{\perp})=S^{\perp}$. Other trivial relations are: $(\mathscr{L}S)^{\perp}=S^{\perp}$; $S_1\subseteq S_2 \Rightarrow S_1^{\perp}\supseteq S_2^{\perp}$; $S\subseteq S^{\perp\perp}$; and for any family of sets $S_i\subseteq R^n$:

(2.6.2) $$\left(\bigcup S_i\right)^{\perp}=\bigcap S_i^{\perp}.$$

For an arbitrary point $X\in R^n$ and a subspace L we define the

(2.6.3) *orthogonal projection*, briefly *projection*

of X into L as the point $\bar{X} \in L$ for which

$$\|X-\bar{X}\|^2 := (X-\bar{X})^T(X-\bar{X})$$

assumes its minimal value.

(2.6.4) **Lemma.** *There can be at most one projection of a point $X \in R^n$ into a subspace $L \subseteq R^n$*

Proof. Assume that $\bar{X}_1 \neq \bar{X}_2$ are two points minimizing $\|X-\bar{X}\|$. Then the relation

$$\left\|X - \frac{\bar{X}_1+\bar{X}_2}{2}\right\|^2 = \frac{1}{2}\|X-\bar{X}_1\|^2 + \frac{1}{2}\|X-\bar{X}_2\|^2 - \left\|\frac{\bar{X}_1-\bar{X}_2}{2}\right\|^2$$

would disclose $\dfrac{\bar{X}_1+\bar{X}_2}{2}$ to have an even shorter distance from X. □

(2.6.5) **Lemma.** *Let L be a subspace of R^n. If $X \in R^n$ admits an orthogonal decomposition $X = Y + Z$ with $Y \in L$ and $Z \in L^\perp$, then Y and Z are the projections of X into L and $L^\perp$, respectively.*

Proof. For any $U \in L$ and $V \in L^\perp$, we have

$$\|X-U\|^2 = \|X-Y\|^2 + \|Y-U\|^2 \geqslant \|X-Y\|^2,$$

$$\|X-V\|^2 = \|X-Z\|^2 + \|Z-V\|^2 \geqslant \|X-Z\|^2. \quad \square$$

(2.6.6) **Lemma.** *If Y is the projection of X into L, then $X-Y$ is the projection of X into $L^\perp$.*

Proof. First we show that $X - Y \in L^\perp$. Since Y is the projection of X into L, we have, for any nonzero $U \in L$,

$$\left\|X - Y - \frac{U^T(X-Y)}{U^T U} U\right\|^2 - \|X-Y\|^2 = -\left[\frac{U^T(X-Y)}{U^T U}\right]^2 \geqslant 0$$

and therefore $U^T(X-Y)=0$. Hence $X-Y$ lies in $L^\perp$ and is the projection of X into $L^\perp$ according to Lemma (2.6.5). □

The proof outlined for the following theorem may be extended to a proof of the theorem of Weyl (2.8.8) (compare section 2.9).

(2.6.7) **Theorem.** *Every subspace of the R^n is an intersection of finitely many planes.*

Now we use the fact that every subspace L of the R^n has a finite set of generators (theorem (2.2.5)):

$$L=\mathscr{L}\{A_1,\dots,A_k\}, \qquad A_i\neq 0.$$

Consider the plane $E:=\{X\mid A_1^T X=0\}$. Define a map $p: R^n\to E$ by putting

$$p(X):=X-\frac{A_1^T X}{A_1^T A_1}A_1.$$

Of course, $p(X)$ is the orthogonal projection of X into E. But we do not need this information. Abbreviating $\bar{A}_i:=p(A_i)$ we have

(2.6.8) $\qquad L=\mathscr{L}\{A_1\}\oplus\bar{L}, \quad$ *with* $\quad \bar{L}:=\mathscr{L}\{\bar{A}_2,\dots,\bar{A}_k\}.$

Proof of (2.6.8). For any $X\in L=\mathscr{L}\{A_1,\dots,A_k\}$ we have

$$X=\sum_{i=1}^{k}u_i A_i=\left(u_1+\sum_{i=2}^{k}u_i\frac{A_1^T A_i}{A_1^T A_1}\right)A_1+\sum_{i=2}^{k}u_i\left(A_i-\frac{A_1^T A_i}{A_1^T A_1}A_1\right),$$

whence there exists α such that

$$X=\alpha A_1+\sum_{i=2}^{k}u_i\bar{A}_i.$$

On the other hand, any expression $\alpha A_1+\sum_{i=2}^{k}u_i\bar{A}_i$ is a linear combination of $A_1,\dots,A_k$ and therefore an element of L. Clearly, $A_1^T\bar{A}_i=0$.

As a consequence of (2.6.8), we note that

(2.6.9) $\qquad X\in L \quad$ *if and only if* $\quad \bar{X}:=p(X)\in\bar{L}.$

Proof of (2.6.9). Each $X\in L$ has a decomposition $X=\alpha A_1+Y$ with $Y\in\bar{L}$. Obviously, $p(A_1)=0$ and $p(Y)=Y$. Hence $\bar{X}=p(X)=\alpha p(A_1)+p(Y)=Y\in\bar{L}$. On the other hand, if $\bar{X}\in\bar{L}$ then $X=(A_1^T X/A_1^T A_1)A_1+\bar{X}\in\mathscr{L}\{A_1\}+\bar{L}=L$ according to (2.6.8).

We are now able to prove that if $\bar{L}$ is an intersection of finitely many planes, then so is L. More precisely,

(2.6.10) $\quad \bar{L}=E_1\cap\cdots\cap E_l,\quad$ *where* $\quad E_j:=\{X\mid B_j^T X=0\},\quad$ *implies* $L=\bar{E}_1\cap\cdots\cap\bar{E}_l$ *where* $\bar{E}_j:=\{X\mid\bar{B}_j^T X=0\}$ *and* $\bar{B}_j:=p(B_j)$.

Proof of (2.6.10). $X\in L$ is equivalent to $\bar{X}\in\bar{L}$ by (2.6.9). By hypothesis $\bar{X}\in\bar{L}$ is equivalent to $B_j^T\bar{X}=0$ for all j. But $B_j^T\bar{X}=\bar{B}_j^T\bar{X}=\bar{B}_j^T X$. Therefore $X\in L$ if and only if $\bar{B}_j^T X=0$ for all j.

The result (2.6.10) makes it possible to prove theorem (2.6.7) by induction over the number of generators. The induction may be initiated by expressing $\{0\}$ as an intersection of coordinate planes. ☐

(2.6.11) **Theorem.** *For every subspace $L \in R^n$ one has*

(i) $L = L^{\perp\perp}$,
(ii) $R^n = L \oplus L^{\perp}$,
(iii) *every $X \in R^n$ has a projection into L.*

Proof. By theorem (2.6.7) we may write $L = \bigcap_{i=1}^{i=s} E_i$, where $E_i := \{X \mid A_i^T X = 0\}$ and $A_i \neq 0$. Let

$$p^{(i)}(X) := X - \frac{\bar{A}_i^T X}{\bar{A}_i^T \bar{A}_i} \bar{A}_i,$$

where $\bar{A}_i := p^{(i-1)} \circ \cdots \circ p^{(1)}(A_i)$. Then

$$\bar{X} = p^{(1)} \circ p^{(2)} \circ \cdots \circ p^{(s)}(X)$$ [1]

is the projection of X into L, as $A_1 = \bar{A}_1$ and by (2.6.8),

$$L = \mathscr{L}\{\bar{A}_1\} \oplus \cdots \oplus \mathscr{L}\{\bar{A}_s\}.$$

This proves (iii). (iii) implies (ii) via lemmas (2.6.4) and (2.6.6). (ii) implies (i): Every $X \in L^{\perp\perp}$ admits a decomposition $X = Y + Z$ with $Y \in L$ and $Z \in L^{\perp}$. Then $X^T Z = 0$ and $Y^T Z = 0$. It follows that $Z^T Z = Y^T Z + Z^T Z = X^T Z = 0$. Therefore $Z = 0$ and $X = Y \in L$. □

For any two linear subspaces L and K we have

(2.6.12)
(i) $(L + K)^{\perp} = (L \cup K)^{\perp} = L^{\perp} \cap K^{\perp}$,
(ii) $(L \cap K)^{\perp} = \mathscr{L}(L^{\perp} \cup K^{\perp}) = L^{\perp} + K^{\perp}$.

Proof. (i) Since $L \cup K \supseteq L$, $(L \cup K)^{\perp} \subseteq L^{\perp}$, and similarly $(L \cup K)^{\perp} \subseteq K^{\perp}$, whence $(L \cup K)^{\perp} \subseteq L^{\perp} \cap K^{\perp}$. To prove the other direction, suppose $X \in L^{\perp} \cap K^{\perp}$. Then $Y^T X = 0$ for all $Y \in L$ as well as for all $Y \in K$. Thus $(L \cup K)^{\perp} \supseteq L^{\perp} \cap K^{\perp}$. From (i) we obtain with the help of theorem (2.6.11): $(L \cap K)^{\perp} = (L^{\perp\perp} \cap K^{\perp\perp})^{\perp} = (L^{\perp} \cup K^{\perp})^{\perp\perp} = (\mathscr{L}(L^{\perp} \cup K^{\perp}))^{\perp\perp} = \mathscr{L}(L^{\perp} \cup K^{\perp})$. □

There are important relations between the dimension of a linear subspace L and its

(2.6.13) *codimension* $(= \operatorname{codim} L)$,

i.e. the minimal number of planes with intersection L. If $L = E_1 \cap \cdots \cap E_k$ is such a minimal intersection of planes with $E_i = \{X \mid A_i^T X = 0\}$ then $\{A_1, \ldots, A_k\}$ is a minimal spanning set of $L^{\perp}$ by (2.6.12.ii). Hence

(2.6.14) $$\operatorname{codim} L = \dim L^{\perp}.$$

---

[1] If the vectors A_i are linearly independent, then $\bar{X} = X - A(A^T A)^{-1} A^T X$, where A denotes the matrix $(A_1, \ldots, A_s)$.

Since $R^n = L \oplus L^\perp$ also by theorem (2.6.11), we have

(2.6.15) $$n = \dim L + \dim L^\perp = \dim L + \operatorname{codim} L.$$

2.7. Cones. Polarity

In the theory of cones, polar sets play the role which orthogonal complements play in the theory of subspaces. For an arbitrary subset $S \subseteq R^n$ we define its

(2.7.1) *polar cone* $S^p := \{Y \in R^n | Y^T X \leqslant 0 \text{ for all } X \in S\}$.

S^p is a cone: $\mathscr{C}(S^p) = S^p$. Other trivial relations are: $(\mathscr{C}(S))^p = S^p$; $S_1 \subseteq S_2 \Rightarrow S_1^p \supseteq S_2^p$; $S \subseteq S^{pp}$; and for any family of sets $S_i \subseteq R^n$:

(2.7.2) $$(\bigcup S_i)^p = \bigcap S_i^p.$$

The concept of orthogonal projections is extended to cones in a straightforward way. The

(2.7.3) *orthogonal projection*, briefly *projection*

of $X \in R^n$ into a cone $C \subseteq R^n$ is the point $\bar{X} \in C$ for which $\|X - \bar{X}\|^2 := (X - \bar{X})^T (X - \bar{X})$ assumes its minimal value. The same argument as was used for the proof of (2.6.4) shows that the projection into a cone is unique. Finally, we introduce the

(2.7.4) *orthogonal sum* $S_1 \boxplus S_2$

of two sets S_1 and S_2. A set S is the orthogonal sum of S_1 and S_2 if every element X of S admits a unique orthogonal decomposition $X = Y + Z$ with $Y \in S_1$, $Z \in S_2$, and $Y^T Z = 0$. The direct sum $L_1 \oplus L_2$ of two orthogonal subspaces is also an orthogonal sum: $L_1 \oplus L_2 = L_1 \boxplus L_2$.

In analogy to lemmas (2.6.5) and (2.6.6) we have:

(2.7.5) **Lemma.** *Let $C \subseteq R^n$ be a cone. If $X \in R^n$ admits an orthogonal decomposition $X = Y + Z$, with $Y \in C$, $Z \in C^p$ and $Y^T Z = 0$, then Y and Z are the projections of X into C and C^p. Conversely, if $X \in R^n$ has a projection into C, then it also has a projection into C^p and both projections constitute an orthogonal decomposition of X.*

Proof. Let $X = Y + Z$ be an orthogonal decomposition as specified in the lemma. For any $U \in C$ and $V \in C^p$, we then have $U^T Z \leqslant 0$ and $V^T Y \leqslant 0$. Therefore

$$\|X - U\|^2 - \|X - Y\|^2 = \|U - Y\|^2 - 2U^T Z \geqslant 0,$$
$$\|X - V\|^2 - \|X - Z\|^2 = \|V - Z\|^2 - 2V^T Y \geqslant 0.$$

In order to prove the converse, we shall show that if Y is the projection of X into C, then $X-Y \in C^p$ and $Y^T(X-Y)=0$. This means that Y and $X-Y$ form an orthogonal decomposition. But we have already proved that in this case $X-Y$ is the projection into C^p.

We first show that $X-Y \in C^p$. Consider the relation

$$\left\| X - Y - \frac{U^T(X-Y)}{\|U\|^2} U \right\|^2 - \|X-Y\|^2 = -\left[\frac{U^T(X-Y)}{\|U\|}\right]^2.$$

It implies either $U^T(X-Y)=0$, or that $V := Y + \dfrac{U^T(X-Y)}{\|U\|^2} U$ has a shorter distance from X than Y. Hence V cannot be an element of C, which it would be if $U^T(X-Y)>0$ and $U \in C$. Thus $U^T(X-Y) \leqslant 0$ for all $U \in C$, which was to be shown.

Next it is to be verified that Y and $X-Y$ are orthogonal. If $Y \neq 0$, then $\alpha=1$ is a double zero of the quadratic function

$$\|X-\alpha Y\|^2 - \|X-Y\|^2 = \alpha^2 Y^T Y - 2\alpha X^T Y + 2 Y^T X - Y^T Y.$$

Therefore the discriminant $(Y^T(X-Y))^2$ equals zero. □

The definition (2.7.1) may be written as

$$S^p = \bigcap_{X \in S} \{Y \mid X^T Y \leqslant 0\}$$

manifesting S^p as an intersection of homogeneous halfspaces[1]. By definition $Y \in S^p$ holds if and only if $\{X \mid Y^T X \leqslant 0\} \supseteq S$. Hence

$$S^{pp} = \bigcap_{Y \in S^p} \{X \mid Y^T X \leqslant 0\}$$

shows that S^{pp} is the intersection of all homogeneous halfspaces which contain S. We formulate these results as

(2.7.6) **Lemma.** *S^{pp} is the intersection of all homogeneous halfspaces which contain S. Every polar set S^p is an intersection of halfspaces. Hence*

$$S^{ppp} = S^p$$

holds for every subset $S \subseteq R^n$.

[1] A homogeneous halfspace is a halfspace of the form $\{X \mid Y^T X \leqslant b\}$ with $b=0$.

(2.7.7) **Theorem.** *Let* $C \subseteq R^n$ *be a cone. Between the statements*

(i) C *is the intersection of halfspaces,*
(ii) $C = C^{pp}$,
(iii) $R^n = C \boxplus C^p$ (Fig. 5),
(iv) *every* $X \in R^n$ *has a projection into* C,

hold the following logical relations: (i) *is equivalent to* (ii); (iii) *implies* (ii); *and* (iii) *is equivalent to* (iv).

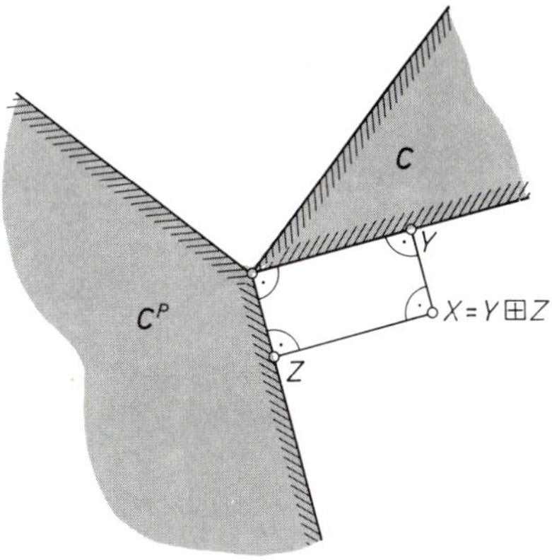

Fig. 5. $R^n = C \boxplus C^p$

Proof. The equivalence of (i) and (ii) is an immediate consequence of lemma (2.7.6). (iii) $\Rightarrow$ (ii): Since $R^n = C \boxplus C^p$, each element $X \in C^{pp}$ admits an orthogonal decomposition $X = Y + Z$ with $Y \in C$, $Z \in C^p$ and $Y^T Z = 0$. Further we have $X^T Z \leqslant 0$ since, by hypothesis, $X \in C^{pp}$ and $Z \in C^p$. Hence $Z^T Z = (Y+Z)^T Z = X^T Z \leqslant 0$. This implies $Z = 0$, and $X = Y \in C$. Thus we have established that $C^{pp} \subseteq C$. (iii) $\Leftrightarrow$ (iv): This is an immediate consequence of lemma (2.7.5) and the uniqueness of projections. □

We now focus our attention on those cones which have a finite set of generators. For these cones all of the four statements of theorem (2.7.7) are true as a consequence of our next theorem. That every finitely generated cone is intersection of *finitely* many halfspaces is the content of Weyl's theorem, which we shall prove in section 2.8 as well as in section 2.9.

(2.7.8) **Theorem.** *If* C *is the conical hull of a finite set then every point* $X \in R^n$ *has a projection into* C.

Proof. According to theorem (2.6.11) each point $X \in R^n$ has a projection $\bar{X}$ into $\mathscr{L}C$. If $\bar{X}$ has a projection $\bar{\bar{X}}$ into C, then $\bar{\bar{X}}$ is also the projection of X into C. Indeed, we have

$$\|X-U\|^2=\|X-\bar{X}\|^2+\|\bar{X}-U\|^2 \geqslant \|X-\bar{X}\|^2+\|\bar{X}-\bar{\bar{X}}\|^2=\|X-\bar{\bar{X}}\|^2$$

for all $U \in C$. Thus it suffices to prove the existence of a projection for every $X \in \mathscr{L}(C)$.

Let $S:=\{A_1, \ldots, A_k\}$, $A_i \neq 0$, be the set of generators of C. Then every $X \in \mathscr{L}C$ may be written as a linear combination of the generators A_i:

$$X=w_1 A_1+\cdots+w_k A_k.$$

If all coefficients w_i are nonnegative then $X \in C$, and there is nothing to prove. Assume therefore that $w_j<0$ holds for some j. Finally, let Y be any point of C:

$$Y=v_1 A_1+\cdots+v_k A_k, \qquad v_i \geqslant 0 \quad \text{for all} \quad i.$$

Then the segment $\mathscr{H}\{X, Y\}$ contains a point Z which belongs to one of the cones $\mathscr{C}(S \sim \{A_i\})$, $i=1, \ldots, k$. Indeed, $Z:=Y+\alpha(X-Y)$ with

$$\alpha:=\min \left\{\frac{v_i}{v_i-w_i} \,\middle|\, w_i-v_i<0\right\}$$

in such a point. Now, we are able to proceed by induction. Theorem (2.7.8) is true for $k=0$. Assume that it holds for $k-1$. Then X has a projection X_i into each of the cones $\mathscr{C}(S \sim \{A_i\})$, and we have

$$\|X-Y\| \geqslant \|X-Z\| \geqslant \min_i \|X-X_i\|$$

for every $Y \in C$. Hence that one among the X_i which is closest to X is the projection of X into C. □

As a consequence of theorems (2.7.7) and (2.7.8) we note

(2.7.9) **Lemma.** *For every finitely generated cone* C,

$$C^{pp}=C.$$

Finally we list some relations linking orthogonal complements and polar sets. Clearly, $S^p \supseteq S^{\perp}$ for every set $S \subseteq R^n$. The relation

(2.7.10) $$L^{\perp}=L^p$$

is true of every subspace $L \subseteq R^n$. More general, we have

(2.7.11) $$S^{\perp}=S^p \cap(-S)^p$$

for arbitary $S \subseteq R^n$. $S^p \supseteq S^\perp$ and (2.7.10) give $S^{pp} \subseteq S^{\perp\perp} = \mathscr{L}S$. Hence $\mathscr{L}(S^{pp}) \subseteq \mathscr{L}S$. On the other hand, $\mathscr{L}(S^{pp}) \supseteq \mathscr{L}S$ since $S^{pp} \supseteq S$. Thus

(2.7.12) $$\mathscr{L}(S^{pp}) = \mathscr{L}S.$$

2.8. Polyhedral Cones

A cone which is a polyhedron (2.3.2) is called a

(2.8.1) *polyhedral cone.*

Clearly, the solution set of a homogeneous system of inequalities

$$A^T X \leqslant 0$$

is a polyhedral cone. On the other hand,

(2.8.2) *every polyhedral cone is the solution set of a homogeneous system of inequalities.*

Proof. Assume that $C := \{X \mid A^T X \leqslant B\}$ is a cone. Since every cone contains the origin 0, we must have $B \geqslant 0$. Homogeneity implies $A^T X \leqslant \alpha B$ for every $X \in C$ and $\alpha > 0$. Hence $A^T X \leqslant 0$ for all $X \in C$. □

As a consequence of (2.8.2), every nonvoid face of a polyhedral cone is again a polyhedral cone.

By A_i, $i = 1, \ldots, m$, we denote the m columns of a matrix A. According to the trivial formula (2.7.2), we have

(2.8.3) $$\{X \mid A^T X \leqslant 0\} = (\mathscr{C}\{A_1, \ldots, A_m\})^p.$$

Applying lemma (2.7.9), we get $(\mathscr{C}\{A_1, \ldots, A_m\})^{pp} = \mathscr{C}\{A_1, \ldots, A_m\}$. Therefore we can "dualize" formula (2.8.3) and get an expression for the polar of a polyhedral cone:

(2.8.4) $$\{X \mid A^T X \leqslant 0\}^p = \mathscr{C}\{A_1, \ldots, A_m\}.$$

Formula (2.8.4) is equivalent to the fundamental

(2.8.5) **Theorem (Farkas Lemma).** *The halfspace* $H = \{X \mid A_0^T X \leqslant 0\}$ *contains the polyhedral cone* $C = \{X \mid A^T X \leqslant 0\}$ *if and only if* $A_0 \in \mathscr{C}\{A_1, \ldots, A_m\}$, *where* A_i *denotes the* i-th *column of* A (*compare* (1.4.8)).

Proof. $H = \{X \mid A_0^T X \leqslant 0\} \supseteq C$ is equivalent to $A_0 \in C^p$. But formula (2.8.4) gives $C^p = \mathscr{C}\{A_1, \ldots, A_m\}$. □

The finite-basis theorem (2.5.8) specializes to the

(2.8.6) **Theorem of Minkowski** [4]. *Every polyhedral cone has a finite set of generators.*

The theorem of Minkowski enables us to apply theorems (2.7.7) and (2.7.8) to polyhedral cones.

(2.8.7) *For every polyhedral cone* C *one has*

(i) $C = C^{pp}$,

(ii) $R^n = C \boxplus C^p$,

(iii) *Each* $X \in R^n$ *has a projection into* C.

Combined with lemma (2.7.9), the theorem of Minkowski yields the

(2.8.8) **Theorem of Weyl** [*1*]. *Every cone with finitely many generators is a polyhedral cone.*

Proof. The polar of a finitely generated cone $C = \mathscr{C}\{A_1, \ldots, A_m\}$ is a polyhedral cone. This follows from formula (2.8.3). Hence C^p, too, has a finite set $\{B_1, \ldots, B_l\}$ of generators (theorem of Minkowski), and its polar C^{pp} is again a polyhedral cone. But we have $C = C^{pp}$ according to lemma (2.7.9). □

Formula (2.8.4), the Farkas lemma, and lemma (2.7.9) express essentially the same fact. We encountered this fact in chapter 1 in the form of the Kuhn-Fourier theorem (1.1.9), the transposition theorem of Gordan (1.6.3), and the duality theorem of linear programming (1.7.13). All theorems mentioned above belong to a class of theorems which express the "key-fact" of the theory of linear inequalities. The "key-fact", together with the theorem of Minkowski, was used for deriving the theorem of Weyl. In turn, the theorem of Weyl implies $C^{pp} = C$ by virtue of theorem (2.7.7). Minkowski's theorem is also a consequence of the theorem of Weyl. The latter implies formula (2.8.4) via lemma (2.7.9). Let then $C = \{X \mid A^T X \leqslant 0\}$ be a polyhedral cone. The—trivial—formula (2.8.3) yields $C = \mathscr{C}\{A_1, \ldots, A_k\}^p$. But $\mathscr{C}\{A_1, \ldots, A_k\} = \{X \mid B^T X \leqslant 0\}$ according to the theorem of Weyl. Hence formula (2.8.4) gives

$$C = \{X \mid B^T X \leqslant 0\}^p = \mathscr{C}\{B_1, \ldots, B_l\},$$

where B_i denotes the i-th column of the matrix B.

The theorems of Weyl and Minkowski show that the polyhedral cones form a sublattice of the lattice of all cones (compare section 2.1). Polarity constitutes an involutorical anti-automorphism of this lattice. By replacing each set by its polar and interchanging intersections and conical hulls we may translate every statement into a

(2.8.9) *dual statement,*

e.g. the formula (2.8.3) into formula (2.8.4).

The theorem of Weyl is the dual of the theorem of Minkowski. From this one might conclude that it was not necessary to prove the theorem of Weyl in addition to the theorem of Minkowski. This would have been correct, if the lattice character of the set of all polyhedral cones would have been established in advance. However, in our approach, the theorem of Weyl was needed in order to establish this lattice character.

The set of inner points C^I of a polyhedral cone C is in general not a cone, since it need not contain the origin. Indeed

(2.8.10) $$0 \in C^I \quad \textit{if and only if} \quad C = \mathscr{L} C.$$

Proof. If $0 \in C^I$ and $X \in \mathscr{L} C$, then by lemma (2.3.11) there exists $\varepsilon > 0$ such that $\varepsilon X \in C$. Hence $X \in C$. ▯

Another important property of C^I is expressed by formula

(2.8.11) $$(C^I)^p = C^p.$$

Proof. Clearly $(C^I)^p \supseteq C^p$. Assume $Y \in (C^I)^p \sim C^p$. Then $Y^T X_0 > 0$ for some $X_0 \in C$. Let $U \in C^I$. Then $Y^T U \leqslant 0$. Now consider any point $X = X_0 + \theta(U - X_0)$ where

$$0 < \theta < \frac{Y^T X_0}{Y^T X_0 - Y^T U}.$$

We have $Y^T X > 0$ by definition of θ, as well as $X \in C^I$ by lemma (2.3.13). This contradicts $Y \in (C^I)^p$. ▯

2.9. A Direct Proof of the Theorem of Weyl

We summarize the results of the previous section: The theorems of Minkowski and Farkas may be derived from the theorem of Weyl. However, the theorems of Farkas and Minkowski must be combined in order to yield the theorem of Weyl. Therefore, it is desirable to have a short direct proof of the theorem of Weyl.

Let

$$S := \{A_1, \ldots, A_k\}, \qquad A_i \neq 0 \quad \text{for} \quad i = 1, \ldots, k$$

be a finite point set generating the cone $\mathscr{C} S$. Such a finitely generated cone may be described by an intersection formula. It will be on this formula that we shall base a proof of the theorem of Weyl (2.8.8), following the lines of the proof given for the analogous theorem (2.6.7).

(2.9.1) **Lemma.** *$\mathscr{C}S$ is either a halfline, or the formula*

$$\mathscr{C}S = \bigcap_i \mathscr{C}(S \cup \{-A_i\})$$

holds.

Proof. Clearly $\mathscr{C}S \subseteq \bigcap_i \mathscr{C}(S \cup \{-A_i\})$. The problem is to demonstrate the converse inclusion $\mathscr{C}S \supseteq \bigcap_i \mathscr{C}(S \cup \{-A_i\})$. Let X be any point of $\bigcap_i \mathscr{C}(S \cup \{-A_i\})$. We want to show $X \in \mathscr{C}S$. Since X is contained in every one of the k cones $\mathscr{C}(S \cup \{-A_i\})$, it can be written in the following k ways:

$$\begin{aligned} X &= \rho_{11} A_1 + \alpha_{12} A_2 + \cdots + \alpha_{1k} A_k, \\ X &= \alpha_{21} A_1 + \rho_{22} A_2 + \cdots + \alpha_{2k} A_k, \\ &\cdots \qquad\qquad \cdots \\ X &= \alpha_{k1} A_1 + \alpha_{k2} A_2 + \cdots + \rho_{kk} A_k. \end{aligned} \tag{2.9.2}$$

Here the coefficients α_{ij} are nonnegative, whereas the sign of the coefficients ρ_{ii} along the main diagonal remains undetermined.

If ρ_{11} happens to be nonnegative, then X is recognized as an element of $\mathscr{C}S$, and we have reached our goal. If $\rho_{11} < 0$, the second expression in (2.9.2) may be replaced by a mean of the first and the second such that first coefficient $\alpha_{21}^{(1)}$ of the new expression vanishes. The third and the following expression can be handled in the same manner. We wind up with the following new expressions (the first one is still the same):

$$\begin{aligned} X &= \rho_{11}^{(1)} A_1 + \alpha_{12}^{(1)} A_2 + \cdots + \alpha_{1k}^{(1)} A_k, \\ X &= \qquad\quad\; \rho_{22}^{(1)} A_2 + \cdots + \alpha_{2k}^{(1)} A_k, \\ &\cdots \qquad\qquad \cdots \\ X &= \qquad\quad\; \alpha_{k2}^{(1)} A_2 + \cdots + \rho_{kk}^{(1)} A_k. \end{aligned}$$

All but the first expression do not contain the generator A_1 anymore.

If $\rho_{22}^{(1)} \geqslant 0$ then, again, X is established as an element of $\mathscr{C}S$. If $\rho_{22}^{(1)} < 0$, we repeat the above procedure with the second column, and so on. If all $\rho_{ii}^{(i-1)}$ turn out to be negative, then the process ends up with

$$\begin{aligned} X &= \rho_{11}^{(k)} A_1, \\ X &= \qquad \rho_{22}^{(k)} A_2, \\ &\cdots \qquad\qquad \cdots \\ X &= \qquad\qquad\qquad \rho_{kk}^{(k)} A_k, \end{aligned}$$

where $\rho_{ii}^{(k)} < 0$ for all i. Hence all A_i are positive multiples of one another, and $\mathscr{C}S$ is a halfline. This concludes the proof of lemma (2.9.1). □

Now we may proceed in complete analogy to the proof of theorem (2.6.7). We introduce the map

$$p(X) := X - \frac{A_1^T X}{A_1^T A_1} A_1,$$

abbreviate $\bar{A}_i := p(A_i)$, and show

$$\mathscr{C}(S \cup \{-A_1\}) = \mathscr{L}\{A_1\} + \bar{C}, \quad \textit{where} \quad \bar{C} := \mathscr{C}\{\bar{A}_2, \ldots, \bar{A}_k\}.$$

It follows that

$$X \in \mathscr{C}(S \cup \{-A_1\}) \quad \textit{if and only if} \quad \bar{X} := p(X) \in \bar{C}.$$

This is used to prove that if $\bar{C}$ is an intersection of finitely many half-spaces, then so is $\mathscr{C}(S \cup \{-A_1\})$. More precisely,

(2.9.3) $\bar{C} = H_1 \cap \cdots \cap H_l$, *where* $H_j := \{X \mid B_j^T X \leqslant 0\}$, *implies* $\mathscr{C}(S \cup \{-A_1\}) = \bar{H}_1 \cap \cdots \cap \bar{H}_l$, *where* $\bar{H}_j := \{X \mid \bar{B}_j^T X \leqslant 0\}$ *and* $\bar{B}_j := p(B_j)$.

Again, this result enables us to complete the proof of the theorem of Weyl by induction over the number of generators: If the theorem is proved for $k-1$, then it is also valid for the cones $\mathscr{C}(S \cup \{-A_i\})$ and, by the intersection formula (2.9.1), for the cone $\mathscr{C} S$ itself.

2.10. Lineality Spaces

At the beginning of section 2.8 we observed that as a consequence of (2.8.2) the nonvoid faces of a polyhedral cone were again cones. For an arbitrary convex cone C in R^n we have correspondingly:

(2.10.1) *every nonvoid extreme subset W of C is again a convex cone.*

Proof. Suppose $Z \in W$. Then $Z \in (\alpha Z, \beta Z)$ whenever $0 \leqslant \alpha < 1$ and $1 < \beta$. Since $\alpha Z, \beta Z \in C$, we conclude $\alpha Z, \beta Z \in W$ according to (2.4.2). In other words, if $Z \in W$ then $\lambda Z \in W$ for all $\lambda \geqslant 0$. ▯

The set

$$L := C \cap (-C)$$

is clearly a linear subspace. It is called the

(2.10.2) *lineality space*

of C. The lineality space is the largest linear subspace contained in C. Indeed, if X belongs to any subspace, then so must $-X$.

(2.10.3) *The lineality space L is the smallest nonvoid extreme subset of C.*

Proof. Suppose $X, Y \in C$ and $Z \in (X, Y) \cap L$. In other words, $Z = \lambda X + \mu Y$ for some $\lambda, \mu > 0$. By definition of L, $-Z \in C$. Hence $-X = \frac{1}{\lambda}(-Z) + \frac{\mu}{\lambda} Y$ and $-Y = \frac{\lambda}{\mu} X + \frac{1}{\mu}(-Z) \in C$. We conclude that $X, Y \in C \cap (-C) = L$, and the implication (2.4.2), which characterizes extreme subsets, has thus been established.

The minimality property of L remains to be shown. Suppose $W \neq \emptyset$ is extreme, and let $Z \in L$. Then $0 = \frac{1}{2}(Z) + \frac{1}{2}(-Z)$. Since Z and $-Z$ are both in C, and since $0 \in W$ by (2.10.1), (2.4.2) gives $Z \in W$. Thus each nonvoid extreme subset W of C contains L. □

If $L = C \cap (-C) = \{0\}$, that is, if the origin is an extreme point of C, then C is called a

(2.10.4) *pointed cone.*

The following theorem will be generalized in section 2.12:

(2.10.5) **Theorem.** *Every cone C is the direct sum of its lineality space L and the pointed cone $\bar{C} := L^{\perp} \cap C$:*

$$C = L \oplus \bar{C}.$$

Proof. We have $R^n = L \oplus L^{\perp}$ (theorem (2.6.11)). Hence every $X \in R^n$ may be decomposed uniquely:

$$X = Y + Z \quad \text{with} \quad Y \in L \quad \text{and} \quad Z \in L^{\perp}.$$

If $Z \in C$, then $X \in C$ since $Y \in L \subseteq C$. But the converse holds, too: If $X \in C$, then $Z \in C$ since $-Y \in L \subseteq C$ and $Z = X - Y$. Thus

$$C = L \oplus (L^{\perp} \cap C) = L \oplus \bar{C}.$$

It remains to be shown that $\bar{C}$ is pointed. To this end we determine the lineality space $\bar{C} \cap (-\bar{C})$ of $\bar{C}$:

$$\bar{C} \cap (-\bar{C}) = L^{\perp} \cap C \cap (-L^{\perp}) \cap (-C) = L^{\perp} \cap C \cap (-C) = L^{\perp} \cap L = \{0\}. \ \square$$

A cone C is called

(2.10.6) *blunt*

if $\mathscr{L} C = R^n$. By (2.7.11) we have $(\mathscr{L} C)^{\perp} = C^{\perp} = C^p \cap (-C)^p = \tilde{L}$ where $\tilde{L}$ denotes the lineality space of C^p. This yields at once:

(2.10.7) **Theorem.** *The polar of a pointed cone C is blunt and vice versa, provided $C = C^{pp}$.*

Considering polyhedral cones only, theorem (2.10.7) is a geometrical formulation of the *transposition theorem of Gordan* (compare (1.6.3)):

The system

$$AX=0, \quad X \geqslant 0$$

has a nontrivial solution if (and trivially only if) there is no U such that

$$A^T U < 0.$$

Indeed, we have the following two lemmas:

(2.10.8) **Lemma.** *The cone $C := \{Y \mid A^T Y \leqslant 0\}$ is blunt if and only if a point U exists with $A^T U < 0$.*

Proof. A point U with $A^T U < 0$ exists if and only if C has no singular boundary planes. Theorem (2.4.8) then yields the lemma. □

(2.10.9) **Lemma.** *The cone $C := \mathscr{C}\{A_1, \ldots, A_k\}$, where $A_i \neq 0$, for $i=1, \ldots, k$, is pointed if and only if there is no $X \neq 0$ solving the system*

(2.10.10) $$AX=0, \quad X \geqslant 0,$$

in which $A=(A_1, \ldots, A_k)$ denotes the matrix with the columns A_i.

Proof. Let X be a solution of (2.10.10). If a component x_i of X does not vanish, then $-A_i \in \mathscr{C}\{A_1, \ldots, A_k\}$ and $0 \neq A_i \in C \cap (-C)$. Conversely, if the lineality space $C \cap (-C)$ contains a nonzero point B then there are vectors $Y, Z \geqslant 0$ such that $B = AY$ and $-B = AZ$. Consequently $X := Y + Z$ is a nonzero solution of (2.10.10). □

The lineality space of a polyhedral cone $C = \{X \mid A^T X \leqslant 0\}$ is $L = \{X \mid A^T X = 0\}$. For a cone C given by generators the lineality space L is spanned by those generators of C which happen to lie in L. We shall prove a more general result. We partition a finite family of generators $S := \{A_1, \ldots, A_k\}$ into its

(2.10.11) *lineal* and *conical parts,*

depending on whether or not A_i belongs to the lineality space of $\mathscr{C} S$. A subset $T \subseteq S$ is said to

(2.10.12) *determine*

a face F of $\mathscr{C} S$ if $T = S \cap F$. The lineal part of S thus determines the lineality space of $\mathscr{C} S$. Suppose $X = \lambda_1 A_1 + \cdots + \lambda_k A_k \in F$, where F is a face of $\mathscr{C} S$. Since F is an extreme subset, $\lambda_i > 0$ implies $A_i \in F$. This shows that

(2.10.13) *if $T \subseteq S$ determines a face F of $\mathscr{C} S$, then $F = \mathscr{C} T$.*

(2.10.14) **Theorem.** *$T \subseteq S$ determines a face F of $\mathscr{C} S$ if and only if $S \sim T$ is the conical part of $S \cup -T$.*

Proof. We first show that $F := \mathscr{C}T$ is a face of $\mathscr{C}S$ if $S \sim T$ is the conical part of $S \cup -T$. Let $X \in \mathscr{C}T$, $X = Y + Z$, $Y, Z \in \mathscr{C}S$. Then there exist nonnegative coefficients μ_i, ν_i, λ_j so that

$$Y = \sum_{A_i \in S} \mu_i A_i, \qquad Z = \sum_{A_i \in S} \nu_i A_i, \qquad X = \sum_{A_j \in T} \lambda_j A_j = \sum_{A_i \in S} (\mu_i + \nu_i) A_i.$$

Thus

$$0 = \sum_{A_i \in S} (\mu_i + \nu_i) A_i + \sum_{A_j \in T} \lambda_j (-A_j).$$

Now $\mu_i + \nu_i = 0$ must hold if $A_i \in S \sim T$, since otherwise $-A_i \in \mathscr{C}(S \cup -T)$ could be deduced from the above relation, contradicting the hypothesis that $S \sim T$ forms the conical part of $S \cup -T$. As $\mu_i, \nu_i \geqslant 0$, $\mu_i = \nu_i = 0$. It follows that $Y, Z \in \mathscr{C}T$, which shows that $F = \mathscr{C}T$ is a face.

If $A_i \in \mathscr{C}T$, then $-A_i \in \mathscr{C}(S \cup -T)$ hence A_i cannot belong to $S \sim T$, since the latter constitutes the conical part of $S \cup -T$. Hence $A_i \in \mathscr{C}T$ implies $A_i \in T$, and T indeed determines the face $\mathscr{C}T$.

To prove the "only if" part, suppose $T = F \cap S$ for some face F of $\mathscr{C}S$, and assume $-A_i \in \mathscr{C}(S \cup -T)$. Then there exist $Y \in \mathscr{C}S$ and $X \in \mathscr{C}T$ such that $-A_i = Y - X$. Since $X \in F$, this implies $Y, A_i \in F$. In particular $A_i \in T$. Thus $S \sim T$ is the conical part of $S \cup -T$. ◻

For related characterizations of faces and lineality spaces see Wets and Witzgall [*1*], [*2*].

2.11. Homogenization

In order to extend our results on polyhedral cones to general polyhedra, we associate with an arbitrary set $S \subseteq R^n$ the cone

(2.11.1) $$\mathscr{G}S := \left\{ \begin{pmatrix} X \\ 1 \end{pmatrix} \in R^{n+1} \,\middle|\, X \in S \right\}^{pp} \qquad \text{(Fig. 6)}.$$

In the other direction, we define the map

(2.11.2) $$\mathscr{D}\tilde{S} := \left\{ X \in R^n \,\middle|\, \begin{pmatrix} X \\ 1 \end{pmatrix} \in \tilde{S} \right\}, \qquad \tilde{S} \subseteq R^{n+1}.$$

The map $\mathscr{D}$ is intersection preserving:

(2.11.3) $$\mathscr{D}(\tilde{S}_1 \cap \tilde{S}_2) = \mathscr{D}\tilde{S}_1 \cap \mathscr{D}\tilde{S}_2.$$

This simple fact allows an extension of Weyl's theorem (2.8.8) to general polyhedra (see Goldman [*1*]). The theorem gained will be the important converse of the finite basis theorem (2.5.8).

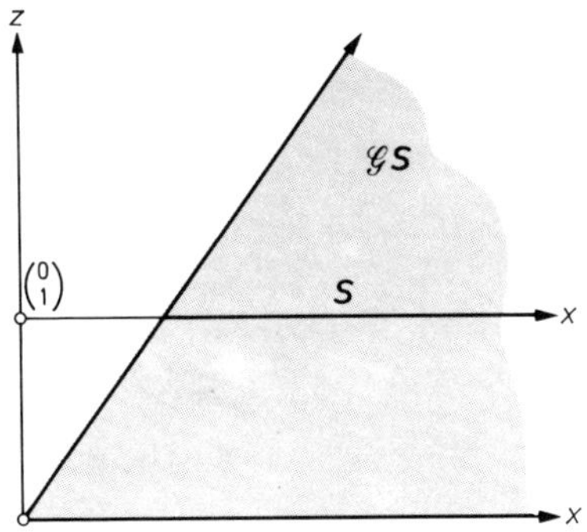

Fig. 6. Homogenization $\mathscr{G}S$ of a halfline S

(2.11.4) **Theorem.** *If* $G=\{Y_1,\dots,Y_k\}$ *and* $H=\{Z_1,\dots,Z_l\}$ *are finite point sets, then*

$$P:=\mathscr{H}G+\mathscr{C}H$$

is a polyhedron.

Proof. Consider the cone

$$\tilde{C}:=\mathscr{C}\left\{\begin{pmatrix}Y_1\\1\end{pmatrix},\dots,\begin{pmatrix}Y_k\\1\end{pmatrix},\begin{pmatrix}Z_1\\0\end{pmatrix},\dots,\begin{pmatrix}Z_l\\0\end{pmatrix}\right\}.$$

We have

$$\mathscr{D}\tilde{C}=P. \tag{2.11.5}$$

Indeed, $X\in\mathscr{D}\tilde{C}$ holds if and only if $\begin{pmatrix}X\\1\end{pmatrix}\in\tilde{C}$, that is, if $\begin{pmatrix}X\\1\end{pmatrix}=\Sigma\lambda_i\begin{pmatrix}Y_i\\1\end{pmatrix}+\Sigma\mu_j\begin{pmatrix}Z_j\\0\end{pmatrix}$ holds for suitable nonnegative λ_i,μ_j. This is only possible if, in addition, $\Sigma\lambda_i=1$ holds. But this is the necessary and sufficient condition for $X=\Sigma\lambda_i Y_i+\Sigma\mu_j Z_j$ to lie in P. This proves (2.11.5).

According to the theorem of Weyl (2.8.8), the cone $\tilde{C}$ is an intersection of finitely many homogeneous halfspaces $\tilde{H}_i\subseteq R^{n+1}$:

$$\tilde{C}=\bigcap_i\tilde{H}_i.$$

By (2.11.3), we have $P=\mathscr{D}\tilde{C}=\cap\mathscr{D}\tilde{H}_i$. But $\mathscr{D}\tilde{H}_i$ is a—possibly degenerated—halfspace $H_i\subseteq R^n$. This proves the theorem. □

For latter applications, particularly the derivation of separation theorems, we need some additional properties of the homogenization operators $\mathscr{G}$ and $\mathscr{D}$. Let then $\tilde{C}\subseteq R^{n+1}$ be any cone, and put

$$\tilde{H}_z:=\left\{\begin{pmatrix}X\\z\end{pmatrix}\middle|\,z\geqslant 0\right\}.$$

Then $\tilde{H}_z^I = \left\{ \begin{pmatrix} X \\ z \end{pmatrix} \middle| z > 0 \right\}$ and $\mathscr{C}\left\{ \begin{pmatrix} X \\ 1 \end{pmatrix} \middle| X \in \mathscr{D}\tilde{C} \right\} = \mathscr{C}(\tilde{C} \cap \tilde{H}_z^I)$. Definition (2.11.1) then gives

$$(2.11.6) \qquad \mathscr{G}\mathscr{D}\tilde{C} = (\tilde{C} \cap \tilde{H}_z^I)^{pp}.$$

Let $G = \{Y_1, \ldots, Y_k\}$ and $H = \{Z_1, \ldots, Z_l\}$ be the basis (2.5.7) of a polyhedron P, and consider, as in the proof of theorem (2.11.4), the cone $\tilde{C} := \mathscr{C}\left\{ \begin{pmatrix} Y_1 \\ 1 \end{pmatrix}, \ldots, \begin{pmatrix} Z_l \\ 0 \end{pmatrix} \right\}$. Since $\tilde{C}$ is a polyhedral cone and by (2.8.11), we have $\tilde{C} = (\tilde{C}^I)^{pp} \subseteq (\tilde{C} \cap H_z^I)^{pp} \subseteq \tilde{C}^{pp} = \tilde{C}$. From (2.11.5) and (2.11.6) then follows

(2.11.7) *If* $P = \mathscr{H}\{Y_1, \ldots, Y_k\} + \mathscr{C}\{Z_1, \ldots, Z_l\}$ *then*

$$\mathscr{G}P = \mathscr{C}\left\{ \begin{pmatrix} Y_1 \\ 1 \end{pmatrix}, \ldots, \begin{pmatrix} Y_k \\ 1 \end{pmatrix}, \begin{pmatrix} Z_1 \\ 0 \end{pmatrix}, \ldots, \begin{pmatrix} Z_l \\ 0 \end{pmatrix} \right\}.$$

If a nonvoid polyhedron is represented by an inequality system, then $\mathscr{G}P$ may be represented in terms of this system. More precisely,

(2.11.8) *if* $P = \{X \mid A^T X \leqslant B\} \neq \emptyset$ *then*

$$\mathscr{G}P = \left\{ \begin{pmatrix} X \\ z \end{pmatrix} \middle| (A^T, -B)\begin{pmatrix} X \\ z \end{pmatrix} \leqslant 0,\ z \geqslant 0 \right\}.$$

Proof. For abbreviation, we put $\tilde{C} := \left\{ \begin{pmatrix} X \\ z \end{pmatrix} \middle| (A^T, -B)\begin{pmatrix} X \\ z \end{pmatrix} \leqslant 0,\ z \geqslant 0 \right\}$. Since the halfspace $H_z = \left\{ \begin{pmatrix} X \\ z \end{pmatrix} \middle| z \geqslant 0 \right\}$ is one of the halfspaces which make up the representation of $\tilde{C}$ by inequalities, we have $\tilde{C}^I \subseteq \tilde{C} \cap \tilde{H}_z^I \subseteq \tilde{C}$ by the definition (2.3.14) of $\tilde{C}^I$, and consequently $(\tilde{C}^I)^{pp} \subseteq (\tilde{C} \cap \tilde{H}_z^I)^{pp} \subseteq \tilde{C}^{pp}$. Since $\tilde{C}$ is a polyhedral cone, we have by (2.8.7) and (2.8.11): $\tilde{C} = \tilde{C}^{pp} = (\tilde{C}^I)^{pp}$. Thus $\tilde{C} = (\tilde{C} \cap H_z^I)^{pp} = \mathscr{G}\mathscr{D}\tilde{C} = \mathscr{G}P$ by (2.11.6) and $\mathscr{D}\tilde{C} = P$. ▯

Note that (2.11.8) (and also (2.11.7)) implies

$$(2.11.9) \qquad \mathscr{D}\mathscr{G}P = P \quad \textit{for every polyhedron} \quad P \subseteq R^n.$$

There are some irregularities in the behaviour of the operator $\mathscr{G}$. It is not a straightforward inverse of the operator $\mathscr{D}$, and neither is it intersection preserving as is the latter. These irregularities are caused by the occurence of points $\begin{pmatrix} Y \\ 0 \end{pmatrix} \in \mathscr{G}P$ with $Y \neq 0$. Such points are of geometrical significance:

(2.11.10) *If* $\begin{pmatrix} Y \\ 0 \end{pmatrix} \in \mathscr{G}P$ *with* $Y \neq 0$ *then* Y *is a direction of infinity of* P, i.e., $X + \theta Y \in P$ *holds for all* $X \in P$ *and* $\theta \geqslant 0$.

Proof. Suppose $X \in P$ and $\theta \geqslant 0$. Then $\begin{pmatrix} X \\ 1 \end{pmatrix} + \theta \begin{pmatrix} Y \\ 0 \end{pmatrix} = \begin{pmatrix} X + \theta Y \\ 1 \end{pmatrix} \in \mathscr{G}P$. By (2.11.9), $X + \theta Y \in P$. ▯

We derive now a result which may be regarded as the "dual" of (2.11.6). Let S be any subset of the R^n. Then

(2.11.11) $\mathscr{D}\mathscr{G}S$ *is the intersection of all halfspaces which contain* S.

Proof. For each halfspace $H_i \supseteq S$, we have $\mathscr{D}\mathscr{G}S \subseteq \mathscr{D}\mathscr{G}H_i = H_i$ by (2.11.9). Hence $\mathscr{D}\mathscr{G}S \subseteq \cap H_i$. But $\mathscr{G}S$ is defined as a polar set, and therefore it is an intersection of halfspaces (2.7.6). Since $\mathscr{D}$ is intersection preserving, $\mathscr{D}\mathscr{G}S$ is an intersection of halfspaces containing S. This implies $\mathscr{D}\mathscr{G}S \supseteq \cap H_i$. ▯

2.12. Decomposition and Separation of Polyhedra

We now have the tools for analyzing the structure of general polyhedra. We shall distinguish "bounded" and "pointed" polyhedra. A polyhedron is called

(2.12.1) *bounded*, or a *polytope*

if an $\alpha \in R$ exists with $|x_i| \leqslant \alpha$ for every component of each point $X \in P$. The faces of a bounded polyhedron are bounded, hence the primitive faces (2.5.1) of a bounded polyhedron are vertices, and (2.5.2) yields the following special case of the well known theorem of Krein and Milman [*1*]:

(2.12.2) **Theorem.** *Every bounded polyhedron is the convex hull of its finitely many vertices.*

We call a polyhedron

(2.12.3) *pointed*

if it has vertices. Every nonvacuous bounded polyhedron is pointed. Again we employ (2.5.2) to prove

(2.12.4) **Theorem.** (i) *The polyhedron* $P = \{X \mid A^T X \leqslant B\}$ *is bounded if and only if it contains no halfline. If* $P \neq \emptyset$ *the latter statement holds if and only if the associated homogeneous system of inequalities*

$$A^T Z \leqslant 0$$

admits no nonzero solution. (ii) *The polyhedron P is pointed if and only if P contains no line, i.e., if the associated homogeneous system of equations*

$$A^T W = 0$$

admits only the trivial solution $W = 0$.

Proof. (i) If all primitive faces of P are bounded, then all primitive faces are vertices, and P is the convex hull of finitely many points. Such a set is clearly bounded. Hence a nonbounded polyhedron has at least one nonbounded primitive face, which, of course, contains a half line (see lemma (2.5.6)). The remaining statements of (i) are trivial.

(ii) If P contains a line $\mathcal{M}\{X_1, X_2\}$ then $W := X_1 - X_2$ is a nontrivial solution of $A^T W = 0$. Every face of P is characterized by an inequality system

$$A^T X \leqslant B,$$
$$-(A_I)^T X \leqslant -({}_I B),$$

where the index vector I characterizes corresponding submatrices of A and B. This system leads to the same homogeneous linear system $A^T W = 0$ as does the original system $A^T X \leqslant B$. Hence P contains a line if and only if every face of P contains a line. If P is pointed, then it has vertices, that is, faces which do not contain a line. Hence P does not contain a line, either. Conversely, if P contains no line, then the same is true for its primitive faces. But the only primitive polyhedra containing no lines are points and halflines. Hence P has vertices. □

Let $G = \{Y_1, \ldots, Y_k\}$, $H = \{Z_1, \ldots, Z_l\}$ be a finite basis (2.5.7) of a polyhedron $P = \mathcal{H} G + \mathcal{C} H$. According to the converse (2.11.4) of the finite basis theorem, the set $\mathcal{H} G$ is a polyhedron and $\mathcal{C} H$ a polyhedral cone. Note that $\mathcal{C} H$ is uniquely determined by P. Indeed

(2.12.5) *if* $P = \{X \mid A^T X \leqslant B\}$ *then* $\mathcal{C} H = \{Z \mid A^T Z \leqslant 0\}$.

Proof. We use the homogenization technique described in the preceding section. By (2.11.7) and (2.11.8), we have

$$\mathcal{G} P = \mathcal{C}\left\{\begin{pmatrix} Y_1 \\ 1 \end{pmatrix}, \ldots, \begin{pmatrix} Y_k \\ 1 \end{pmatrix}, \begin{pmatrix} Z_1 \\ 0 \end{pmatrix}, \ldots, \begin{pmatrix} Z_l \\ 0 \end{pmatrix}\right\}$$
$$= \left\{\begin{pmatrix} X \\ z \end{pmatrix} \,\middle|\, (A^T, -B)\begin{pmatrix} X \\ z \end{pmatrix} \leqslant 0,\ z \geqslant 0\right\}.$$

Consider those points of $\mathcal{G} P$ for which the last coordinate vanishes. Clearly

$$\mathcal{C} H = \left\{X \,\middle|\, \begin{pmatrix} X \\ 0 \end{pmatrix} \in \mathcal{G} P\right\} = \{X \mid A^T X \leqslant 0\}. \quad \square$$

By theorem (2.10.5), $\mathscr{C}H$ can be written as a direct sum of a pointed cone C and the lineality L of $\mathscr{C}H$. Hence every polyhedron P can be written in the form $P=Q+(C\oplus L)$, where Q is a bounded polyhedron, C a pointed cone, and L a subspace. However, a triple Q, C, L satisfying the above decomposition formula need not satisfy the formula $P=(Q+C)\oplus L$. We shall prove that also a decomposition of this stronger type exists (Motzkin [*2*]).

(2.12.6) **Decomposition Theorem.** *Every polyhedron P can be decomposed in the following manner*

$$P=(Q+C)\oplus L,$$

where Q is a bounded polyhedron, C a pointed cone, and L a linear subspace. If $P=\{X \mid A^T X\leqslant B\}$, then $L=\{W \mid A^T W=0\}$ and $C\oplus L=\{Z \mid A^T Z\leqslant 0\}$. $Q+C$ is a pointed polyhedron.

Proof. Again we use homogenization techniques. Let $P=\mathscr{H}G+\mathscr{C}H$. Then the polyhedral cone $\mathscr{C}H$ has essentially the same lineality space as the cone $\mathscr{G}P$. More precisely:

(2.12.7) *If L is the lineality space of $\mathscr{C}H$, and $\tilde{L}$ the lineality space of $\mathscr{G}P$, then* $\tilde{L}=\left\{\begin{pmatrix}X\\0\end{pmatrix} \,\middle|\, X\in L\right\}$.

Indeed, if $P=\{X \mid A^T X\leqslant B\}$, then $\mathscr{C}H=\{Z \mid A^T Z\leqslant 0\}$ by (2.12.5), and $\mathscr{G}P=\left\{\begin{pmatrix}X\\z\end{pmatrix} \,\middle|\, (A^T, -B)\begin{pmatrix}X\\z\end{pmatrix}\leqslant 0,\ z\geqslant 0\right\}$ by (2.11.8). From this $\tilde{L}=\mathscr{G}P\cap(-\mathscr{G}P)=\left\{\begin{pmatrix}W\\0\end{pmatrix} \,\middle|\, A^T W=0\right\}$.

By theorem (2.10.5), there exists a pointed polyhedral cone $\tilde{C}$ such that $\mathscr{G}P=\tilde{C}\oplus\tilde{L}$. Now the following statements are equivalent

$$\begin{pmatrix}X\\1\end{pmatrix}\in\tilde{C}\oplus\tilde{L} \iff \left\{\begin{array}{l}\begin{pmatrix}X\\1\end{pmatrix}=\begin{pmatrix}Y\\1\end{pmatrix}+\begin{pmatrix}W\\0\end{pmatrix}\\ \begin{pmatrix}Y\\1\end{pmatrix}\in\tilde{C},\quad \begin{pmatrix}W\\0\end{pmatrix}\in\tilde{L}\end{array}\right\} \iff \left\{\begin{array}{l}X=Y+W,\\ Y\in\mathscr{D}\tilde{C},\\ W\in L.\end{array}\right.$$

As a consequence $P=\mathscr{D}\mathscr{G}P=\mathscr{D}\tilde{C}\oplus L$. The cone $\tilde{C}$ has a finite basis: $\tilde{C}=\mathscr{C}\left\{\begin{pmatrix}Y_1\\1\end{pmatrix},\dots,\begin{pmatrix}Z_l\\0\end{pmatrix}\right\}$. Thus

$$\mathscr{D}\tilde{C}=\mathscr{H}\{Y_1,\dots,Y_k\}+\mathscr{C}\{Z_1,\dots,Z_l\}=Q+C.$$

Since $\tilde{C}$ is pointed its lineality space reduces to the origin, and so does the lineality space of C according to (2.12.7). Hence C is pointed. By theorem (2.12.4) it follows that $\mathscr{D}\tilde{C}=Q+C$ is a pointed polyhedron.

Obviously, $C \oplus L = \mathscr{C} H$. Therefore $C \oplus L = \{Z \mid A^T Z \leqslant 0\}$ by (2.12.5), and consequently $L = \{W \mid A^T W = 0\}$. This completes the proof of the theorem. ▯

One can use the converse (2.11.4) of the finite basis theorem to prove a separation theorem for polyhedra. This separation theorem —again a formulation of the "key fact"—was derived in chapter 1 from the theorem of Kuhn und Fourier (see (1.1.10)). A plane $E = \{X \mid A^T X = b\}$

(2.12.8) *separates*

two convex sets K_1 and K_2 if K_1 and K_2 are, respectively, included in the corresponding halfspaces

$$K_1 \subseteq H^+ = \{X \mid A^T X \leqslant b\} \quad \text{and} \quad K_2 \subseteq H^- = \{X \mid A^T X \geqslant b\}.$$

(2.12.9) **Separation Theorem for Polyhedra.** *Two nonvoid polyhedra P_1 and P_2 with $P_1 \cap P_2 = \emptyset$ can be separated by a plane E such that*

$$E \cap P_1 = E \cap P_2 = \emptyset.$$

Proof. The standard trick for dealing with the separation of two convex sets K_1 and K_2 consists of reducing this problem to the problem of separating the origin 0 from the convex set $K_1 - K_2$. So we are led to examine $P_1 - P_2$.

By virtue of the finite basis theorem (2.5.8) and its converse (2.11.4), the set $P_1 - P_2$ is again a polyhedron. Indeed if

$$P_1 = \mathscr{H} G_1 + \mathscr{C} H_1, \qquad P_2 = \mathscr{H} G_2 + \mathscr{C} H_2,$$

then

$$P_1 - P_2 = (\mathscr{H} G_1 - \mathscr{H} G_2) + (\mathscr{C} H_1 - \mathscr{C} H_2).$$

Plainly, $\mathscr{C} H_1 - \mathscr{C} H_2 = \mathscr{C}(H_1 \cup (-H_2))$. Now let $X = X^{(1)} - X^{(2)}$, where

$$X^{(1)} = \sum_{i=1}^{k} \lambda_i X_i^{(1)}, \qquad \sum_{i=1}^{k} \lambda_i = 1, \qquad \lambda_i \geqslant 0, \qquad X_i^{(1)} \in G_1,$$

and

$$X^{(2)} = \sum_{j=1}^{l} \mu_j X_j^{(2)}, \qquad \sum_{j=1}^{l} \mu_j = 1, \qquad \mu_j \geqslant 0, \qquad X_j^{(2)} \in G_2.$$

Then

$$X = \left(\sum_{j=1}^{l} \mu_j\right) X^{(1)} - \left(\sum_{i=1}^{k} \lambda_i\right) X^{(2)} = \sum_{i=1}^{k} \sum_{j=1}^{l} \lambda_i \mu_j (X_i^{(1)} - X_j^{(2)}),$$

and one concludes

$$\mathscr{H}G_1 - \mathscr{H}G_2 = \mathscr{H}(G_1 - G_2)$$

and a finite basis has been established for $P_1 - P_2$. Since $P_1 - P_2$ does not contain the origin 0, there is a boundary plane $E_0 = \{X \mid A^T X = b_0\}$ which separates $P_1 - P_2$ and 0 without containing the latter. For every $X_1 \in P_1$ and $X_2 \in P_2$ we then have $A^T(X_1 - X_2) \leqslant b_0 < 0$ therefore

$$A^T X_1 < A^T X_2. \tag{2.12.10}$$

Since P_2 is nonvoid, the linear function $A^T X$ is bounded above on P_1, and theorem (2.5.9) ensures the existence of the maximum. Analogously it is seen that $\{A^T X \mid X \in P_2\}$ has a minimum. By (2.12.10), we then have

$$\max\{A^T X \mid X \in P_1\} < \min\{A^T X \mid X \in P_2\}.$$

Hence there exists a number b in between. The plane $E := \{X \mid A^T X = b\}$ then separates P_1 and P_2 in the sense of the theorem. □

2.13. Face Lattices of Polyhedral Cones

For a polyhedral cone C the set

$$\mathscr{F}(C) := \{F \neq \emptyset \mid F \text{ face of } C\}$$

is a finite complete lattice (see Birkhoff [*1*], Hermes [*1*]) under set inclusion, since for every subset $\mathscr{S} \subseteq \mathscr{F}(C)$ there exists a "greatest lower bound"

$$\operatorname{glb}\mathscr{S} = \bigcap_{F \in \mathscr{S}} F$$

and a "least upper bound"

$$\operatorname{lub}\mathscr{S} = \bigcap_{H \in \mathscr{T}} H, \qquad \mathscr{T} := \{H \in \mathscr{F}(C) \mid H \supseteq \bigcup_{F \in \mathscr{S}} F\}.$$

This follows immediately from the fact that the intersection of arbitrary many faces is again a face. For the greatest lower bound and the least upper bound of two faces F and G, we use the familiar notation

$$\begin{aligned} F \wedge G &:= \operatorname{glb}\{F, G\} = F \cap G, \\ F \vee G &:= \operatorname{lub}\{F, G\}. \end{aligned} \tag{2.13.1}$$

There is a close relationship between the face lattice $\mathscr{F}(C)$ and the face lattice $\mathscr{F}(C^p)$ of the polar cone. Consider the map $\mathscr{P} := \mathscr{F}(C) \to \mathscr{F}(C^p)$ which is defined by

$$\mathscr{P}F := F^{\perp} \cap C^p.$$

Obviously,

$$F \subseteq G \Rightarrow \mathscr{P}F \supseteq \mathscr{P}G.$$

In order to verify that $\mathscr{P}F$ is a face, we remember that according to Minkowski's theorem (2.8.6) the polyhedral cone F is spanned by finitely many generators X_i. Then $F = \mathscr{C}\{X_1, \ldots, X_k\}$ implies $F^\perp = \bigcap_i E_i$ where $E_i = \{Y \mid X_i^T Y = 0\}$, and the halfspaces $H_i := \{Y \mid X_i^T Y \leqslant 0\}$ contain C^p. In other words, we may add these halfspaces to those constraints which already characterize the polyhedral cone C^p. With respect to this enlarged representation of C^p by linear inequalities, the planes E_i are boundary planes. Therefore $\mathscr{P}F = \bigcap E_i \cap C^p$ which proves that $\mathscr{P}F$ is a face of C^p.

By (2.7.11), we have $F^\perp = F^p \cap (-F^p)$, and therefore $\mathscr{P}F = (-F^p) \cap C^p$. Then $\mathscr{P}\mathscr{P}F = (-((-F^p) \cap C^p)^p) \cap C = (-(-F + C)) \cap C = (F - C) \cap C$. Obviously, $F \subseteq (F - C) \cap C$. But the converse inclusion holds, too. Indeed, if $X \in (F - C) \cap C$ then $X = Y - Z$ with $X, Z \in C$ and $Y \in F$. Since F is extreme, one has $X, Z \in F$ by (2.4.2). This proves

$$\mathscr{P}\mathscr{P}F = F,$$

and we have established the

(2.13.2) **Theorem.** *The map $\mathscr{P}$ which is defined by $\mathscr{P}F := F^\perp \cap C^p$ is an involution as well as an anti-isomorphism of the lattice of faces of a polyhedral cone C onto the lattice of faces of C^p.*

The formula $\mathscr{P}F = F^\perp \cap C^p$ gives rise to $(\mathscr{P}F)^p = \mathscr{L}F + C$. Hence by (2.7.11)

$$(\mathscr{P}F)^\perp = (\mathscr{P}F)^p \cap (-\mathscr{P}F)^p = (\mathscr{L}F + C) \cap (\mathscr{L}F - C) \supseteq \mathscr{L}F.$$

Suppose $X \in (\mathscr{L}F + C) \cap (\mathscr{L}F - C)$. Then there exist points $Y, Z \in \mathscr{L}F$ and $U, V \in C$ with $X = Y + U = Z - V$. Consequently, $Z - Y = U + V \in \mathscr{L}F \cap C$. Since F is an extreme set of C, we have by (2.4.4) that $F = \mathscr{L}F \cap C$. Thus $U + V \in F$. Again, since F is extreme, both U and V must lie in F, and we obtain $X \in \mathscr{L}F$. Thus $(\mathscr{L}F + C) \cap (\mathscr{L}F - C) = \mathscr{L}F$, and we have shown that

(2.13.3) $$\mathscr{L}\mathscr{P}F = (\mathscr{L}F)^\perp.$$

A face F of the cone C

(2.13.4) *covers*

the face G of C if $F \supset G$, but there exists no other face $H \in \mathscr{F}$ such that $F \supset H \supset G$. We also say in this case that G is a

(2.13.5) *facet*

of F. It is then a tautology that each proper face, and consequently each boundary point of F, is contained in at least one facet of F.

Now consider facets of the cone C itself. They correspond via the polarity operator $\mathscr{P}$ to those faces H which cover the lineality space of the polar cone C^p. The lineality space of C^p is the only nonempty face of each such face H. The faces H are therefore primitive faces (2.5.1), and it follows from (2.5.2) that

(2.13.6) *each face F of C^p is a conical hull of faces H which cover the lineality space of C^p.*

By theorem (2.13.2), this statement may be dualized:

(2.13.7) *each face F of C is an intersection of facets G of C.*

The following theorem links the dimension of faces to lattice properties

(2.13.8) **Theorem.** *Let F and G be faces of a polyhedral cone C. If F covers G, then*

$$\dim G = \dim F - 1 .$$

Proof. Without restriction of generality, we assume $F = C$. Then any facet G of C corresponds to a face H of C^p which covers the lineality space:

$$\mathscr{P} G = H .$$

Now H was seen to be primitive, and by theorem (2.5.4), it is therefore either a subspace or a halfsubspace. H cannot be a subspace, because then it would be contained in the lineality space of C^p instead of covering it. Thus H is a halfsubspace whose only face is the lineality space $C^p \cap (-C^p) = C^\perp = (\mathscr{L} C)^\perp$. Clearly, $\dim H = \dim(\mathscr{L} C)^\perp + 1$, and by (2.6.14),

$$\dim H = n - \dim C + 1 .$$

Since $\mathscr{L} H = (\mathscr{L} G)^\perp$ by (2.13.3), we have similarly

$$\dim H = n - \dim G ,$$

which proves $\dim G = \dim C - 1$. □

The face lattice $\mathscr{F}(C)$ has none of the strong lattice properties like distributivity or semimodularity. However, one has:

(2.13.9) **Theorem.** *The face lattice $\mathscr{F}(C)$ of a polyhedral cone C is relatively complemented, i.e. if $H, F, G \in \mathscr{F}(C)$ and*

$$H \subseteq F \subseteq G ,$$

then there exists an $\bar{F} \in \mathscr{F}(C)$ *such that*

$$F \wedge \bar{F} = H \quad \textit{and} \quad F \vee \bar{F} = G .$$

Moreover, it can be assumed, that

$$\dim F + \dim \bar{F} = \dim H + \dim G .$$

Proof. We consider at first the special case in which G covers F, and F covers H. By (2.13.7), H is the intersection of facets $F_1, \ldots, F_k$ of G. Thus

$$H = F_1 \cap \cdots \cap F_k = F \cap F_1 \cap \cdots \cap F_k .$$

The deletion of F_k may or may not change the value of the latter expression. In the first case, $F \cap F_1 \cap \cdots \cap F_{k-1} = F$ because F covers H. But then $H = F \cap F_k$. In the second case, F_k was superfluous and can be deleted. If $k=2$, then $H = F \cap F_1$. If $k>2$, the above argument may be repeated. In any case there exists a second facet $\bar{F}$ of G such that $F \wedge \bar{F} = H$ and, plainly, $F \vee \bar{F} = G$.

The cases $F=G$ and $F=H$ are trivial. In order to prove relative complementary for the general case $G \supset F \supset H$ we select two sequences of faces

$$G = F_0^0 \supset F_1^0 \supset \cdots \supset F_l^0 = F ,$$
$$F = F_l^0 \supset F_l^1 \supset \cdots \supset F_l^h = H ,$$

such that each face covers its successor. It is left to the reader to establish the existence of such sequences by arguments similar to the one used in the above paragraph.

In particular, F_{l-1}^0 covers F_l^0 which covers F_l^1. Hence there exists a facet F_{l-1}^1 of F_{l-1}^0 such that $F_{l-1}^1 \wedge F_l^0 = F_l^1$ and $F_{l-1}^1 \vee F_l^0 = F_{l-1}^0$. Similarly a face F_{l-2}^1 is found by complementing $F_{l-2}^0 \supset F_{l-1}^0 \supset F_{l-1}^1$, and so on. This process can be continued until the sequence $G = F_0^0 \supset F_1^0 \supset F_1^1$ has been complemented by a facet F_0^1 of G. Then we start again with $F_{l-1}^1 \supset F_l^1 \supset F_l^2$, complementing it by F_{l-1}^2 (Fig. 7), and so on until a face F_0^2 of G has been determined. This procedure can be repeated h times, yielding the faces F_s^r, $0 < r \leqslant h$, $0 \leqslant s < l$. We maintain that $\bar{F} := F_0^h$ is the relative complement to $F = F_l^0$ in the sequence $G \supset F \supset H$.

This is a special case ($r := h$, $s := 0$, $r' := 0$, $s' := l$) of the relations

$$F_s^r \wedge F_{s'}^{r'} = F_{s'}^r , \qquad F_s^r \vee F_{s'}^{r'} = F_s^{r'} \tag{2.13.10}$$

holding for $0 \leqslant r' \leqslant r \leqslant h$, $0 \leqslant s \leqslant s' \leqslant l$. These relations can be proved by induction on the number $k := r - r' + s' - s$. Indeed, (2.13.10) is trivially true for $k=0$. Now, suppose that (2.13.10) is true for $k \geqslant 0$, and suppose that r, r', s, s' are such that $0 \leqslant r' < r \leqslant h$, $0 \leqslant s < s' \leqslant l$, $r - r' + s' - s = k+1$. As $F_s^r = F_s^r \wedge F_s^{r-1}$, $F_{s'}^{r-1} = F_{s+1}^{r-1} \wedge F_{s'}^{r-1}$ by construction, the induction hypothesis yields

$$F_s^r \wedge F_{s'}^{r'} = F_s^r \wedge (F_s^{r-1} \wedge F_{s'}^{r'}) = F_s^r \wedge F_{s'}^{r-1}$$
$$= (F_s^r \wedge F_{s+1}^{r-1}) \wedge F_{s'}^{r-1}$$
$$= F_{s+1}^r \wedge F_{s'}^{r-1} = F_{s'}^r .$$

The second identity of (2.13.10) is proved analogously.

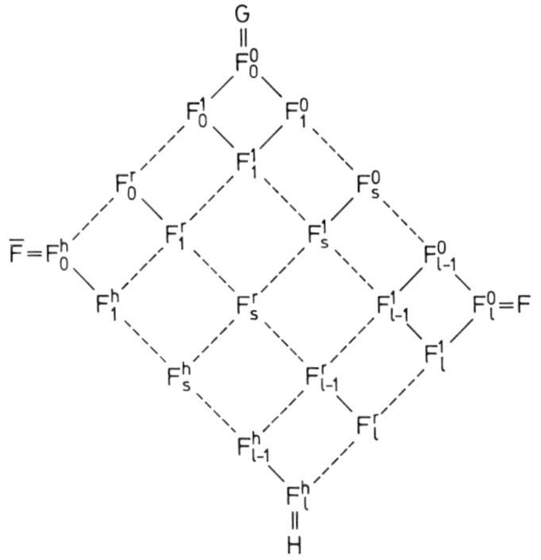

Fig. 7. Construction of a relative complement $\bar{F}$ of F in $G \supset F \supset H$

The dimensionality statement for $\bar{F} = F_l^0$, $\dim \bar{F} = \dim G + \dim H - \dim F$ follows at once from (2.13.8). ▯

Note that the construction of the relative complement $\bar{F}$ has been achieved by purely lattice-theoretical means based only on application of (2.13.7) to all levels.

2.14. Polar and Dual Polyhedra

The faces of a polyhedron $P = \{X \mid A^T X \leqslant B\}$ in R^n form also a lattice $\mathscr{F}(P)$ under set inclusion, provided one includes the empty face. In section 2.11 we described a homogenization operator $\mathscr{G}$ which assigns to $P \neq \phi$ the polyhedral cone

$$\mathscr{G}P = \left\{ \begin{pmatrix} X \\ 1 \end{pmatrix} \;\middle|\; X \in P \right\}^{pp} = \left\{ \begin{pmatrix} X \\ z \end{pmatrix} \;\middle|\; A^T X - Bz \leqslant 0,\; z \geqslant 0 \right\}.$$

Each face F of P can be written (see (2.3.6))

$$F = P \cap \bigcap_{i \in I} H_i^-, \qquad H_i^- := \{X \mid -A_i^T X \leqslant -b_i\},$$

where I consists of some column indices of A. Then

$$\begin{aligned} \mathscr{G} F &= \left\{ \begin{pmatrix} X \\ z \end{pmatrix} \,\middle|\, A^T X - Bz \leqslant 0,\ -A_I^T X + {}_I B z \leqslant 0,\ z \geqslant 0 \right\} \\ &= \mathscr{G} P \cap \bigcap_{i \in I} \left\{ \begin{pmatrix} X \\ z \end{pmatrix} \,\middle|\, A_i^T X - b_i z = 0 \right\}, \end{aligned}$$

which shows that $\mathscr{G} F$ is again a face.

Similarly, if $\tilde{F}$ is a face of $\mathscr{G} P$, then

$$F := \left\{ X \,\middle|\, \begin{pmatrix} X \\ 1 \end{pmatrix} \in \tilde{F} \right\} = \mathscr{D} \tilde{F}$$

is a face of P. Indeed, utilizing the extremeness of faces (theorem (2.4.7)), we get

$$\left.\begin{matrix} X, Y \in P \\ \\ Z \in (X\ Y) \cap \mathscr{D} \tilde{F} \end{matrix}\right\} \Rightarrow \left\{\begin{matrix} \begin{pmatrix} X \\ 1 \end{pmatrix}, \begin{pmatrix} Y \\ 1 \end{pmatrix} \in \mathscr{G} P \\ \\ \begin{pmatrix} Z \\ 1 \end{pmatrix} \in \tilde{F} \end{matrix}\right\} \Rightarrow \begin{pmatrix} X \\ 1 \end{pmatrix}, \begin{pmatrix} Y \\ 1 \end{pmatrix} \in \tilde{F} \Rightarrow X, Y \in \mathscr{D} \tilde{F}.$$

In section 2.11, the operators $\mathscr{G}$ and $\mathscr{D}$ were seen to be essentially inverse to each other. Indeed (2.11.9) states $\mathscr{D} \mathscr{G} P = P$ for polyhedra.

As both maps $\mathscr{G}$ and $\mathscr{D}$ preserve set-theoretical inclusion, we find that

(2.14.1) *$\mathscr{G}$ is an isomorphism of $\mathscr{F}(P)$ onto a sublattice of $\mathscr{F}(\mathscr{G} P)$, and $\mathscr{D}$ is a homomorphism of $\mathscr{F}(\mathscr{G} P)$ onto $\mathscr{F}(P)$.*

Moreover,

(2.14.2) *$\tilde{F} \in \mathscr{F}(\mathscr{G} P)$ is the image $\mathscr{G} F$ of some $F \in \mathscr{F}(P)$ precisely if either $\tilde{F} = \left\{ \begin{pmatrix} 0 \\ 0 \end{pmatrix} \right\}$ or $\mathscr{D} \tilde{F} \neq \phi$. In this case*

$$\mathscr{G} \mathscr{D} \tilde{F} = \tilde{F}.$$

Proof: This is trivial if $\tilde{F} = \left\{ \begin{pmatrix} 0 \\ 0 \end{pmatrix} \right\}$, since then $\tilde{F} = \mathscr{G} \phi$ by definition of $\mathscr{G}$. To complete the proof of the remaining case, note that the plane $E_z := \left\{ \begin{pmatrix} X \\ z \end{pmatrix} \,\middle|\, z = 0 \right\}$ is a boundary plane of $\mathscr{G} P$ by (2.11.8), and therefore also a boundary plane of $\tilde{F} \in \mathscr{F}(\mathscr{G} P)$. If $\mathscr{D} \tilde{F} \neq \phi$, then E_z is a nonsingular boundary plane, and consequently $\tilde{F}^I \subseteq H_z^I$ where $H_z^I = \left\{ \begin{pmatrix} X \\ z \end{pmatrix} \;\middle|\; z > 0 \right\}$. Then (2.11.6) and (2.8.11) yield

$$\mathscr{G}\mathscr{D}\tilde{F} = (\tilde{F} \cap H_z^I)^{pp} \supseteq (\tilde{F}^I \cap H_z^I)^{pp} = (\tilde{F}^I)^{pp} = \tilde{F}.$$

Trivially $\mathscr{G}\mathscr{D}\tilde{F} \subseteq \tilde{F}$. ▯

As a first application, we extend theorem (2.13.8) to polyhedra:

(2.14.3) **Theorem.** *If $F \neq \phi$ is a facet of a polyhedron P, then* $\dim F = \dim P - 1$.

Proof. A facet is understood, as in definition (2.13.5), to be a face "covered" by P in $\mathscr{F}(P)$. It follows readily from (2.14.1) and (2.14.2) that $\mathscr{G}P$ covers $\mathscr{G}F$. Thus $\dim \mathscr{G}F = \dim \mathscr{G}P - 1$. Now, $\{X_1, \ldots, X_k\} \subseteq S$ is a minimal spanning set of $\mathscr{M}S$ precisely if $\left\{ \begin{pmatrix} X_1 \\ 1 \end{pmatrix}, \ldots, \begin{pmatrix} X_k \\ 1 \end{pmatrix} \right\}$ is a basis of $\mathscr{L}\mathscr{G}S$. Thus $\dim \mathscr{G}S = \dim \mathscr{M}S + 1$ for any set $S \subseteq R_n$. This proves the theorem. ▯

(2.14.4) *If P is a polytope, i.e. a bounded polyhedron, then the lattices $\mathscr{F}(\mathscr{G}P)$ and $\mathscr{F}(P)$ are isomorphic.*

Proof. We recall from section 2.11 that $\begin{pmatrix} Y \\ 0 \end{pmatrix} \in \mathscr{G}P$ if and only if Y is a "direction of infinity" of P, that is, if $X + \Theta Y \in P$ for all $X \in P$ and $\Theta \geqslant 0$. Since P is bounded, $Y = 0$ is its only direction of infinity. In other words, $\tilde{F} = \left\{ \begin{pmatrix} 0 \\ 0 \end{pmatrix} \right\}$ is the only element of $\mathscr{F}(\mathscr{G}P)$ for which $\mathscr{D}\tilde{F} = \phi$. Thus (2.14.3) follows from (2.14.2). ▯

As a consequence of (2.14.4), theorem (2.13.9) extends to polytopes.

(2.14.5) **Theorem.** *The face lattice $\mathscr{F}(P)$ of a polytope P is relatively complemented.*

This theorem is false for halfmanifolds, since they have only one proper nonempty face. Since every unbounded polyhedron contains faces which are half manifolds, the theorem is false for every polyhedron not a polytope.

Theorem (2.13.2) shows that if $\mathscr{F}$ is the face lattice of a polyhedral cone C, then the lattice which the lattice-theoretical dual to $\mathscr{F}$ is also realizable as the face lattice of a polyhedral cone, namely of C^p. In order to prove a similar result for polytopes, we need polarity for sets.

For an arbitrary set $S \subseteq R^n$, one defines its

(2.14.6) *polar set* $S^\pi := \{Y \in R^n \mid Y^T X \leqslant 1 \text{ for all } X \in S\}$.

S^π is convex: $\mathscr{H}(S^\pi) = S^\pi$. Other trivial relations are: $0 \in S^\pi$; $(\mathscr{H} S)^\pi = S^\pi$; $S_1 \subseteq S_2 \Rightarrow S_1^\pi \supseteq S_2^\pi$; $S \subseteq S^{\pi\pi}$; and for any family of sets $S_i \subseteq R^n$:

$$\left(\bigcup S_i\right)^\pi = \bigcap S_i^\pi.$$

The concept of polar sets is not as straightforward conceptually as is the concept of polar cones. Every argument, moreover, based on polar sets can also be based on the more fundamental notion of polar cones by means of homogenization techniques. To this end, let $\sigma: R^{n+1} \to R^{n+1}$ be the map which reflects R^{n+1} through the coordinate plane characterized by $z=0$:

$$\sigma\begin{pmatrix} X \\ z \end{pmatrix} := \begin{pmatrix} X \\ -z \end{pmatrix}.$$

Then

(2.14.7) $$S^\pi = \mathscr{D}\sigma(\mathscr{G} S)^p.$$

Proof. Clearly

$$Y^T X \leqslant 1 \text{ for all } X \in S \iff \begin{pmatrix} Y \\ -1 \end{pmatrix}^T \begin{pmatrix} X \\ 1 \end{pmatrix} \leqslant 0 \quad \text{for all } X \in S.$$

Thus by definitions (2.14.6) and (2.11.1)

$$Y \in S^\pi \iff \begin{pmatrix} Y \\ -1 \end{pmatrix} \in \left\{\begin{pmatrix} X \\ 1 \end{pmatrix} \;\middle|\; X \in S\right\}^p = \left\{\begin{pmatrix} X \\ 1 \end{pmatrix} \;\middle|\; X \in S\right\}^{ppp} = (\mathscr{G} S)^p. \quad \square$$

If P is a polyhedron, then so is $\mathscr{G} P$ by (2.11.8) and therefore P^π by (2.14.7). The latter is called

(2.14.8) *polar*

to P. One has

(2.14.9) $P^{\pi\pi} = P$ *if and only if* $0 \in P$.

Proof. We first show that

(2.14.10) $$0 \in P \Rightarrow (\sigma \mathscr{G} P)^p = \mathscr{G}(P^\pi).$$

Clearly, $\begin{pmatrix} 0 \\ 1 \end{pmatrix} \in \mathscr{G} P$. Hence $\begin{pmatrix} Y \\ w \end{pmatrix}^T \begin{pmatrix} 0 \\ 1 \end{pmatrix} = w \leqslant 0$ for all $\begin{pmatrix} Y \\ w \end{pmatrix} \in (\mathscr{G} P)^p$ by definition of polarity, and $w=0$ characterizes a face $\tilde{F}$ of $(\mathscr{G} P)^p$. $\tilde{F} \neq (\mathscr{G} P)^p$ since $\begin{pmatrix} 0 \\ -1 \end{pmatrix} \in (\mathscr{G} P)^p$. Hence $((\mathscr{G} P)^p)^I \subseteq \sigma(H_z^I)$ or $(\sigma \mathscr{G} P)^{pI} \subseteq H_z^I$. By (2.11.6) and (2.11.8), we then have $\mathcal{G}(P^\pi)=\mathcal{G}\,\mathcal{D}((\sigma\mathcal{G}P)^p)=((\sigma\mathcal{G}P)^p\cap H_z^I)^{pp}$ $\supseteq((\sigma\mathcal{G}P)^{pI})^{pp}=(\sigma\mathcal{G}P)^p$. Plainly, $((\sigma\mathcal{G}P)^p\cap H_z^I)^{pp}\subseteq(\sigma\mathcal{G}P)^p$, and (2.14.10) is proved.

From (2.14.10) and (2.11.9),

$$P^{\pi\pi}=\mathcal{D}((\sigma\mathcal{G}(P^\pi))^p)=\mathcal{D}((\sigma(\sigma\mathcal{G}P)^p)^p)=\mathcal{D}(\mathcal{G}P)^{pp}=\mathcal{D}\mathcal{G}P=P.$$

This proves that $0\in P$ is sufficient for $P^{\pi\pi}=P$ to hold; its necessity had been seen before. □

That the shape of the polar polyhedron depends rather drastically on the location of P is also seen in the following statement:

(2.14.11) *P^π is bounded if and only if $0\in P^I$ and $\mathcal{L}P=R^n$.*

Proof. Consider the n-cube

$$W_\rho:=\{X\in R^n \mid \max\{|x_1|,\dots,|x_n|\}\leqslant\rho\}.$$

Then

$$(W_\rho)^\pi:=\left\{Y\in R^n \;\middle|\; |y_1|+\cdots+|y_n|\leqslant\frac{1}{\rho}\right\}.$$

Suppose $0\in P^I$ and $\mathcal{L}P=R^n$. By theorem (2.12.2), $(W_1)^\pi$ is the convex hull of its vertices $Y_1,\dots,Y_k$. Since $Y_i\in\mathcal{L}P$ for $i=1,\dots,k$, we conclude from lemma (2.3.12) that there exists $\rho>0$ such that $\rho Y_i\in P$ for all i. Thus $(W_\rho)^\pi\subseteq P$, and by (2.14.9), $W_\rho=W_\rho^{\pi\pi}$ contains P^π, which is therefore bounded.

Conversely, if $0\notin P^I$, then there exists a boundary plane E_i $=\{X\mid A_i^TX=b_i\}$ of P with $b_i\leqslant 0$, so that $A_i\neq 0$ and $A_i^TX\leqslant b_i$ for all $X\in P$. Clearly A_i is a direction of infinity of $P^\pi=\{Y\mid Y^TX\leqslant 1$ for all $X\in P\}$. Hence P^π is not bounded. If $0\in P^I$, but $\mathcal{L}P\neq R^n$, then there exists by theorem (2.4.8) a singular boundary plane $E_i=\{X\mid A_i^TX=0\}\supseteq P$ with $A_i\neq 0$, and the argument proceeds as above. □

By (2.14.10), if P is a polytope of full dimension n and $0\in P^I$, then the same is true for P^π. Moreover, the face lattices $\mathcal{F}(P)$ and $\mathcal{F}(P^\pi)$ are isomorphic respectively to the face lattices $\mathcal{F}(\mathcal{G}P)$ and $\mathcal{F}(\mathcal{G}(P^\pi))$ of the cones $\mathcal{G}P$ and $\mathcal{G}(P^\pi)$ (2.14.4). Finally (2.14.10) shows that $(\mathcal{G}P)^p$ and $\mathcal{G}(P^\pi)$ have isomorphic face lattices. By theorem (2.13.2), the lattices $\mathcal{F}(\mathcal{G}P)$ and $\mathcal{F}((\mathcal{G}P)^p)$ are duals of each other. The following theorem now follows readily:

(2.14.12) **Theorem.** *For every full dimensional polytope $P\subseteq R^n$ there exists a polytope P whose face lattice is dual to the one of P. Such a polytope is called*

dual

to P.

The problem of structurally characterizing face lattices of polyhedra is unsolved. Partial results for structurally characterizing the graphs formed by the edges of polytopes, and more generally the cell-complexes formed by the faces of polytopes have been obtained by Whitney [*1*], Balinski [*1*], Grünbaum and Motzkin [*1*], Klee [*31*], Klee and Walkup [*1*]. In the case of "simplicial" polytopes, i.e. polytopes whose facets are simplices, the Euler-Poincaré relations

$$f_0 - f_1 + \cdots + (-1)^{d-1} f_{d-1} = 1 - (-1)^d$$

between the numbers f_k, $k=0,\ldots,d-1$, of k-dimensional faces of any given d-polytope can be supplemented by additional linear relations between the quantities f_k (Sommerville [*1*], Klee [*27*]).

2.15. Gale Diagrams

We conclude this chapter with a short outline of Gale diagrams. This beautiful and powerful technique has been described by Grünbaum [*6*] following private communication from M. A. Perles. It has been extracted from the pioneering work of Gale [*4*], [*6*], [*7*] and some aspects of the work of Motzkin [*2*], [*6*], [*8*].

In what follows, we shall consider families $Z=\{Z_1,\ldots,Z_n\}$ rather than sets of points, the difference being that some or all points Z_i in Z may be equal. Every subset I of $N:=\{1,\ldots,n\}$ can be written as $I=\{i_1,\ldots,i_k\}$ with $1\leqslant i_1<\cdots<i_k\leqslant n$, and corresponds uniquely to a subfamily

$$Z_I := \{Z_{i_1},\ldots,Z_{i_m}\}$$

of Z. Operations between subfamilies are defined via operations between respective index sets, e.g., $Z\sim Z_I := Z_{N\sim I}$, $Z_{I_1}\cap Z_{I_2} := Z_{I_1\cap I_2}$.

Now consider a family $Z=\{Z_1,\ldots,Z_n\}$ of points in R^l. The technique to be described rests on the following observation: the collection of all subfamilies T of Z for which

$$0\in(\mathscr{C}(Z\sim T))^I, \tag{2.15.1}$$

i.e. for which $\mathscr{C}(Z\sim T)$ is a linear subspace (2.8.10), form a lattice under set inclusion with the intersection as cap operation. Indeed, if T_1 and T_2 are subfamilies of Z, then

$$\mathscr{C}(Z\sim(T_1\cap T_2)) = \mathscr{C}(Z\sim T_1) + \mathscr{C}(Z\sim T_2).$$

If therefore $\mathscr{C}(Z\sim T_1)$ and $\mathscr{C}(Z\sim T_2)$ are subspaces, then so is $\mathscr{C}(Z\sim(T_1\cap T_2))$. Note that $T:=Z$ always satisfies (2.15.1) since

$(\mathscr{C}\emptyset)^I = \{0\}^I = \{0\}$. Note further that condition (2.15.1) is equivalent to

(2.15.2) $\qquad$ *either* $\quad 0 \in (\mathscr{H}(Z \sim T))^I \quad$ *or* $\quad T = Z$.

This latter condition is somewhat easier to visualize. We call the lattice we just defined the

(2.15.3) $\qquad$ *Gale lattice*

of the family Z.

(2.15.4) **Theorem.** *For every family $X = \{X_1, \ldots, X_n\}$ of points there exists a family $Z = \{Z_1, \ldots, Z_n\}$ of points, such that the face lattice of the polyhedral cone $\mathscr{C}X$ is isomorphic to the Gale lattice of Z, and conversely, for every family $Z = \{Z_1, \ldots, Z_n\}$ there exists a family $X = \{X_1, \ldots, X_n\}$ such that the Gale lattice of Z and the face lattice of $\mathscr{C}X$ are isomorphic. In both cases, the linear dimensions of the families X and Z relate to their common cardinality as follows:*

$$\dim \mathscr{C} X + \dim \mathscr{L} Z = n.$$

Proof. The proof will be based on the notion of a "Gale transformation". Every linear subspace L of R^n can be represented either as intersection of l hyperplanes, or as spanned by k generators. Both descriptions involve families $X = \{X_1, \ldots, X_n\}$ of points $X_i \in R^k$ and $Z = \{Z_1, \ldots, Z_n\}$ of points $Z_i \in R^l$, from which matrices $(X_1, \ldots, X_n)$ and $(Z_1, \ldots, Z_n)$ are formed:

$$(2.15.5) \qquad \begin{aligned} L &= \{Y = (X_1, \ldots, X_n)^T W \mid W \in R^k\}, \\ L &= \{Y \mid (Z_1, \ldots, Z_n) Y = 0\}. \end{aligned}$$

Associating Z_i with X_i for $i = 1, \ldots, n$ is called a

(2.15.6) $\qquad$ *Gale transformation.*

Neither of the representations (2.15.5) is unique. We call two families $X = \{X_1, \ldots, X_n\}$, $\bar{X} = \{\bar{X}_1, \ldots, \bar{X}_n\}$

(2.15.7) $\qquad$ *linearly equivalent*

if there exist matrices T, $\bar{T}$ such that $\bar{X}_i = T X_i$ and $X_i = \bar{T} \bar{X}_i$ for $i = 1, \ldots, n$. With this definition, families of points from spaces of different dimension may be linearly equivalent. We state without proof, that two families X, $\bar{X}$ define the same subspace L precisely if they are linearly equivalent. The same holds for families Z in (2.15.5). A Gale transformation is therefore a mapping of classes of linearly equivalent families of points. Note that both Gale lattices and face lattices are invariant under transformations which preserve linear equivalence.

We proceed to show that if the family Z arises from family X by Gale transformation, then the face lattice of $\mathscr{C}X$ and the Gale lattice of Z are isomorphic. To this end we consider those subfamilies X_I which determine (2.10.12) faces F of $\mathscr{C}X$, i.e. for which $X_I = F \cap X$ and consequently $F = \mathscr{C}X_I$ (2.10.13). The natural lattice formed by all such subfamilies is plainly isomorphic to the face lattice of $\mathscr{C}X$. By theorem (2.10.14), the subfamilies $X_I = F \cap X$ are characterized by the fact that $X_{N\sim I}$ forms the conical part (2.10.11) of the family $X \cup -X_I$, i.e. none of the generators X_h in $X_{N\sim I}$ belongs to the lineality space of the cone $\mathscr{C}(X \cup -X_I)$:

(2.15.8) $$-X_h \notin \mathscr{C}(X \cup -X_I) \quad \text{for} \quad h \in N \sim I.$$

The Gale lattice of the family $Z = \{Z_1, \ldots, Z_n\}$ is isomorphic to the natural lattice of those subfamilies Z_I for which $0 \in \mathscr{C}(Z_{N\sim I})^I$ or, equivalently, $\mathscr{C}(Z_{N\sim I})$ is a linear subspace:

(2.15.9) $$-Z_h \in \mathscr{C}(Z_{N\sim I}) \quad \text{for} \quad h \in N \sim I.$$

The face lattice of $\mathscr{C}X$ and the Gale lattice of Z are therefore isomorphic if (2.15.8) and (2.15.9) are equivalent. By the Farkas' lemma (2.8.5) and in view of (2.8.3), (2.15.8) amounts to the existence of

$$U \in \mathscr{C}(X \cup -X_I)^p = \{U \mid X_i^T U \leq 0 \text{ for } i \in N, \ -X_j^T U \leq 0 \text{ for } j \in I\}$$

such that $-X_h^T U > 0$. Then $Y := -(X_1, \ldots, X_n)^T U \in L$ satisfies $Y \geq 0$, $Y_I = 0$, $y_h > 0$, and by (2.15.5), $(Z_1, \ldots, Z_n) Y = 0$. One readily concludes that the existence of such a vector $Y \in L$ is necessary and sufficient for $-Z_h \in \mathscr{C}(Z_{N\sim I})$ to hold.

We have proved that every face lattice is a Gale lattice. The converse follows immediately from the fact, that the Gale transformation (2.15.6) can be inverted up to linear equivalence. In other words, to each family Z there exists a family X such that Z can be obtained from X by Gale transformation.

The dimensional relation finally follows immediately from the fact that

$$\dim \mathscr{L}X = \dim L = \operatorname{codim} \mathscr{L}Z. \quad \square$$

The Gale lattice of a family $Z = \{Z_1, \ldots, Z_n\}$ is not affected if one multiplies the points Z_i with any number $\lambda_i > 0$. In particular, Z can be normalized so that all nonzero points Z_i have euclidean length 1. A family so normalized is called a

(2.15.10) *Gale diagram*

if in addition $\mathscr{C}Z = R^l$, or equivalently, $0 \in (\mathscr{H}Z)^I$. In this case, the empty subfamily of Z belongs to the Gale lattice, and consequently the lineality space of the corresponding cone $\mathscr{C}X$ has dimension 0 and $X_i \neq 0$ for all $i=1,\ldots,n$. In other words, there exists a family $\hat{X}=\{\hat{X}_1,\ldots,\hat{X}_n\}$ so that the face lattice of $\mathscr{C}\hat{X}$ is isomorphic to the face lattice of the convex hull $\mathscr{H}\hat{X}$. Thus

(2.15.11) *the Gale lattice of a Gale diagramm is isomorphic to the face lattice of a polytope, and vice versa.*

Consider for instance a polytope of dimension d with $d+3$ vertices $V_1,\ldots,V_{d+3}$. Its face lattice is isomorphic to the face lattice of the cone spanned by the $(d+1)$-dimensional points

$$X_i := \begin{pmatrix} V_i \\ 1 \end{pmatrix}, \qquad i=1,\ldots,d+3.$$

According to theorem (2.15.4), a corresponding Gale diagram can be realized in R^2.

Polytopes with isomorphic face lattices are called

(2.15.12) *combinatorially equivalent.*

A class of combinatorially equivalent polyhedra constitutes a

(2.15.13) *combinatorial type.*

M. A. Perles has characterized and counted all combinatorial types of polyhedra of dimension d having $d+3$ vertices. See Grünbaum [6] for this and other applications of Gale diagrams. Simplices of dimension d have as Gale diagram the family consisting of $d+1$ copies of the origin. The following family in R^1 is interesting

$$Z=\{+1,\ -1,\ +1,\ -1,\ +1,\ -1\}.$$

It is the Gale diagram of some polytope in R^4. This polytope has 6 vertices. Each pair of points is connected by an edge, since any two elements of Z can be removed and the remainder still spans the interval $\mathscr{H}\{-1,+1\}$, which contains 0 as an inner point. We have therefore found an example of a "neighborly" polytope not a simplex. The faces of the polytope are readily determined: they correspond to those sub-families of Z that arise from Z by removing an arbitrary element of value $+1$ and an arbitrary element of value -1.

CHAPTER 3

Convex Sets

This chapter surveys the basic properties of convex sets as they will be needed to develop the duality theory of convex programming and to derive saddle point theorems. The material presented is mostly classical: The important separation theorems for convex sets, the existence of supports, Helly's theorem, and Brouwer's fixed point theorem (see Bonnesen and Fenchel [*1*], Eggleston [*1*], and Fan [*7*]). In addition, extreme sets are discussed in detail.

All considerations are restricted to finite dimensional linear spaces over the field R of real numbers. The linear space R^n is considered as a normed space. The fact that the maximum norm and the euclidean norm generate identical topologies in R^n is instrumental for the developments of this chapter: The maximum norm generates neighborhoods which are bounded polyhedra. These can be described by their finite sets of vertices. The euclidean norm, on the other hand, permits application of the results on orthogonal projections derived in chapter 2.

The reader should be familiar with the elementary concepts of topology including topological spaces, neighborhoods, open and closed sets, convergence and continuity (Alexandroff and Hopf [*1*], Bourbaki [*2*], Kelley [*2*], Moore [*1*]).

Klee [*34*] lists the known separation and support properties of convex sets.

3.1. The Normed Linear Space R^n

One of the characterizing properties of the field R of real numbers (Aumann [*1*], McShane and Botts [*1*]) is the following: Each non empty subset S which is bounded above has a least upper bound, called the *supremum* ($\sup S$) of S, and each non-empty subset T which is bounded below has a greatest lower bound, called the *infimum* ($\inf T$).

The field R of real numbers admits a "natural" topology, which can be generated for instance by the neighborhood system consisting of intervals

$$U_\varepsilon(x) := \{y \in R \mid |y-x| \leq \varepsilon\}, \qquad \varepsilon > 0.$$

In R^n the "natural" topology is the product topology which arises from the natural topology in R. It can be generated by choosing n-cubes as neighborhoods of a point X:

$$U_\varepsilon(X) := \{Y \in R^n \mid |y_i - x_i| \leq \varepsilon,\ i=1,\dots,n\}, \qquad \varepsilon > 0.$$

Putting

$$p_0(Z) := \max_i |z_i| \tag{3.1.1}$$

we may write $U_\varepsilon(X) := \{Y \mid p_0(Y-X) \leq \varepsilon\}$. The function p_0 is called the

maximum norm

in R^n. In general, a

norm $\quad p\colon R^n \to R$

in R^n is defined by the properties (Minkowski [*3*]):

(3.1.2)
- (i) $p(X) \geq 0$ *for all* $X \in R^n$,
- (ii) $p(X) = 0$ *if and only if* $X = 0$,
- (iii) $p(\lambda X) = \lambda p(X)$ *for all* $X \in R^n$ *and* $\lambda \geq 0$,
- (iv) $p(X+Y) \leq p(X) + p(Y)$ *for all* $X, Y \in R^n$.

A linear space R^n together with a norm forms a (compare for instance Day [*1*])

(3.1.3) *normed linear space.*

Any norm p on R^n can be used to define a neighborhood system

$$U_\varepsilon^p(X) := \{Y \in R^n \mid p(Y-X) \leq \varepsilon\}, \qquad \varepsilon > 0$$

and thereby a Hausdorff topology on R^n.

A norm provides R^n with more structure than just a topology. For instance we may define a function $f\colon R^n \to R$ to be

(3.1.4) *uniformly continuous*

with respect to p if for every $\varepsilon > 0$ there exists a number $\delta(\varepsilon) > 0$ such that $|f(X) - f(Y)| \leq \varepsilon$ if $p(X-Y) \leq \delta(\varepsilon)$. Also we may define a sequence $\{X_i\}_{i=1,2,\dots}$ of points in R^n to be a

(3.1.5) *Cauchy sequence*

if for every $\varepsilon>0$ there exists an integer $N(\varepsilon)$ such that $p(X_i-X_j)\leqslant\varepsilon$ whenever $i,j\geqslant N(\varepsilon)$. Let us call two norms p_1 and p_2

(3.1.6) *equivalent*

if they not only generate the same topologies, but also give rise to identical sets of uniformly continuous functions and Cauchy sequences.

Norms p_1 and p_2 generate the same topology if and only if for each $\varepsilon>0$ there exists a $\delta=\delta(\varepsilon)>0$ such that both

$$U_\varepsilon^{p_1}(X)\supseteq U_\delta^{p_2}(X) \quad \text{and} \quad U_\varepsilon^{p_2}(X)\supseteq U_\delta^{p_1}(X),$$

in other words if and only if

(3.1.7) *for each* $\varepsilon>0$ *there exists* $\delta=\delta(\varepsilon)$ *such that* $p_1(Y)\leqslant\varepsilon$ *if* $p_2(Y)\leqslant\delta$ *and* $p_2(Y)\leqslant\varepsilon$ *if* $p_1(Y)\leqslant\delta$.

Clearly (3.1.7) also ensures that p_1 and p_2 define the same uniformly continuous functions and the same Cauchy sequences. For two norms to be equivalent (3.1.6) it is therefore sufficient that they generate the same topology.

A space with uniform continuity and Cauchy sequences is called a

(3.1.8) *uniform space*

(Weil [*1*], Bourbaki [*2*]). A uniform space over a set S is characterized by a system $\mathscr{U}$ of sets $U\subseteq S\times S$ satisfying the axioms:

(i) $U\supseteq\Delta$,
(ii) $U\in\mathscr{U}\Rightarrow U^{-1}\in\mathscr{U}$,
(iii) $U\in\mathscr{U}\,\exists\, V\in\mathscr{U}$: $V\circ V\subseteq U$,
(iv) $U,V\in\mathscr{U}\Rightarrow U\cap V\in\mathscr{U}$,
(v) $U\in\mathscr{U}, V\supseteq U\Rightarrow V\in\mathscr{U}$.

Here Δ denotes the "diagonal" $\{(X,X)\mid X\in S\}$ of $S\times S$, U^{-1} stands for $\{(X,Y)\mid (Y,X)\in U\}$, and $V\circ V$ denotes the set of all pairs (X,Z) such that a $Y\in S$ can be found with (X,Y) and (Y,Z) both in V. The system $\mathscr{U}$ is called a "uniformity".

If p is a norm in R^n, then all sets containing a set $U_\varepsilon:=\{(X,Y)\mid p(Y-X)\leqslant\varepsilon\}$ with $\varepsilon>0$ constitute a uniformity in R^n. Two norms are equivalent (3.1.6) if they give rise to the same uniformity. As we have seen, two norms whose topologies are identical generate identical uniformities. In general, however, two uniformities with the same topology need not coincide.

Our main theorem can now be stated:

(3.1.9) **Theorem.** *All norms on* R^n *are equivalent in the sense* (3.1.6).

The *proof* follows Ostrowski [*1*], and illustrates some important properties of norms. We first note that

(3.1.10) *two norms p_1 and p_2 are equivalent if and only if there exists a number $m>0$ such that*

$$\frac{1}{m} \leqslant \frac{p_1(X)}{p_2(X)} \leqslant m$$

holds for all points $X \neq 0$ in R^n.

Proof of (3.1.10). If the inequalities in (3.1.10) are satisfied, then $p_1(X) \leqslant m p_2(X)$ and $p_2(X) \leqslant m p_1(X)$, and (3.1.7) follows if one puts $\delta(\varepsilon) := \varepsilon/m$. On the other hand, if (3.1.7) is satisfied, one has in particular $p_1(X) \leqslant 1$ if $p_2(X) \leqslant \delta := \delta(1)$. Thus $p_1\left(\frac{\delta}{p_2(X)} X\right) \leqslant 1$, since $p_2\left(\frac{\delta}{p_2(X)} X\right) = \delta$ by the homogeneity of p_2. The homogeneity of p_1 then gives

$$\frac{p_1(X)}{p_2(X)} \leqslant \frac{1}{\delta},$$

which proves one side of the inequality in (3.1.10). The other side follows by a symmetrical argument.

Next we single out the maximum norm $p_0(X) = \max_i |x_i|$ and prove:

(3.1.11) *Every norm p is uniformly continuous on R^n with respect to the maximum norm.*

Proof of (3.1.11). The unit cube

$$W := \{Y \mid |y_i| \leqslant 1,\ i = 1, \ldots, n\}$$

is the convex hull of its vertices $Z_1, \ldots, Z_r$, $r = 2^n$. Hence every point $Y \in W$ can be written as $Y = u_1 Z_1 + \cdots + u_r Z_r$ with $u_k \geqslant 0$ and $\Sigma u_k = 1$. For the norm p we then have

$$p(Y) \leqslant u_1 p(Z_1) + \cdots + u_r p(Z_r) \leqslant \max p(Z_i) =: m.$$

This gives $p(Y) \leqslant \varepsilon$ for all $Y \in \frac{\varepsilon}{m} W$. Further, $p(X+Y) \leqslant p(X) + \varepsilon$ for all $Y \in \frac{\varepsilon}{m} W$, and $p(X) \leqslant p(X+Y) + p(-Y) \leqslant p(X+Y) + \varepsilon$, since $Y \in \frac{\varepsilon}{m} W$ implies $-Y \in \frac{\varepsilon}{m} W$. Noting that $Y \in \frac{\varepsilon}{m} W$ precisely when $p_0(Y) \leqslant \frac{\varepsilon}{m}$, we have:

$$|p(X+Y) - p(X)| \leqslant \varepsilon \quad \text{if} \quad p_0(Y) \leqslant \frac{\varepsilon}{m}.$$

This establishes the uniform continuity of p, proving (3.1.11).

As a final step, we note that norm p assumes its infimum on the set

$$A := \{X \mid p_0(X) = 1\}$$

which is the boundary of the unit n-cube. This is a consequence of the continuity (3.1.11) of p and the well known compactness of A.

The reader may want to ascertain that the proof of the compactness of A does not presuppose the quite elementary result we are trying to prove. We note therefore that the compactness of A rests on the following two facts: (i) Every closed finite interval $I := \{x \mid a \leqslant x \leqslant b\} \subseteq R$ is compact. (ii) A sequence $\{X_i\}_{i=1,2,\ldots}$ of points in R^n is convergent with respect to the maximum norm p_0 if it is convergent in each component separately. With the help of (ii) one derives from (i) that every cartesian product $I_1 \times \cdots \times I_n \subseteq R^n$, where the I_i are either single points or closed finite intervals, is also compact. Clearly, A is a union of finitely many sets of this kind. Statement (i) is an immediate consequence of the existence of the supremum and the infimum in R. Statement (ii) follows from the definition of the maximum norm.

Let then

$$l := \inf\{p(X) \mid X \in A\} = \min\{p(X) \mid X \in A\} > 0.$$

The norm p has an upper bound u on A. This can be concluded from its continuity and the compactness of A, but has been also established directly in the proof of (3.1.11). Putting $m := \max\left\{u, \frac{1}{l}\right\}$, we have

$$\frac{1}{m} \leqslant \frac{p(X)}{p_0(X)} \leqslant m \quad \text{for all} \quad X \quad \text{with} \quad p_0(X) = 1.$$

Homogeneity extends this result immediately to all $X \neq 0$. According to (3.1.10), every norm p is thus equivalent to p_0. This proves theorem (3.1.9). □

In infinite dimensional spaces, different norms may create different topologies. Consider the linear space of all sequences $X = (x_1, x_2, \ldots)$ with $x_i \in R$ and $\|X\|^2 := \sum_{i=1}^{\infty} |x_i|^2 < \infty$. Then $\|X\|$ is a norm and the space becomes a Hilbert space with this norm. But $p(X) := \sum_{i=1}^{\infty} \frac{1}{i}|x_i|$ is also a norm on the same space, and while the sequence of unit vectors

$$E_i := (0, \ldots, 0, \underbrace{1}_{i\text{-th entry}}, 0, \ldots)$$

diverges in Hilbert space, it converges in p-space.

We are now able to prove the important

(3.1.12) **Cauchy Criterion.** *A sequence $\{X_i\}_{i=1,2,\ldots}$ of points in R^n converges if and only of it is a Cauchy sequence.*

Proof. Since it makes no difference which norm is chosen to define Cauchy sequences, we select the maximum norm $p_0(X) = \max_i |x_i|$. We then see immediately that a sequence is Cauchy if and only if it is Cauchy in each component separately, just as a sequence converges if and only if it converges in each component separately. It suffices therefore to prove (3.1.12) for the 1-dimensional case. But any Cauchy sequence $\{x_i\}_{i=1,2,\ldots}$ in R converges towards $x := \sup_k \inf\{x_i \mid i \geqslant k\}$. □

The natural topology of R^n is compatible with the linear operations $(X, Y) \to X + Y$ and $(\lambda, X) \to \lambda X$. In other words

(3.1.13) **Theorem.** *The normed space R^n is a topological linear space. Affine maps $f: R^n \to R^m$ defined by $m \times n$-matrices A and m-vectors B,*

$$f(X) := A X + B,$$

are continuous.

Proof. The continuity of the linear operations follows from the inequalities

$$p_0(X + Y - X_i - Y_i) \leqslant p_0(X - X_i) + p_0(Y - Y_i),$$

$$p_0(\lambda X - \lambda_i X_i) \leqslant |\lambda| \, p_0(X - X_i) + |\lambda - \lambda_i| \; |p_0(X) - p_0(X_i)| + |\lambda - \lambda_i| \, p_0(X).$$

Each component of $A X_i + B$ is a linear expression involving the components x_k^i of X_i. If $X_i \to X$, then $x_k^i \to x_k$ for $k = 1, \ldots, n$. The continuity of the linear operations in R^1 then yields $A X_i + B \to A X + B$. □

3.2. Closure and Relative Interior of Convex Sets

In the previous section, the algebraic operations and, in particular linear maps arising from matrix multiplication

$$X \to A^T X$$

were seen to be continuous. Sets of the form $\{X \mid A^T X \leqslant B\}$ and $\{X \mid A^T X = B\}$ are therefore closed. In other words, all polyhedra are closed sets. It was shown in section 2.6 that in R^n all linear subspaces, and therefore all linear manifolds, are solution sets of finite systems of linear equations and thus sets of the form $\{X \mid A^T X = B\}$. Linear subspaces and manifolds are therefore also closed sets. This is not true for infinite dimensional spaces.

Let S be any subset of R^n. Then we denote by

(3.2.1) $$\bar{S}$$

its topological closure, that is, the smallest closed set containing S. Since all manifolds are closed sets, one has $\mathcal{M}S \subseteq \mathcal{M}\bar{S} \subseteq \mathcal{M}\overline{\mathcal{M}S} = \mathcal{M}\mathcal{M}S = \mathcal{M}S$. Therefore

(3.2.2) $$\mathcal{M}\bar{S} = \mathcal{M}S \quad \textit{and} \quad \mathcal{L}\bar{S} = \mathcal{L}S.$$

Each point $Y \in \bar{S}$ is the limit of a sequence $\{X_i\}_{i=1,2,\dots}$ of points in S: choose $X_i \in U_i \cap S$ where $U_i := \left\{X \in R^n \;\middle|\; p(X-Y) \leqslant \frac{1}{i}\right\}$ for some fixed norm p. If two sequences $\{X_i^{(1)}\}$, $\{X_i^{(2)}\}$ converge to Y_1 and Y_2, respectively, then by virtue of the homogeneity (3.1.2.(iii)) and the subadditivity (3.1.2.(iv)) of an arbitrary norm the sequence $\{\lambda X_i^{(1)} + \mu X_i^{(2)}\}_{i=1,2,\dots}$ converges to $\lambda Y_1 + \mu Y_2$. Hence

(3.2.3) *the closure of a convex set K is convex.*

The closure $\bar{S}$ of S with respect to the natural topology in R^n is identical with the closure of S in the relative topology induced on $\mathcal{M}S$. The closure of S is therefore the same regardless of the dimension of the space in which it is embedded.

One needs a distinction between the "boundary" and the "interior" of sets. It is advantageous to define these concepts in a manner that does not depend on the embedding of the sets. A point X is called a

(3.2.4) *relative interior point* or *inner point*

of S if X is an interior point of S with respect to the relative topology induced in $\mathcal{M}S$. We denote the set of inner points of S by

(3.2.5) $$S^I.$$

S^I is open in $\mathcal{M}S$, and since $\mathcal{M}(S^I) \subseteq \mathcal{M}S$, S^I is open in $\mathcal{M}(S^I)$. Hence $S^{II} = S^I$.

The reader is warned, that the formation of the relative interior is not a monotonic operation: In general, $S_1 \subseteq S_2$ does not imply $S_1^I \subseteq S_2^I$, as any polyhedron with facets shows. However, the implication is clearly justified whenever $\mathcal{M}S_1 = \mathcal{M}S_2$.

We call a point X which is a boundary point of S relative to $\mathcal{M}S$ simply a

(3.2.6) *boundary point*

of S. This definition deviates from the common use of language, but it is consistent with our definition (2.3.9) of "boundary point" for polyhedra. Indeed we have:

(3.2.7) *If X is an inner (boundary) point of polyhedron* $P = \{X \mid A^T X \geqslant B\}$ *in the algebraic sense* (2.3.9), *then it is also an inner (boundary) point in the topological sense* (3.2.4), *and vice versa.*

Proof. An inequality $A_i^T X \geqslant b_i$ is called a singular inequality (2.3.5) of the system $A^T X \geqslant B$, if $A_i^T X = b_i$ for all solutions of the system. In the algebraic sense (2.3.9) an inner point X is then defined as one with $A_i^T X > b_i$ for all nonsingular inequalities. Since linear functions are continuous, there exists in R^n a neighborhood U of X such that $A_i^T Y > b_i$ for all $Y \in U$ and all nonsingular inequalities. X will be an inner point in the topological sense if

$$U \cap \mathscr{M} P \subseteq P$$

because $U \cap \mathscr{M} P$ is a neighborhood of X in $\mathscr{M} P$ by definition of the relative topology. Let then $Y \in U \cap \mathscr{M} P$. According to theorem (2.4.8), $\mathscr{M} P$ is the solution set of all equations $A_i^T X = b_i$ that arise from singular inequalities of the system $A^T X \geqslant B$, and U was chosen so as to satisfy the nonsingular inequalities. Hence Y, belonging both to $\mathscr{M} P$ and U, satisfies all inequalities that characterize P.

Now let X be a boundary point and Z an inner point of P, both in the algebraic sense (2.3.9). Then clearly $X(\varepsilon) := X + \varepsilon(X - Z) \notin P$ for all $\varepsilon > 0$, while $X(\varepsilon) \to X$ if $\varepsilon \to 0$. This shows that X is also a boundary point in the topological sense (3.2.6). Since every point of P is either an inner point or a boundary point, both directions of (3.2.7) are herewith proved. □

The existence of an algebraic inner point Z of a given polyhedron P, assumed in the above argument, is guaranteed by lemma (2.3.10). A corresponding lemma holds for topological inner points and arbitrary convex sets.

(3.2.8) **Lemma.** *Every nonvoid convex set K has inner points.*

Proof. There exist a finite number of points $X_1, \ldots, X_s \in K$ such that $\mathscr{M}\{X_1, \ldots, X_s\} = \mathscr{M} K$. By lemma (2.3.10), the polyhedron

$$P := \mathscr{H}\{X_1, \ldots, X_s\} \subseteq K$$

(theorem (2.11.4)) has algebraic inner points, which by lemma (3.2.7) are topological inner points of P with respect to $\mathscr{M} P = \mathscr{M} K$. These points are a fortiori inner points of K. □

The analogy between polyhedra and general convex sets in R^n extends further. The following two lemmas are perfect analogues of lemmas (2.3.11) and (2.3.13).

(3.2.9) **Lemma.** *Let K be a convex set. If $X \in K^I$, then for each $Y \in \mathscr{M}K$ there exists an $\varepsilon > 0$ such that*

$$X + \varepsilon(Y - X) \in K \quad \text{and} \quad X - \varepsilon(Y - X) \in K.$$

Proof. If $X \in K^I$, then there exists a polyhedral relative neighborhood U of X which contains X as an inner point in the topological sense (3.2.4). According to (3.2.7), X is an inner point also in the algebraical sense (2.3.9) and the existence of a suitable ε follows from lemma (2.3.11). □

The above lemma yields immediately that

(3.2.10) $$\mathscr{M}(K^I) = \mathscr{M}K \quad \textit{and} \quad \mathscr{L}(K^I) = \mathscr{L}K$$

for all convex sets K in R^n.

Translations $t(X) \to X + Y$ and scalings $s(X) \to \mu X$ with $\mu \neq 0$, are clearly homeomorphisms of R^n into itself, and therefore these maps take open sets into open sets. The same is true—with respect to the relative topology—for translations and scalings which map a linear manifold onto itself. The upshot of these facts is the important and useful (compare Klee [*5*], Eggleston [*1*]):

(3.2.11) **Accessibility Lemma.** *If K is convex, $X \in K^I$ and $Y \in \bar{K}$, then K^I contains the entire open segment (X, Y).*

Proof. Let $Z = \lambda X + \mu Y$, $\lambda, \mu > 0$, $\lambda + \mu = 1$. We first show that there exist $\bar{X} \in K^I$ and $\bar{Y} \in K$ such that $Z = \lambda \bar{X} + \mu \bar{Y}$. In other words, we show that it suffices to prove the lemma under the stronger hypothesis $Y \in K$. To this end, let $U \subseteq K$ be an open neighborhood of X in $\mathscr{M}K$. Then $V := \frac{1}{\mu}(Z - \lambda U)$ is an open neighborhood in $\mathscr{M}K$ of $\frac{1}{\mu}(Z - \lambda X) = Y$. Since $Y \in \bar{K}$ there exists in V a point $\bar{Y} \in K$. The point $\bar{X} := \frac{1}{\lambda}(Z - \mu \bar{Y})$ then lies in U and therefore in K^I. Clearly $Z = \lambda \bar{X} + \mu \bar{Y}$.

We assume now that $Y \in K$, and that X, Z, U are as before. Then $W := \lambda U + \mu Y \subseteq K$ is a relative neighborhood of Z. Hence $Z \in W \subseteq K$, which shows that Z is an inner point of K. □

With the help of lemma (3.2.11) one obtains easily (compare Eggleston [*1*]):

(3.2.12) *The relative interior K^I of a convex set is convex.*

(3.2.13) *If K is convex, then $\overline{K^I} = \bar{K}$ and $(\bar{K})^I = K^I$.*

The accessibility lemma provides a method for characterizing inner points. Indeed $X \in K^I$ whenever there exist $Y \in K^I$ and $Z \in \bar{K}$ such that

$X \in (Y, Z)$. In view of the existence (3.2.8) of inner points, this leads immediately to the following algebraic rather than topological criterion for inner points:

(3.2.14) *Let K be a convex set. Then the following statements are equivalent:*

(i) $X \in K^I$,
(ii) *for each* $Y \in K$ *there exists* $Z \in K$ *such that* $X \in (Y, Z)$,
(iii) $(X, Y) \cap K \neq \emptyset$ *for all* $Y \in \mathcal{M} K$.

Note that all subsets K of R^n with the property that for $X, Y \in K$ there exists $Z \in K$ with $X \in (Y, Z)$ are convex and relatively open in their affine hull. However, this sufficient condition can be further weakened. To this end we call a set $S \subseteq R^n$

(3.2.15) *star open*

with respect to a point V if S is the union of a family of open line segments (X_i, Y_i), $i \in I$, whose intersection contains V. Every relatively open convex set $K = K^I$ in R^n is clearly star open with respect to anyone of its points. Conversely

(3.2.16) **Lemma** (Dubins [*1*]). *If a convex set $K \subset R^n$ is star open with respect to some point $V \in K$, then $K = K^I$.*

Proof. Let $X, Y \in K$. By star openness there exist $\bar{X}, Y' \in K$ such that $X \in (\bar{X}, V)$, $V \in (Y, Y')$. The line segment $(\bar{X}, Y')$ then intersects the line $\mathcal{M}\{X, Y\}$ in a point $Z \in K$ with $X \in (Y, Z)$. Thus condition (3.2.14.(ii)) is met. □

For general linear topological spaces, the algebraic characterization (3.2.14) and the topological characterization (3.2.4) of the relative interior of a convex set need not coincide. The set of algebraic inner points of a convex set is also called the

(3.2.17) *core*

of K (see Klee [*5*]). The core of a convex set K contains the relative interior of K with respect to any topology which makes the space a linear topological space. Lemma (3.2.8) fails in spaces of infinite dimension. The convex set of all finite sequences of real nonnegative numbers in Hilbert space has no algebraically interior points.

We shall now examine intersections, unions, convex hulls, as well as Minkowski sums

$$K_1 + K_2 := \{X_1 + X_2 \mid X_1 \in K_1, X_2 \in K_2\}$$

and their relationship to closure and relative interior. K_1 and K_2 will always denote convex sets, and S any set in R^n.

The convex hull of a closed set need not be closed. For an example consider a line and a point not on it. However

(3.2.18) *The convex hull of a compact set S in R^n is compact.*

Proof. Consider the compact set $I:=\left\{U\in R_+^{n+1}\;\middle|\;\sum_{i=1}^{n+1}u_i=1\right\}$ and let S^{n+1} denote the cartesian product of $n+1$ copies of S. As the cartesian product of compact sets is again compact, so is the set $S^{n+1}\times I$. Each element Z of S^{n+1} can be interpreted as a matrix $Z=(Z_1,\ldots,Z_{n+1})$ with columns in S. The mapping $\phi: S^{n+1}\times I\to\mathscr{H}S$ defined by $\phi(Z,U):=ZU$ is continuous (theorem (3.1.13)). By Carathéodory's theorem (2.2.12), $\phi(S^{n+1}\times I)=\mathscr{H}(S)$. Thus $\mathscr{H}(S)$ is the continuous image of a compact set, and therefore it is compact also. □

Proposition (3.2.18) has been proved for Banach spaces by Mazur [*2*]. In general linear topological spaces, the convex hull of a compact set need not be closed. As an example, consider the set $C[0,1]$ of all continuous real valued functions on the unit interval $[0,1]\subset R$. All real valued functions $\phi: C[0,1]\to R$ with the topology of pointwise convergence form a linear topological space E. For every $x\in[0,1]$, let $\phi_x\in E$ be the mapping defined by $\phi_x(f):=f(x)$ where $f\in C[0,1]$. The set $\phi_{[0,1]}$ of all these mappings ϕ_x is homeomorphic to $[0,1]$, hence compact. But $\mathscr{H}\phi_{[0,1]}$ is not closed, as the Riemann integral $\rho(f):=\int_0^1 f(x)\,dx$ belongs to $\overline{\mathscr{H}\phi_{[0,1]}}$ but not to $\mathscr{H}\phi_{[0,1]}$. Moreover, $\overline{\mathscr{H}\phi_{[0,1]}}$ is not compact. Additional conditions which ensure that the closed convex hull of a compact set is compact are stated in a theorem of Krein [*1*] (see Köthe [*1*]). Note that the convex hull of a compact set which is the union of a finite number of convex compact sets is again compact. The proof is essentially the one of (3.2.18).

For general sets S we assert

(3.2.19)
$$\overline{\mathscr{H}S}\supseteq\mathscr{H}\bar{S}.$$

Proof. Let $X\in\mathscr{H}\bar{S}$. Then X is of the form $\sum_{i=1}^{k}\lambda_i X^i$ with $\Sigma\lambda_i=1$, $\lambda_i\geqslant 0$ and $X^i\in\bar{S}$ for $i=1,\ldots,k$, and there exist sequences $\{X_j^i\}_{j=1,2,\ldots}$ with $X_j^i\in S$ for all j and $X_j^i\to X^i$ as $j\to\infty$ for $i=1,\ldots,k$. Since linear operations are continuous (3.1.13), $\sum_{i=1}^{k}\lambda_i X_j^i\to\sum_{i=1}^{k}\lambda_i X^i=X$. □

(3.2.20)
$$\text{(i)}\quad \bar{K}_1\cap\bar{K}_2\supseteq\overline{K_1\cap K_2}.$$
$$\text{(ii)}\quad \mathscr{H}(\bar{K}_1\cup\bar{K}_2)\subseteq\overline{\mathscr{H}(K_1\cup K_2)}.$$

Proof. (i) is a general relationship for arbitrary closed sets in any topological space. (ii) follows from (3.2.19) since $\bar{K}_1\cup\bar{K}_2=\overline{K_1\cup K_2}$ for arbitrary sets.

(3.2.21) *If* $K_1^I \cap K_2^I \neq \emptyset$ *then,*

(i) $$(K_1 \cap K_2)^I = K_1^I \cap K_2^I,$$

(ii) $$(\mathscr{H}(K_1 \cup K_2))^I = \mathscr{H}(K_1^I \cup K_2^I).$$

Proof. (i) Let $X \in K_1^I \cap K_2^I$ and let $Y \in K_1 \cap K_2$. Applying (3.2.14) to K_i yields the existence of $Z_i \in K_i$ with $X \in (Y, Z_i)$ for $i=1,2$. Clearly $Z_i \in K_1 \cap K_2$ for at least one $i=1,2$. Thus (3.2.14.(ii)) is satisfied and $X \in (K_1 \cap K_2)^I$ must hold. To prove the converse inclusion, let $X \in (K_1 \cap K_2)^I$. By hypothesis, there exists $V \in K_1^I \cap K_2^I$, and by (3.2.14), there exists $Z \in K_1 \cap K_2$ with $X \in (V, Z)$. The accessibility lemma (3.2.11) applied to K_i yields immediately $X \in K_i^I$ for $i=1,2$.

(ii) Observe that by (3.2.10)

$$\mathscr{M}(K_1^I \cup K_2^I) = \mathscr{M}(\mathscr{M} K_1^I \cup \mathscr{M} K_2^I) = \mathscr{M}(\mathscr{M} K_1 \cup \mathscr{M} K_2) = \mathscr{M}(K_1 \cup K_2)$$

and therefore

(3.2.22) $$\mathscr{M}\mathscr{H}(K_1^I \cup K_2^I) = \mathscr{M}(K_1^I \cup K_2^I) = \mathscr{M}(K_1 \cup K_2) = \mathscr{M}\mathscr{H}(K_1 \cup K_2).$$

We use this to prove that in general

(3.2.23) $$(\mathscr{H}(K_1 \cup K_2))^I = (\mathscr{H}(K_1^I \cup K_2^I))^I.$$

The "$\supseteq$" inclusion follows immediately as (3.2.22) insures that the above formation of the relative interior is a monotonic operation. As to the remaining inclusion, note that by (3.2.13) and (3.2.20.(ii)),

$$\mathscr{H}(K_1 \cup K_2) \subseteq \mathscr{H}(\bar{K}_1 \cup \bar{K}_2) = \mathscr{H}(\overline{K_1^I} \cup \overline{K_2^I}) \subseteq \overline{\mathscr{H}(K_1^I \cup K_2^I)},$$

and therefore again in view of (3.2.22), $(\mathscr{H}(K_1 \cup K_2))^I \subseteq (\overline{\mathscr{H}(K_1^I \cup K_2^I)})^I = (\mathscr{H}(K_1^I \cup K_2^I))^I$. This proves (3.2.23).

It remains to be shown that $\mathscr{H}(K_1^I \cup K_2^I)$ is relatively open provided there exists $V \in K_1^I \cap K_2^I$. Let $X \in \mathscr{H}(K_1^I \cup K_2^I)$. Then $X = \lambda X_1 + \mu X_2$, where $X_i \in K_i^I$ for $i=1,2$. By (3.2.14) there exist $Z_i \in K_i^I$ with $X_i \in (V, Z_i)$ for $i=1,2$. Then $Z := \lambda Z_1 + \mu Z_2 \in \mathscr{H}(K_1^I \cup K_2^I)$ and $X \in (V, Z)$. Thus $\mathscr{H}(K_1^I \cup K_2^I)$ is star open with respect to V, whence $\mathscr{H}(K_1^I \cup K_2^I)^I = \mathscr{H}(K_1^I \cup K_2^I)$ by lemma (3.2.16). □

If K_1 and K_2 are convex, then so is $K_1 + K_2$. We examine the topological properties of $K_1 + K_2$. It is easy to see that $K_1 + K_2 \subseteq \bar{K}_1 + \bar{K}_2 \subseteq \overline{K_1 + K_2}$ and therefore

(3.2.24) $$\overline{\bar{K}_1 + \bar{K}_2} = \overline{K_1 + K_2}.$$

$\bar{K}_1 + \bar{K}_2$ need not be closed, and therefore $\bar{K}_1 + \bar{K}_2 = \overline{K_1 + K_2}$ does not hold in general.

Consider, for instance, in R^2 the sets

$$\left\{\begin{pmatrix} x_1 \\ x_2 \end{pmatrix} \;\middle|\; x_1 \geqslant 0,\ x_2 \leqslant 0\right\}, \quad \left\{\begin{pmatrix} x_1 \\ x_2 \end{pmatrix} \;\middle|\; x_1 x_2 \geqslant 1,\ x_1 \geqslant 0\right\}.$$

For $n=1,2,\ldots$, the points $\begin{pmatrix} 1/n \\ 0 \end{pmatrix} = \begin{pmatrix} 1/n \\ n \end{pmatrix} + \begin{pmatrix} 0 \\ -n \end{pmatrix}$ belong to the sum of the two sets, but the point $\begin{pmatrix} 0 \\ 0 \end{pmatrix} = \lim_{n\to\infty} \begin{pmatrix} 1/n \\ 0 \end{pmatrix}$ does not. Consider also the following two closed convex cones:

$$C_1 := \left\{\begin{pmatrix} x_1 \\ x_2 \\ x_3 \end{pmatrix} \;\middle|\; x_1 \geqslant 0,\ x_2 \geqslant 0,\ x_3 \geqslant 0,\ x_1 x_2 \geqslant x_3^2\right\},$$

$$C_2 := \left\{\begin{pmatrix} x_1 \\ x_2 \\ x_3 \end{pmatrix} \;\middle|\; x_1 \leqslant 0,\ x_2 \geqslant 0,\ x_3 \leqslant 0,\ -x_1 x_2 \geqslant x_3^2\right\}.$$

The point $(0,0,1)^T$ is then a boundary point of $C_1 + C_2$ without belonging to it. The sum of two closed convex cones is closed if $C_1 \cap (-C_2)$ is a subspace (see for instance Klee [*5*]). The sum of a compact set and a closed set is closed, as is the sum of polyhedra. For more general collections of closed convex sets which are closed under addition see Gale and Klee [*1*].

As a contrast, $K_1^I + K_2^I$ is always open in $\mathscr{M}(K_1 + K_2)$, and one has the following relation (Rockafellar [*1*]), which is important for our derivation of separation theorems in the next section:

$$K_1^I + K_2^I = (K_1 + K_2)^I. \tag{3.2.25}$$

Proof. First note that for arbitrary $S_1, S_2 \subseteq R^n$,

$$\mathscr{M}(S_1 + S_2) = \mathscr{M}(\mathscr{M}\, S_1 + \mathscr{M}\, S_2) = \mathscr{M}\, S_1 + \mathscr{M}\, S_2.$$

Therefore we obtain from (3.2.2) and (3.2.10)

$$\mathscr{M}(K_1 + K_2) = \mathscr{M}(\bar{K}_1 + \bar{K}_2) = \mathscr{M}(K_1^I + K_2^I).$$

Formation of the relative interior of the following sets is therefore monotonic, and (3.2.13) and (3.2.24) yield

$$(K_1 + K_2)^I \subseteq (\bar{K}_1 + \bar{K}_2)^I = (\overline{K_1^I} + \overline{K_2^I})^I \subseteq (\overline{K_1^I + K_2^I})^I = (K_1^I + K_2^I)^I \subseteq K_1^I + K_2^I.$$

This proves one inclusion. Now consider any $X \in K_1^I + K_2^I$, $Y \in K_1 + K_2$. Then $X = X_1 + X_2$, $Y = Y_1 + Y_2$ with $X_i \in K_i^I$, $Y_i \in K_i$ for $i=1,2$. By (3.2.14), there exists $Z_i \in K_i$ with $X_i \in (Y_i, Z_i)$ for $i=1,2$. Thus $X \in (Y, Z_1 + Z_2)$, and $X \in (K_1 + K_2)^I$ follows from (3.2.14). □

3.3. Separation of Convex Set

Let us consider the euclidean norm

$$\|X\| = +\sqrt{X^T X}$$

in R^n. A point Y of a convex set K is the orthogonal projection (2.6.3) of a point X into K if $\|Y-X\| = \inf\{\|Z-X\| \mid Z \in K\}$. In chapter 2 we established the existence of orthogonal projections for linear subspaces and polyhedral cones. We recognized the important polarity relation $C^{pp} = C$ to be a consequence of the existence of orthogonal projections into C (theorem (2.7.7)).

(3.3.1) **Lemma.** *Let K be a closed convex set of R^n. Then each point of R^n has an orthogonal projection into K.*

Proof. Define for any $X \in R^n$

$$\gamma := \inf\{\|Y-X\| \mid Y \in K\} \geqslant 0.$$

There exists a sequence $\{Y_i\}_{i=1,2,\ldots}$ of points $Y_i \in K$ such that $\|Y_i - X\| \to \gamma$. Moreover, the identity

$$\frac{1}{2}\left\|Y_i - Y_k\right\|^2 = \left\|Y_i - X\right\|^2 + \left\|Y_k - X\right\|^2 - 2\left\|X - \frac{Y_i + Y_k}{2}\right\|^2$$

holds. If we choose i and k sufficiently large we have therefore $\|Y_i - X\|^2 \leqslant \gamma^2 + \varepsilon$ and $\|Y_k - X\|^2 \leqslant \gamma^2 + \varepsilon$. Further $\|X - (Y_i + Y_k)/2\| \geqslant \gamma$ since $(Y_i + Y_k)/2 \in K$. Thus $\|Y_i - Y_k\|^2 \leqslant 4\varepsilon$. The Cauchy-criterion (3.1.12) then yields the existence of a limit $Y \in K$ with $\|Y - X\| = \gamma$. □

Closed convex sets in R^n are characterized by the existence and uniqueness of orthogonal projections (Motzkin [*1*]). For a related characterization see Klee [*2*]. In Hilbert space, this is an unsolved problem (for instance Phelps [*1*]).

In particular, each point $X \in R^n$ has a projection (2.6.3) into a closed cone $C \subseteq R^n$. Thus every closed cone satisfies property (iv) of theorem (2.7.7), and we have

(3.3.2) **Theorem.** *Every closed cone in R^n is an intersection of half-spaces $H_i := \{X \mid A_i^T X \leqslant 0\}$. A cone C is closed if and only if $C^{pp} = C$. The cone C^{pp} is the closure $\bar{C}$ of cone C.*

Proof. Only the last statement of the theorem remains to be proved. We have $\bar{C} = \bar{C}^{pp} \supseteq C^{pp} \supseteq C$, which implies $C^{pp} = \bar{C}$ since C^{pp} is an intersection of halfspaces (2.7.6) and therefore closed. □

Note that theorem (3.3.2) is false for incomplete spaces. For instance, let R be the field of rationals. Then the set

$$C:=\left\{\begin{pmatrix} x \\ y \end{pmatrix} \in R^2 \;\middle|\; y \leqslant \sqrt{2}\,x\right\}$$

is a closed cone. However, $C^p = \{0\}$ and $C^{pp} = R^2 \supset C$.

The closed cones form a lattice under set inclusion. However unlike the case of polyhedral cones, the sum of two arbitrary closed cones need not be closed (compare section 3.2). Thus, the closure of the sum plays the role of the "cup" product, and the dual of

$$(C_1 \cup C_2)^p = C_1^p \cap C_2^p.$$

reads as follows

(3.3.3) $$(C_1 \cap C_2)^p = \overline{C_1^p + C_2^p}.$$

Proof. $(C_1 \cap C_2)^p = (C_1^{pp} \cap C_2^{pp})^p = (C_1^p \cup C_2^p)^{pp} = (C_1^p + C_2^p)^{pp} = \overline{C_1^p + C_2^p}$ by theorem (3.3.2).

Formula (3.3.3) does not hold for general cones. The cones $C_1 := \left\{\begin{pmatrix} x_1 \\ x_2 \end{pmatrix} \;\middle|\; x_1 > 0\right\} \cup \left\{\begin{pmatrix} 0 \\ 0 \end{pmatrix}\right\}$ and $C_2 := -C_1$ provide a simple counter example.

We now employ the duality properties to derive a separation theorem for cones. In chapter 2, a plane $E := \{X \mid A^T X = 0\}$ was said to separate (2.12.8) the cones C_1 and C_2 if $C_1 \subseteq \{X \mid A^T X \leqslant 0\}$ and $C_2 \subseteq \{X \mid A^T X \geqslant 0\}$. However, we shall need a slightly stronger concept of separation: We require that the separating plane E does not contain both C_1 and C_2.

(3.3.4) **Separation Theorem for General Cones.** *In R^n, two cones C_1 and C_2 can be separated by a plane E such that $E \not\supseteq C_1 \cup C_2$ if and only if they have no inner points in common, that is, if and only if*

$$C_1^I \cap C_2^I = \emptyset.$$

Proof. By the definition of the polar (2.7.1), the inclusion relations $C_1 \subseteq \{X \mid A^T X \leqslant 0\}$ and $C_2 \subseteq \{X \mid A^T X \geqslant 0\}$ are equivalent to $A \in C_1^p$ and $-A \in C_2^p$, respectively. The requirement that $E := \{X \mid A^T X = 0\}$ should not contain $C_1 \cup C_2$, translates into

$$A \notin (C_1 \cup C_2)^\perp = (\mathscr{L}(C_1 \cup C_2))^p.$$

Thus C_1 and C_2 are nonseparable in the sense of the lemma if and only if

$$C_1^p \cap (-C_2^p) \subseteq (\mathscr{L}(C_1 \cup C_2))^p.$$

Polarizing this relation yields

$$(C_1^p \cap (-C_2^p))^p \supseteq \mathscr{L}(C_1 \cup C_2),$$

and polarizing again yields the original relations, since $C_1^p \cap (-C_2)^p$ is closed. The above two relations are therefore equivalent. By (3.3.3),

$$(C_1^p \cap (-C_2^p))^p = \overline{C_1^{pp} + (-C_2^p)^p} = \overline{C_1^{pp} - C_2^{pp}} = \overline{\bar{C}_1 - \bar{C}_2}.$$

Hence nonseparability holds if and only if

$$\overline{\bar{C}_1 - \bar{C}_2} \supseteq \mathscr{L}(C_1 \cup C_2). \tag{3.3.5}$$

Now

$$\overline{\bar{C}_1 - \bar{C}_2} \subseteq \overline{\mathscr{L}(\bar{C}_1 - \bar{C}_2)} = \mathscr{L}(\bar{C}_1 - \bar{C}_2) \subseteq \mathscr{L}(\overline{C_1 \cup C_2}) = \mathscr{L}(C_1 \cup C_2).$$

Hence (3.3.5) is only possible if equality holds. But then $\overline{\bar{C}_1 - \bar{C}_2}$ is a subspace, and we have

$$\overline{\bar{C}_1 - \bar{C}_2} = (\overline{\bar{C}_1 - \bar{C}_2})^I = (\bar{C}_1 - \bar{C}_2)^I = C_1^I - C_2^I$$

by (3.2.13) and (3.2.25). Since the cone $C_1^I - C_2^I$ consists of inner points exclusively by (3.2.25), it is a subspace if and only if it contains the origin (compare (2.8.10)), $0 \in C_1^I - C_2^I$, which is of course equivalent to $C_1^I \cap C_2^I \neq \emptyset$. □

We proceed to extend our results from cones to general convex sets. We shall use the homogenization scheme outlined in section 2.11. There we defined the two maps

$$\mathscr{G}K := \left\{ \begin{pmatrix} X \\ 1 \end{pmatrix} \in R^{n+1} \,\middle|\, X \in K \right\}^{pp}, \quad \text{where} \quad K \subseteq R^n,$$

$$\mathscr{D}C := \left\{ X \in R^n \,\middle|\, \begin{pmatrix} X \\ 1 \end{pmatrix} \in C \right\}, \quad \text{where} \quad C \subseteq R^{n+1}.$$

For these maps we have

(3.3.6) *If K is convex, then*

(i) $\mathscr{D}\mathscr{G}K = \bar{K}$.
(ii) $(\mathscr{D}\mathscr{G}K)^I = \mathscr{D}(\mathscr{G}K)^I = K^I$.

Proof. It will suffice to prove the inclusions

$$(\mathscr{D}\mathscr{G}K)^I \subseteq \mathscr{D}(\mathscr{G}K)^I \subseteq K^I.$$

Indeed, $\mathscr{D}\mathscr{G}K$ is an intersection of halfspaces by (2.11.11), and therefore closed. The above inclusions then imply $\mathscr{D}\mathscr{G}K \subseteq \bar{K}$ by (3.2.13). As $\mathscr{D}\mathscr{G}K \supseteq K$, (i) must hold. From this then follows again by (3.2.13) that $(\mathscr{D}\mathscr{G}K)^I = K^I$ and therefore (ii).

Let then $X \in (\mathcal{D}\mathcal{G}K)^I$ and $Y \in \mathcal{D}(\mathcal{G}K)^I \subseteq \mathcal{D}\mathcal{G}K$. By (3.2.14.(ii)), there exists $Z \in \mathcal{D}\mathcal{G}K$ with $X \in (Y, Z)$. Then $\begin{pmatrix} Y \\ 1 \end{pmatrix} \in (\mathcal{G}K)^I$ and $\begin{pmatrix} Z \\ 1 \end{pmatrix} \in \mathcal{G}K$ gives $\begin{pmatrix} X \\ 1 \end{pmatrix} \in (\mathcal{G}K)^I$ by the accessibility lemma (3.2.11). Thus $X \in \mathcal{D}(\mathcal{G}K)^I$. Let then $X \in \mathcal{D}(\mathcal{G}K)^I$ and select an arbitrary $Y \in K$. Then $\begin{pmatrix} X \\ 1 \end{pmatrix} \in (\mathcal{G}K)^I$ and $\begin{pmatrix} Y \\ 1 \end{pmatrix} \in \mathcal{G}K$. By (3.2.14), there exists $\begin{pmatrix} Z \\ w \end{pmatrix} \in \mathcal{G}K$ such that $\begin{pmatrix} X \\ 1 \end{pmatrix}$ belongs to the open line segment bracketed by $\begin{pmatrix} Z \\ w \end{pmatrix}$ and $\begin{pmatrix} Y \\ 1 \end{pmatrix}$. Plainly $w = 1$ and $X \in (Y, Z)$. This implies $X \in K^I$ by (3.2.14), and completes the proof of (3.3.6). □

We have seen (2.11.11) that $\mathcal{D}\mathcal{G}K$ is the intersection of all half-spaces containing K. Hence (3.3.6) gives at once the

(3.3.7) **Theorem.** *Every closed convex set* $K \in R^n$ *is an intersection of halfspaces* $H := \{X \mid A^T X \leqslant b\}$.

Note that theorem (3.3.7) may be interpreted as a separation theorem:

(3.3.8) **Corollary to Theorem (3.3.7).** *In* R^n, *each closed convex set* K *and each point* $X \notin K$ *can be separated by a plane* E *which does not contain* X.

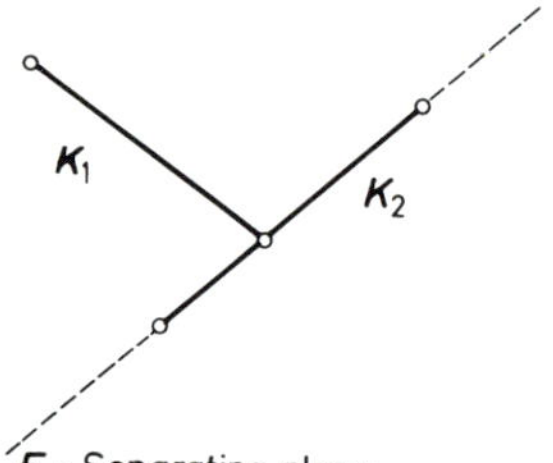

Fig. 8. Separation of two closed line segments without common inner point in R^2

We obtain another separation theorem by extending theorem (3.3.4) (see Fenchel [2]):

(3.3.9) **Separation Theorem for General Convex Sets.** *In* R^n, *two non-void convex sets* K_1 *and* K_2 *can be separated by a plane* E *such that* $E \not\supseteq K_1 \cup K_2$ *if and only if they have no inner points in common, that is, if and only if* (Fig. 8)

$$K_1^I \cap K_2^I = \emptyset.$$

Proof. We note for later use in the proof that $\begin{pmatrix} X \\ z \end{pmatrix} \in (\mathcal{G}K)^I$ implies $z > 0$ provided K is nonvoid. Clearly $z \geqslant 0$ for all $\begin{pmatrix} X \\ z \end{pmatrix} \in \mathcal{G}K$. Assume then that $\begin{pmatrix} X \\ 0 \end{pmatrix} \in (\mathscr{G}K)^I$. Choose any $Y \in K$. Then $\begin{pmatrix} Y \\ 1 \end{pmatrix} \in \mathscr{G}K$ and $\begin{pmatrix} Z \\ w \end{pmatrix} = \begin{pmatrix} X \\ 0 \end{pmatrix} - \alpha\left(\begin{pmatrix} Y \\ 1 \end{pmatrix} - \begin{pmatrix} X \\ 0 \end{pmatrix}\right) \in \mathscr{G}K$ must hold for some $\alpha > 0$ by lemma (3.2.9). But this is impossible since $w = -\alpha < 0$.

We proceed to prove that $(\mathscr{G}K_1)^I \cap (\mathscr{G}K_2)^I = \emptyset$ if and only if $K_1^I \cap K_2^I = \emptyset$, preparing the way for application of theorem (3.3.4). Since the map $\mathscr{D}$ is intersection preserving, we have by (3.3.6(iii))

$$K_1^I \cap K_2^I = \mathscr{D}(\mathscr{G}K_1)^I \cap \mathscr{D}(\mathscr{G}K_2)^I = \mathscr{D}((\mathscr{G}K_1)^I \cap (\mathscr{G}K_2)^I).$$

If $(\mathscr{G}K_1)^I \cap (\mathscr{G}K_2)^I = \emptyset$, then $K_1^I \cap K_2^I = \emptyset$. Suppose then $\begin{pmatrix} X \\ z \end{pmatrix} \in (\mathscr{G}K_1)^I \cap (\mathscr{G}K_2)^I$. According to our preparatory argument, $z > 0$. But then $\begin{pmatrix} Y \\ 1 \end{pmatrix} := \frac{1}{z}\begin{pmatrix} X \\ z \end{pmatrix} \in (\mathscr{G}K_1)^I \cap (\mathscr{G}K_2)^I$ and therefore $Y \in K_1^I \cap K_2^I$.

Now let $K_1^I \cap K_2^I = \emptyset$. Then $(\mathscr{G}K_1)^I \cap (\mathscr{G}K_2)^I = \emptyset$, and by theorem (3.3.4) there exists a plane $\tilde{E}$ which passes through the origin and separates the cones $\mathscr{G}K_1$ and $\mathscr{G}K_2$ without containing both. The plane $E := \mathscr{D}\tilde{E}$, then separates K_1 and K_2, since the map $\mathscr{D}$ is inclusion preserving, and since $K_i \subseteq \mathscr{D}\mathscr{G}K_i$. If E contains K_i, then $\mathscr{G}K_i \subseteq \mathscr{G}E \subseteq \tilde{E}$. Hence $\mathscr{D}\tilde{E}$ does not contain both K_1 and K_2. On the other hand, if a plane E separates K_1 and K_2 without containing both, then the plane $\tilde{E} := \mathscr{L}\mathscr{G}E$ separates $\mathscr{G}K_1$ and $\mathscr{G}K_2$. If $\mathscr{G}K_i \subseteq \mathscr{L}\mathscr{G}E$ then $\mathscr{G}K_i \subseteq \mathscr{G}E$ since $\mathscr{G}E = \mathscr{L}\mathscr{G}E \cap \left\{\begin{pmatrix} X \\ z \end{pmatrix} \,\middle|\, z \geqslant 0\right\}$. Further $\mathscr{G}K_i \subseteq \mathscr{G}E$ implies $K_i \subseteq \mathscr{D}\mathscr{G}K_i \subseteq \mathscr{D}\mathscr{G}E = E$. The plane $\tilde{E} = \mathscr{L}\mathscr{G}E$ thus cannot contain both $\mathscr{G}K_1$ and $\mathscr{G}K_2$, and it therefore separates both cones in the sense of theorem (3.3.4). Hence $(\mathscr{G}K_1)^I \cap (\mathscr{G}K_2)^I = \emptyset$. But then $K_1^I \cap K_2^I = \emptyset$, and both directions of the theorem are proved. □

The separation theorem (3.3.9) does not hold in linear spaces with rational scalars. A simple example are the two sets $K_1 := \{x \in R \mid x < \sqrt{2}\}$, $K_2 := \{x \in R \mid x > \sqrt{2}\}$, where R is the field of rational numbers. However the following separation theorem, sometimes attributed to Stone, holds in full generality (Bourbaki [*3*], Hammer [*2*]):

(3.3.10) *If K_1 and K_2 are convex and disjoint then there exist disjoint convex sets A_1 and A_2 covering the entire space and such that $A_1 \supseteq K_1$ and $A_2 \supseteq K_2$.*

This follows by way of Zorn's lemma from the following elementary observation:

(3.3.11) *If K_1 and K_2 are disjoint convex sets, and if a point X is not in $K_1 \cup K_2$, then $\mathscr{H}(K_1 \cup \{X\})$ and K_2 or $\mathscr{H}(K_2 \cup \{X\})$ and K_1 are again disjoint.*

General separation theorems of the type (3.3.10) exist whenever there is a suitable hull-forming set operation (Ellis [*1*]).

A plane E

(3.3.12) *strictly separates*

two convex sets K_1 and K_2, if the latter belong respectively to the two linearly open halfspaces defined by E. Strict separability of two disjoint linearly open convex sets by a plane is possible in all linear spaces with real scalars (consult Köthe [*1*]). In a linear topological space one has strict separability by closed planes of any pair of disjoint nonvoid open convex sets (Eidelheit [*1*], Kakutani [*1*]).

For a more geometric proof of the separation theorem for convex sets in R^n with R real, see Botts [*1*].

3.4. Supporting Planes and Cones

Probably the best known theorem about convex sets is the supporting plane theorem of Minkowski [*3*], and few theorems rival its range of applicability. The closely related theorem of Hahn-Banach (compare for instance Bourbaki [*3*], Köthe [*1*]) is one of the basic tools of the

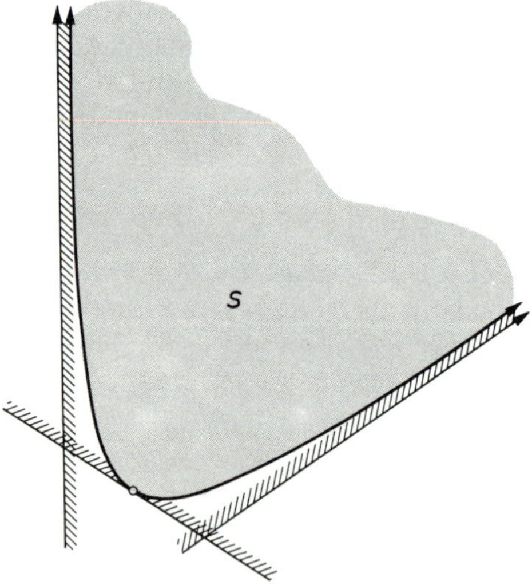

Fig. 9. Supporting halfspaces of a hyperbolic convex set S. Asymptotes support S without meeting it

theory of linear topological spaces. The supporting plane theorem will be seen to be an immediate consequence of the general separation theorem (3.3.9). In addition, we shall study supporting cones.

A halfspace $H := \{X \mid A^T X \leqslant b\}$ is a

(3.4.1) *supporting halfspace*

of a set $S \subseteq R^n$ if $b = \sup\{A^T X \mid X \in S\}$ (Fig. 9). In other words, S is contained in H, but not in any $H_\varepsilon := \{X \mid A^T A \leqslant b - \varepsilon\}$ where $\varepsilon > 0$. The boundary plane $E := \{X \mid A^T X = b\}$ of a supporting halfspace $H = \{X \mid A^T X \leqslant b\}$ of S is a

(3.4.2) *supporting plane*

of S. Notice that the empty convex set S has no supporting halfspace, and thus no supporting plane. Notice also that it is always possible to characterize a supporting plane by an equation $A^T X = b$ in such a manner that the halfspace $H := \{X \mid A^T X \leqslant b\}$ is a supporting halfspace. If both halfspaces $\{X \mid A^T X \leqslant b\}$ and $\{X \mid A^T X \geqslant b\}$ are supporting halfspaces of a set S, then the supporting plane $E := \{X \mid A^T X = b\}$ contains S. We call such a supporting plane a

(3.4.3) *singular supporting plane.*

The role played by singular and nonsingular supporting planes of general convex sets is quite similar to the role played by the singular and nonsingular boundary planes of a polyhedron (section 2.3). In particular one has the

(3.4.4) **Lemma.** *If a supporting plane E of a convex set K contains an inner point of K, then it is a singular supporting plane, that is, it contains K entirely.*

Proof. Let $X \in E \cap K^I$. This means that $\{X\}^I \cap K^I \neq \emptyset$, since $\{X\}^I = \{X\}$. Any supporting plane E of K through X separates the sets K and $\{X\}$. But since $\{X\}^I \cap K^I \neq \emptyset$, any such plane must contain both $\{X\}$ and K by virtue of the general separation theorem (3.3.9). ▯

Furthermore, one has for polyhedra

(3.4.5) **Lemma.** *Any supporting plane of a nonvoid polyhedron P meets P in an entire nonvoid face.*

Proof. Suppose $E = \{X \mid A^T X = b\}$ with $b = \sup\{A^T X \mid X \in K\}$ is a supporting plane of P. Since the linear form $f(X) = A^T X$ is bounded on P, it assumes its supremum on P (theorem (2.5.9)). The exposed set $F := \{X \in P \mid A^T X = b\} = E \cap P$ is therefore nonvoid, and by theorem (2.4.12), F is a face of P. ▯

Since a manifold M possesses only one nonvoid face, namely M itself (compare section 2.3), we conclude from (3.4.5) that

(3.4.6) *all supporting planes of a linear manifold $M \subseteq R^n$ are singular.*

We shall need a slightly different definition of supporting planes. This definition will be based on a notion of "closeness" of sets. Let p be any norm in R^n. Two sets $S, T \subseteq R^n$ are then

(3.4.7) *close*

if $\inf\{p(X-Y) \mid X \in S, Y \in T\} = 0$. No set is close to the empty set. Since all norms p in R^n are equivalent (theorem (3.1.9)), two sets that are close with respect to one norm, are also close with respect to another one.

(3.4.8) *Suppose S is any set in R^n. A plane $E = \{X \mid A^T X = b\}$ with $S \subseteq H := \{X \mid A^T X \leqslant b\}$ is a supporting plane of S if and only if E is close to S.*

Proof. Suppose that E is close to S. We want to show that $S \nsubseteq H_\varepsilon := \{X \mid A^T X \leqslant b - \varepsilon\}$ for all $\varepsilon > 0$. We may assume that closeness is defined with respect to the euclidean norm. The inequality $\|A\| \, \|X-Y\| \geqslant |A^T X - A^T Y|$ then gives $\|X-Y\| \geqslant \dfrac{\varepsilon}{\|A\|}$ for all $X \in E$ and $Y \in H_\varepsilon$, where $\varepsilon > 0$. Hence $S \subseteq H_\varepsilon$ would imply $\inf\{\|X-Y\| \mid X \in E, Y \in S\} > 0$, which contradicts the fact that E and S are close.

To prove the other direction of (3.4.8), let $\{\varepsilon_i\}_{i=1,2,\dots}$ be a sequence of positive numbers converging to zero. If E is a supporting plane, then we may select for each ε_i a point X_i which is in S but not in H_{ε_i}. To each point X_i we assign the point

$$Y_i := X_i + \frac{b - A^T X_i}{A^T A} A .$$

Clearly, $Y_i \in E$ and $\|X_i - Y_i\| \to 0$ as $i \to \infty$. E and S are close, which was to be shown. □

The notion of closeness also permits us to define supporting manifolds of arbitrary dimensions. A manifold $M \subseteq R^n$ is a

(3.4.9) *supporting manifold*

of a set $S \subseteq R^n$, if M is close (3.4.7) to S, and if either $M \cap S^I = \emptyset$ or $M \supseteq S$. In the latter case we speak of a

(3.4.10) *singular supporting manifold.*

Clearly, every (relative) boundary point of S, i.e. every point in $S \sim S^I$, is a nonsingular supporting manifold. It is left to the reader to verify, that supporting planes are indeed instances of supporting manifolds.

We are now able to formulate the

(3.4.11) **General Supporting Plane Theorem.** *Let K be a convex set in R^n. Then for each nonsingular supporting manifold M of K, there exists a nonsingular supporting plane of K containing M.*

Proof. Since the supporting manifold M is nonsingular, we have $M^I \cap K^I = \emptyset$. The general separation theorem (3.3.9) then yields the existence of a plane E which separates M and K without containing both. Since M and K are close to each other, the separating plane E is close to both M and K. Since moreover M and K are in halfspaces determined by E, it follows from (3.4.8) that E is a supporting plane of both M and K. By (3.4.6) E contains M. Recall that E does not contain both M and K. Hence E does not contain K, and is therefore a nonsingular supporting plane of K. □

(3.4.12) **Corollary to Theorem (3.4.11).** *Through each boundary point of a convex set K runs at least one nonsingular supporting plane.*

Let X_0 be any point in R^n, and let again S be any set in R^n. Then the set of all vectors A such that the (possibly degenerate) half-spaces $H := \{X \mid A^T(X - X_0) \leqslant 0\}$ contain S form a cone, the

(3.4.13) *normal cone* $N(S; X_0)$ (Fig. 10)

of S at X_0. If there pass no supporting planes through X_0, then $N(S; X_0)$ is defined to be the cone $\{0\}$.

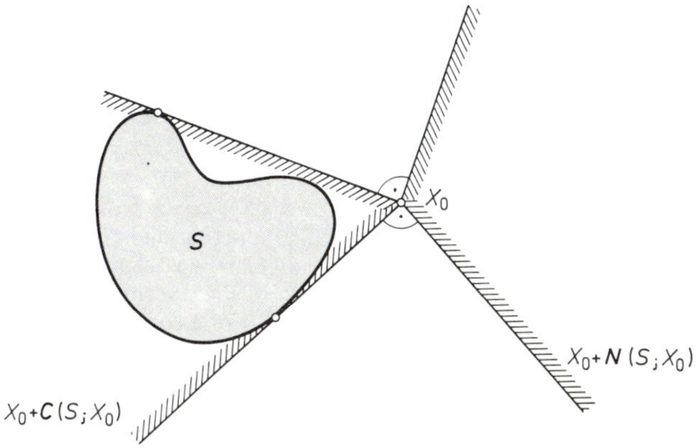

Fig. 10. Supporting and normal cones of a planar set S

Consider now the intersection of all halfspaces $H = \{X \mid A^T(X - X_0) \leqslant 0\} \supseteq S$. This set is obviously the sum of a cone and the point X_0. The cone is called the

(3.4.14) *supporting cone* $C(S; X_0)$ (Fig. 10)

of S at X_0. If the set of halfspaces supporting S at X_0 is empty, then $C(S;X_0)=R^n$.

The cone $C(S;X_0)$ is the intersection of all those halfspaces $\{X \mid A^T X \leqslant 0\}$ for which the halfspaces $H=\{X \mid A^T X \leqslant A^T X_0\}$ support S at X_0. In other words, $C(S;X_0)$ is the intersection of all halfspaces $\{X \mid A^T X \leqslant 0\}$ with $A \in N(S;X_0)$. Hence $C(S;X_0)=N(S;X_0)^p$ by the definition (2.7.1) of the polar. We shall see, that the normal cone $N(S;X_0)$ is in turn the polar of a third cone, the

(3.4.15) *cone of feasible directions* $D(S;X_0)$

of S at X_0. This cone consists of all directions Z with $X_0+\theta Z \in S$ for some $\theta>0$, and of the origin 0.

(3.4.16) **Theorem.** *Suppose* $S \subseteq R^n$ *and* $X_0 \in R^n$. *Between the normal cone* $N=N(S;X_0)$, *the supporting cone* $C=C(S;X_0)$, and *the cone of feasible directions* $D=D(S;X_0)$ *of* S *at* X_0 *holds the following relations*:

(i) $D^p=N$, (ii) $N^p=C$, (iii) $C^p=N$.

Proof. Relation (ii) has been established above. Relation (iii) follows from (i) and (ii) by means of the identity $S^p=S^{ppp}$. This leaves relation (i). Let $A \in N$ and $Z \in D$. Then $X_0+\theta Z \in S$ for some $\theta>0$. For the halfspace $\{X \mid A^T X \leqslant A^T X_0\}$ which supports S at X_0, we thus have $A^T(X_0+\theta Z) \leqslant A^T X_0$ for some $\theta>0$. Hence $A^T Z \leqslant 0$ for all $Z \in D$. This proves $N \subseteq D^p$.

To prove the inclusion $D^p \subseteq N$, suppose $A \in D^p$ and let X be any point of S. Then $Z:=X-X_0 \in D$, and therefore $A^T Z \leqslant 0$. Consequently $A^T X = A^T X_0 + A^T Z \leqslant A^T X_0$ for all $X \in S$, or in other words, $S \subseteq \{X \mid A^T X \leqslant A^T X_0\}$. Hence $A \in N$. □

For almost every boundary point X_0 of a convex set K the supporting cone $C(K;X_0)$ is a halfspace, in other words, the supporting halfspace at X_0 is unique (see for instance Reidemeister [*1*]). This result has been sharpened by Anderson and Klee [*1*]. Busemann and Feller [*1*] established second order smoothness almost everywhere on convex surfaces in 3-space. For a corresponding result about the existence of second derivatives of convex functions of n variables see A. D. Alexandroff [*1*]. The number of boundary points X_0 of a convex set K for which $C(K;X_0)$ is pointed is enumerable (Besicovitch [*1*]).

3.5. Boundedness and Polarity

It was proved in section 2.12 that an unbounded polyhedron P must contain at least one halfline $\{X+\theta Y \mid \theta \geqslant 0\}$. This also follows immediately from the finite basis representation

$$P=\mathscr{H}\{X_1,\ldots,X_k\}+\mathscr{C}\{Y_1,\ldots,Y_l\}$$

of the polyhedron. Any direction Y of a halfline contained in P is a "direction of infinity". Most of the results on directions of infinity of polyhedra carry (e.g. (2.11.10)) over to general convex sets.

(3.5.1) *Every closed nonbounded convex set K possesses nonzero directions of infinity. If Y is any such direction of infinity, then*

$$\{X+\theta Y \mid \theta \geqslant 0\} \subseteq K$$

for all $X \in K$.

Proof. As K is unbounded, it contains a sequence of points $\{X_i\}_{i=1,2,\dots}$ with $\|X_i\| \to \infty$. The sequence

$$\left\{\frac{X_i}{\|X_i\|}\right\}_{i=1,2,\dots}$$

is contained,—and therefore has at least one limit point Y,—in the compact set $\{X \mid \|X\|=1\}$. Clearly $Y \neq 0$. To verify that Y is a direction of infinity, choose any $X \in K$ and $\theta \geqslant 0$. Then $X+\theta Y$ is a limit of the points

$$X + \frac{\theta}{\|X_i\|}(X_i - X) \tag{3.5.2}$$

as $i \to \infty$. Since $0 \leqslant \dfrac{\theta}{\|X_i\|} \leqslant 1$ for sufficiently large i, all but a finite number of the points (3.5.2) are contained in K. Hence $X+\theta Y \in \bar{K} = K$ for all $X \in K$ and all $\theta \geqslant 0$.

This holds for every direction of infinity Y of K with $\|Y\|=1$ since every such direction arises from a sequence of points $X_i \in K$ with $\|X_i\| \to \infty$ in the fashion described above. □

The closedness of K is essential for (3.5.1) to hold, as is seen from the example (Fig. 11)

$$K := \left\{\begin{pmatrix}0\\0\end{pmatrix}\right\} \cup \left\{\begin{pmatrix}x_1\\x_2\end{pmatrix} \;\middle|\; 0 < x_1 \leqslant 1\right\} \subset R^2.$$

For general convex sets, the following weaker statement holds:

(3.5.3) **Theorem.** *Every nonbounded convex set K possesses nonzero directions of infinity. If Y is any such direction of infinity, then*

$$\{X+\theta Y \mid \theta \geqslant 0\} \subseteq K$$

for all $X \in K^I$.

Proof. If K is unbounded, then so is $\bar{K}$. By (3.5.1) there are nonzero directions of infinity of $\bar{K}$. Suppose Y is one of them, and let $X \in K^I$ and $\theta > 0$. Then $X + 2\theta Y \in \bar{K}$ by (3.5.1), and the open segment $(X, X + 2\theta Y)$ is contained in K^I according to the accessibility lemma (3.2.11). In particular $X + \theta Y \in K^I \subseteq K$. □

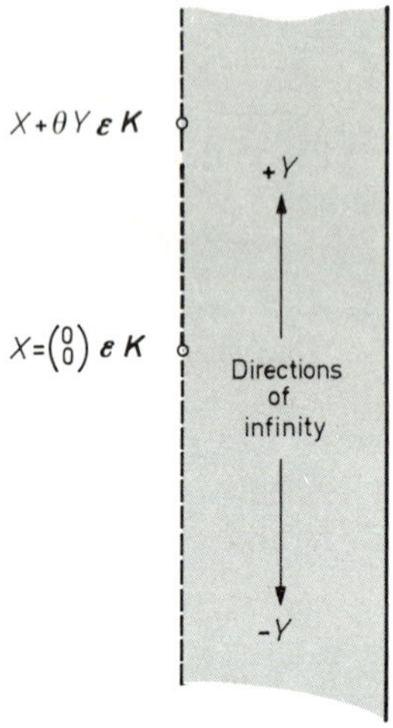

Fig. 11. Example of direction of infinity leading out of set

The preceding proof contains the observation that

(3.5.4) *every direction of infinity of* $\bar{K}$ *is also a direction of infinity of* K^I. *Hence* $\bar{K}$ *is bounded if and only if* K^I *is bounded.*

For polyhedra P, the directions of infinity influence the behavior of the homogenization map $\mathscr{G}$ (2.11.1), which is defined for arbitrary sets S by

$$\mathscr{G} S := \left\{ \begin{pmatrix} X \\ 1 \end{pmatrix} \middle| \; X \in S \right\}^{pp},$$

and which has been employed in section 3.3 to extend the separation theorem (3.3.4) for cones. It was shown that (3.3.6)

$$\mathscr{D} \mathscr{G} K = \bar{K}$$

for arbitrary convex sets K, where

$$\mathscr{D} \tilde{S} := \left\{ X \middle| \begin{pmatrix} X \\ 1 \end{pmatrix} \in \tilde{S} \right\}$$

(see (2.11.2)).

(3.5.5) *The directions of infinity of a convex set* $K \neq \emptyset$ *form a closed cone* C. *The embedding* $\tilde{C} := \left\{ \begin{pmatrix} Y \\ 0 \end{pmatrix} \;\middle|\; Y \in C \right\}$ *of* C *into the horizontal coordinate plane* $\left\{ \begin{pmatrix} X \\ z \end{pmatrix} \;\middle|\; z = 0 \right\}$ *is the intersection of this plane with the closed cone* $\mathscr{G} K$.

Proof. Suppose Y is a direction of infinity of K. Then there exists $X \in K$ such that $X + \theta Y \in K$ for all $\theta \geqslant 0$. Then $\frac{1}{\theta}\begin{pmatrix} X+\theta Y \\ 1 \end{pmatrix} \in \mathscr{G} K$ for all $\theta > 0$, and

$$\begin{pmatrix} Y \\ 0 \end{pmatrix} = \lim_{\theta \to \infty} \frac{1}{\theta} \begin{pmatrix} X+\theta Y \\ 1 \end{pmatrix} \in \overline{\mathscr{G} K} = \mathscr{G} K.$$

This proves one direction of the theorem.

To prove the other direction, suppose $\begin{pmatrix} Y \\ 0 \end{pmatrix} \in \mathscr{G} K$. Then

$$\begin{pmatrix} X \\ 1 \end{pmatrix} + \theta \begin{pmatrix} Y \\ 0 \end{pmatrix} = \begin{pmatrix} X+\theta Y \\ 1 \end{pmatrix} \in \mathscr{G} K$$

for all $X \in K$ and all $\theta \geqslant 0$. Thus $X + \theta Y \in \mathscr{D}\mathscr{G} K = \bar{K}$ establishing Y as a direction of infinity of $\bar{K}$, and therefore of K by (3.5.4). ▯

The closed cone C of directions of infinity of a convex set K is called (Stoker [*1*], Steinitz [*1*]) the

(3.5.6) *characteristic cone*

of K.

We proceed to examine the polar (2.14.6)

$$K^{\pi} := \{ Y \in R^n \mid Y^T X \leqslant 1 \quad \text{for all} \quad X \in K \}$$

of convex sets K. The homogenization operators $\mathscr{G}$ and $\mathscr{D}$ can be used for reducing statements about the polars of sets to statements about the polars of cones. Let $\sigma : R^{n+1} \to R^{n+1}$ be the map which reflects R^{n+1} through the horizontal coordinate plane characterized by $z = 0$:

$$\sigma \begin{pmatrix} X \\ z \end{pmatrix} := \begin{pmatrix} X \\ -z \end{pmatrix}.$$

Then we recall (2.14.7)

$$K^{\pi} = \mathscr{D} \sigma (\mathscr{G} K)^p = \mathscr{D} (\sigma \mathscr{G} K)^p.$$

We use it to prove (see Köthe [*1*] for a direct proof)

(3.5.7) $$K^{\pi\pi} = \overline{\mathscr{H}(K \cup \{0\})} \quad \text{for} \quad K \subseteq R^n.$$

Proof. We show first that $K^{\pi\pi}=\bar{K}$ if $0\in K=\mathscr{H}K$. Let

$$H_z := \left\{\begin{pmatrix} X \\ z \end{pmatrix} \;\middle|\; z\geqslant 0\right\}.$$

Then $H_z^I=\left\{\begin{pmatrix} X \\ z \end{pmatrix} \;\middle|\; z>0\right\}$, and (2.11.6) gives (Fig. 12)

$$\begin{aligned}\mathscr{G}K^{\pi}=\mathscr{G}\mathscr{D}(\sigma\mathscr{G}K)^p=((\sigma\mathscr{G}K)^p\cap H_z^I)^{pp} &= \left[\sigma\mathscr{G}K+\mathscr{C}\left\{\begin{pmatrix} 0 \\ -1 \end{pmatrix}\right\}\right]^p \\ =(\sigma\mathscr{G}K)^p\cap H_z. &\end{aligned}$$

If $0\in K$, then $\begin{pmatrix} 0 \\ -1 \end{pmatrix}\in\sigma\mathscr{G}K$ and $H_z\supseteq(\sigma\mathscr{G}K)^p$. Hence

$$\mathscr{G}K^{\pi}=(\sigma\mathscr{G}K)^p \quad \text{if} \quad 0\in K,$$

or $(\sigma\mathscr{G}K^{\pi})^p=\mathscr{G}K$. From this and (3.3.6)

$$K^{\pi\pi}=\mathscr{D}(\sigma\mathscr{G}K^{\pi})^p=\mathscr{D}\mathscr{G}K=\bar{K}.$$

In order to complete the proof of (3.5.7), we put for arbitrary K

$$\hat{K}:=\overline{\mathscr{H}(K\cup\{0\})}.$$

Since $K^{\pi\pi}$ is closed, convex, and contains 0, $K^{\pi\pi}\supseteq\hat{K}=\hat{K}^{\pi\pi}$. On the other hand, $\hat{K}^{\pi\pi}\supseteq K^{\pi\pi}$. Hence $K^{\pi\pi}=\hat{K}$. □

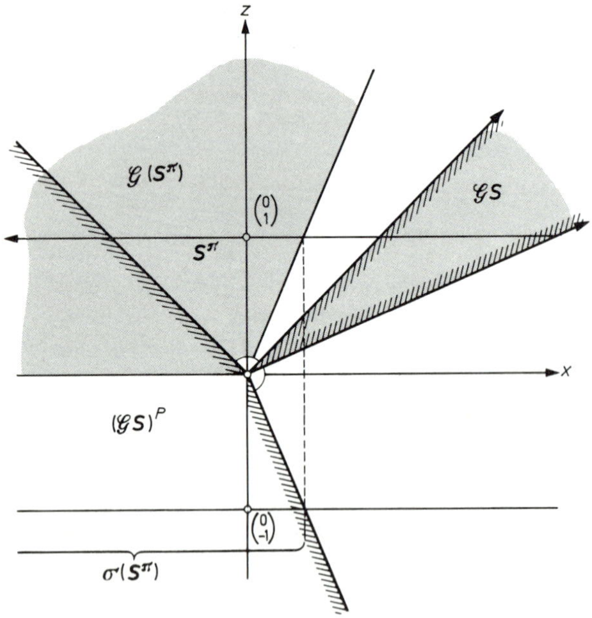

Fig. 12. $\mathscr{G}(S^{\pi})=\sigma((\mathscr{G}S)^p)\cap H_z=(\sigma\mathscr{G}S)^p\cap H_z$

It is readily seen that the polar K^π of a convex set K is bounded if and only if $0 \in K^I$ and $\mathscr{L} K = R^n$ (compare (2.14.11)). This result can be generalized by determining the characteristic cone of the polar S^π of an arbitrary set $S \subseteq R^n$.

(3.5.8) **Theorem.** *The characteristic cone* C *of the polar* S^π *of an arbitrary set* $S \subseteq R^n$ *is the normal cone* $N(S;0)$ (3.4.13) *of* S *at the origin* 0.

Proof. Suppose $A \in N(S;0)$. Then $\{X \mid A^T X \leqslant 0\} \supseteq S$. Hence $(Y+\theta A)^T X \leqslant Y^T X \leqslant 1$ for all $Y \in S^\pi$ and $X \in S$. Thus A is a direction of infinity of S^π, and therefore an element of its characteristic cone. On the other hand, if A is a direction of infinity of S^π, then $(Y+\theta A)^T X \leqslant 1$ for all $\theta \geqslant 0$, $X \in S$ and some $Y \in S^\pi$. This implies $A^T X \leqslant 0$ for all $X \in S$, whence $A \in N(S;0)$. □

3.6. Extremal Properties

In this section, the properties of extreme and exposed subsets, which have been studied in 2.4 for polyhedra, will be investigated for general closed convex sets. We recall the definitions (2.4.1) and (2.4.11) of extreme and exposed subsets, respectively. A convex subset W of a convex set K is called extreme if any representation $Z = \lambda X + (1-\lambda) Y$, $0 < \lambda < 1$, with $X, Y \in K$ of a $Z \in W$ is only possible for $X, Y \in W$. A subset W of K is called exposed if there is a supporting plane E of K such that $W = K \cap E$.

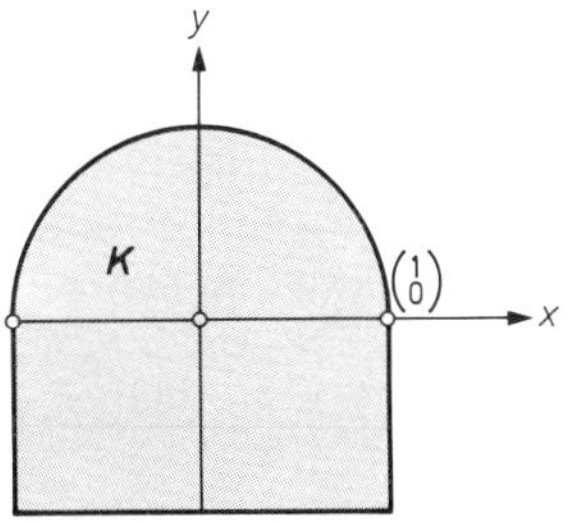

Fig. 13. Extreme point not exposed

Extreme and (exposed) sets of cardinality 1 consist of extreme (exposed) points. The empty set $\emptyset$ and the convex set K itself are extreme as well as exposed subsets of K. Extreme and exposed subsets other than K are called

(3.6.1) *proper.*

In the proof of theorem (2.4.12), every exposed subset was shown to be extreme. The converse holds for polyhedra (theorem (2.4.12)) but not for general closed convex sets. Consider for example the convex set $K \subseteq R^2$ which is the union of the upper half of the unit circle (Fig. 13) $\left\{\begin{pmatrix} x \\ y \end{pmatrix} \in R^2 \;\middle|\; y \geqslant 0,\ x^2+y^2 \leqslant 1\right\}$ and the rectangle $\left\{\begin{pmatrix} x \\ y \end{pmatrix} \;\middle|\; -1 \leqslant y \leqslant 0, -1 \leqslant x \leqslant 1\right\}$. The point $\begin{pmatrix} 1 \\ 0 \end{pmatrix}$ is extreme but not exposed. If V is an extreme subset of an extreme subset W of a convex set K, then V is an extreme subset of K. The above example shows that exposed sets lack this transitivity property. However, if V and W are extreme (exposed) subsets of a closed convex set K with $V \subseteq W$, then V is an extreme (exposed) subset of W. Furthermore,

(3.6.2) *if* $\mathscr{S} = \{W_i \mid i \in I\}$ *is a nonvoid family of extreme (exposed) subsets of a convex set* K, *then the intersection*

$$W := \bigcap_{i \in I} W_i$$

is also an extreme (exposed) subset of K.

Proof. For extreme subsets, (3.6.2) is trivial. Suppose therefore that $\mathscr{S}$ consists of exposed subsets W_i of K, and that $\emptyset, K \notin \mathscr{S}$. Then for each $W_i \in \mathscr{S}$ there exists a supporting plane $E_i := \{X \mid A_i^T X = b_i\}$ such that $W_i = K \cap E_i$ and $A_i^T X \leqslant b_i$ for all $X \in K$.

We proceed to show that without restriction of generality we may assume that $\mathscr{S}$ is finite. By theorem (2.2.5), the set of vectors $\left\{\begin{pmatrix} A_i \\ b_i \end{pmatrix} \;\middle|\; i \in I\right\}$ has a finite maximal linearly independent subset $\left\{\begin{pmatrix} A_j \\ b_j \end{pmatrix} \;\middle|\; j \in J\right\}$, in terms of which one can express every other vector $\begin{pmatrix} A_i \\ b_i \end{pmatrix}$. For the corresponding supporting planes, $\bigcap_{i \in I} E_i = \bigcap_{j \in J} E_j$, whence $\bigcap_{i \in I} W_i = \bigcap_{j \in J} W_j$.

We therefore suppose that $\mathscr{S}$ is finite, and furthermore that $\bigcap_{i \in I} W_i \neq \phi$, the alternative being trivial. For some constants $\lambda_i > 0$, $\sum_{i \in I} \lambda_i = 1$ we define the set

$$E := \{X \mid A^T X = b\}, \quad \begin{pmatrix} A \\ b \end{pmatrix} := \sum_{i \in I} \lambda_i \begin{pmatrix} A_i \\ b_i \end{pmatrix}.$$

If $X \in W \subseteq \bigcap_{i \in I} E_i$, then clearly $X \in E$. Thus $E \cap K \supseteq W$. If $X \in K \sim W$, then there must be one $i \in I$ with $X \notin E_i$. Thus $A_i^T X < b_i$, and therefore $A^T X < b$. Hence $X \notin E$, which gives $E \cap K \subseteq W$. Consequently $E \cap K = W$, which proves W to be exposed. □

This result shows that the extreme (exposed) subsets of a convex set K form a complete lattice under set inclusion with the following lattice operations

$$W_1 \wedge W_2 := W_1 \cap W_2,$$

$$W_1 \vee W_2 := \bigcap \{W \mid W \supseteq W_1 \cup W_2\},$$

where W, W_1, W_2 are extreme (exposed) subsets of K.

(3.6.3) *If W is a proper extreme (exposed) subset of a convex set K, then*

$$\dim W < \dim K.$$

Proof. We recall (2.4.4) that any extreme subset W of K is the intersection of K with the manifold $\mathscr{M}\, W$:

$$W = \mathscr{M}\, W \cap K.$$

Thus $\dim W = \dim K$ would imply $\mathscr{M}\, W = \mathscr{M}\, K$ and therefore $W = K$. This proves (3.6.3) for extreme, and a fortiori, for exposed subsets of K. □

Proper extreme subsets of a convex subset of K are contained in the boundary of K. Indeed, lemma (3.2.9) gives immediately:

(3.6.4) *An extreme subset of a convex set K which meets K^I must coincide with K.*

Since $W = \mathscr{M}\, W \cap K$ for proper extreme subsets of K (2.4.4), we have in view of (3.6.4) $\mathscr{M}\, W \cap K^I = \emptyset$. Thus $\mathscr{M}\, W$ is a nonsingular supporting manifold (3.4.9), (3.4.10) of K. The general supporting plane theorem (3.4.11) then yields the existence of a nonsingular supporting plane E of K with $E \supseteq \mathscr{M}\, W$. In other words,

(3.6.5) *each proper extreme subset V of a convex set K is contained in a proper exposed subset W of K.*

Let us now turn to closed convex sets.

(3.6.6) *Any extreme (exposed) subset W of a closed convex set K in R^n is again closed.*

Proof. Since exposed subsets are extreme, it suffices to consider the latter. Since every extreme subset W of K is convex by definition, we only have to show that every (relative) boundary point X of W belongs to W. If X is such a boundary point of W, then $W \neq \emptyset$ and $X \in K$ by the closedness of K. By lemma (3.2.8) there exists an inner point U of W, and the accessibility lemma (3.2.11) implies

$$Z = \tfrac{1}{2}X + \tfrac{1}{2}U \in W.$$

Then $X \in W$ by definition (2.4.1) of extreme subsets. □

Through each boundary point of a convex set K passes a nonsingular supporting plane of K. Therefore, and in view of (3.6.4),

(3.6.7) *the boundary of a closed convex set K consists of all proper extreme subsets of K.*

In analogy to (2.5.1) we call a closed convex set K (compare Klee [*16*])

(3.6.8) *primitive*

if K is not the convex hull of its (relative) boundary points, that is, if K is not convex hull of its proper extreme subsets. If a closed convex set K is not primitive, then by definition, it is the convex hull of its proper extreme subsets W. By (3.6.6), each such W is closed and convex, and as such either primitive or the convex hull of the proper extreme subsets V of W. This decomposition process terminates, since $\dim V < \dim W < \dim K$ by (3.6.3). Since extreme subsets V of extreme subsets W of K are also extreme subsets of K, we find that

(3.6.9) *every closed convex set $K \neq \emptyset$ is the convex hull of its primitive extreme subsets.*

As in the polyhedral case (section 2.5), primitive closed convex sets can be completely characterized (Klee [*16*]).

(3.6.10) **Theorem.** *A primitive closed convex set K is either a nonvoid linear manifold or half manifold* (2.5.3), *and vice versa.*

Proof. If the closed convex set $K \neq \emptyset$ has no boundary points, then it is easily seen that K is a linear manifold. If K has boundary points, then we prove first that

(3.6.11) *there is a nonsingular supporting plane E of K which contains every boundary point of K.*

By the supporting plane theorem (3.4.12) there are nonsingular planes through each boundary point of K. Assume that there are two different boundary points V_1 and V_2 of K and two different supporting planes $E_1 = \{X \mid A_1^T X = b_1\}$, $E_2 = \{X \mid A_2^T X = b_2\}$ of K with $V_1 \in E_1$, $V_2 \notin E_1$, $V_2 \in E_2$, $V_1 \notin E_2$. More precisely,

$$A_1^T V_1 = b_1, \quad A_1^T V_2 < b_1, \quad A_2^T V_2 = b_2, \quad A_2^T V_1 < b_2.$$

We show that this leads to the contradiction that K is not primitive. Let $U \in K^I$ and choose α_1, α_2 such that

$$\alpha_1 > \frac{b_1 - A_1^T U}{b_1 - A_1^T V_2}, \qquad \alpha_2 > \frac{b_2 - A_2^T U}{b_2 - A_2^T V_1}.$$

Then the points

$$Y_1 := U + \alpha_1(V_1 - V_2), \qquad Y_2 := U + \alpha_2(V_2 - V_1)$$

belong to $\mathscr{M}K$, and satisfy $A_i^T Y_i > b_i$, whence they do not belong to K. Therefore, the line

$$\{U + \lambda(V_1 - V_2) \mid -\infty < \lambda < +\infty\},$$

which passes through U and lies in $\mathscr{M}K$, contains two boundary points X_1 and X_2 of K such that U is a mean of X_1 and X_2. As $U \in K^I$ was arbitrary, it follows that K^I is contained in the convex hull of the boundary of K.

It remains to be shown that K is a halfmanifold, if K has boundary points. By (3.6.11) there is a nonsingular supporting plane $E = \{X \mid A^T X = b\}$ which contains all boundary points of K. Let $H = \{X \mid A^T X \leqslant b\}$ be the corresponding halfspace, $H \supseteq K$. We want to show that

$$K = H \cap \mathscr{M}K,$$

which would prove that K is a halfmanifold. The inclusion $K \subseteq H \cap \mathscr{M}K$ is trivial. To prove the reverse, assume that there is an $X \in H \cap \mathscr{M}K$ with $X \notin K$. Chose $U \in K^I$. Then there is a boundary point Y of K between U and X,

$$Y = \lambda U + \mu X,$$

with $\lambda, \mu > 0$ and $\lambda + \mu = 1$. E is nonsingular, therefore by lemma (3.4.4), $A^T U < b$. Since $X \in K$, $A^T X \leqslant b$. Hence

$$A^T Y = \lambda A^T U + \mu A^T Y < b.$$

The boundary point Y of K would thus not belong to E, contradicting the fact that E contains all boundary points of E. ▯

We note as corollaries to theorem (3.6.10):

(3.6.12) *Every closed convex set K is the convex hull of those extreme subsets which are manifolds and half manifolds.*

(3.6.13) **Theorem.** *Every compact convex set is the convex hull of its extreme points.*

Indeed, the primitive faces of a compact convex set must be compact. But the only compact linear manifolds and half manifolds are points. Similarly, since points and half lines are the only primitives not containing a line:

(3.6.14) **Theorem.** (Klee [*14*]). *Every closed convex set that contains no lines is the convex hull of its set of extreme points and extreme rays.*

Sets which are realizable as sets of extreme points of compact convex sets have been characterized by Björck [*1*].

In general, closed convex sets are not the convex hull of their primitive exposed sets. The latter are, however, distributed densely enough so that the closure of their convex hull indeed yields the original set. We give a proof only in the special case of a compact convex set (Klee [*15*]).

(3.6.15) **Theorem.** *Every compact convex set K in R^n is the closure of the convex hull of its exposed points.*

Proof. As K is compact, there exist for each point $Z \in R^n$ points of K which are farthest from Z in euclidean distance. Any such farthest point X_0 from Z is an exposed point of K since $\{X_0\} = E \cap K$, where E is the supporting plane $\{X \mid (X_0 - Z)^T (X - Z) = \|X_0 - Z\|^2\}$. Consider the set $\overline{F}$ of all points in K which are farthest from some point in R^n. Then $\mathcal{H} F$ is a closed subset of the compact set K, and therefore it is compact, too.

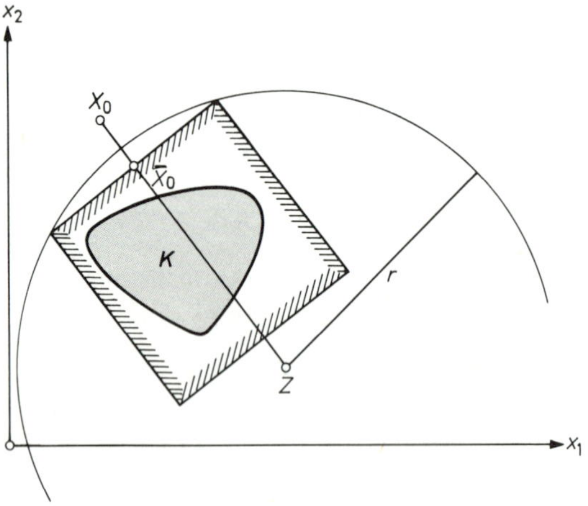

Fig. 14. Separation of X_0 and K by a ball $B(Z,r)$

We claim that $\overline{\mathscr{H} F}=K$. Assume that there exists $X_0 \in K \sim \overline{\mathscr{H} F}$. Applying the separation theorems of section 3.3, one readily constructs an n-cube which contains $\overline{\mathscr{H} F}$ but not X_0, and which is positioned in such a way that the point $\bar{X}_0$ of the cube which is closest to X_0 in euclidean distance is the center of a facet of the n-cube (Fig. 14). It is then easy to find a ball $\{X \mid \|X-Z\| \leqslant r\}$ which contains $\overline{\mathscr{H} F}$ but not X_0. This entails that the point X_0 is farther away from Z than any "farthest" point: an obvious contradiction. Thus $K=\overline{\mathscr{H} F}$ must hold. □

The set of exposed points of a compact convex set K is dense in the set of extreme points of K, the topology being that of R^n. This follows immediately from the preceding theorem and the observation that

(3.6.16) *if* $\overline{\mathscr{H} S}$ *is compact, then* $\bar{S}$ *contains all extreme points of* $\overline{\mathscr{H} S}$.

Proof. Let X be an extreme point. There exists a sequence $\{X_i\}_{i=1,2,\dots}$ of points $X_i \in \mathscr{H} S$ such that $X_i \to X$. Each X_i is a mean of points Z_{ij} in S:

$$X_i = \alpha_{i1} Z_{i1} + \alpha_{i2} Z_{i2} + \cdots + \alpha_{i,k_i} Z_{i,k_i}.$$

Let A denote the set of accumulation points of all Z_{ij}. A is closed. Since $\overline{\mathscr{H} S}$ is compact, A is nonvoid and compact. By theorem (3.2.18), $\mathscr{H} A$ is also compact. Clearly $X \in \overline{\mathscr{H} A} = \mathscr{H} A \subseteq \overline{\mathscr{H} S}$. But X is extreme, and it follows from definition (2.4.1) that $A=\{X\}$. Therefore $\{X\}=A \subseteq \bar{S}$. □

Theorem (3.6.13) fails in the Hilbert space $H=\left\{(x_1,x_2,\dots) \,\middle|\, \sum_k x_k^2 < \infty\right\}$. The set

$$K := \left\{(x_1,x_2,\dots) \in H \,\middle|\, x_k \geqslant 0,\ \sum_k k x_k \leqslant 1\right\}$$

is compact but not convex hull of its extreme points, which are of the form $\left(0,\dots,0,x_k=+\frac{1}{k},0,\dots\right)$. However, it can be asserted under rather general conditions that a compact convex subset S of a linear topological space is the closed convex hull of the extreme points of S. This result is commonly known as the Krein-Milman theorem. It has been proved with various degrees of generality and refinement by Price [*1*], Krein and Milman [*1*], Kelley [*1*], Fan [*9*], Choquet [*2*], [*3*], Pettis [*1*] (consult Bourbaki [*2*], Kelley and Namioka [*1*], Köthe [*1*]).

The notion of extreme points goes back to Minkowski [*3*], who also proved theorem (3.6.13) for $n=3$. Exposed points were introduced by Straszewicz [*1*].

Since the intersection of extreme sets of a convex set K is again extreme, there exists for each point X in K a unique smallest extreme set containing X.

(3.6.17) $W \subseteq K$ *is the smallest extreme set containing* $X \in K$ *if and only if* $X \in W^I$.

Proof. Suppose $X \in W \sim W^I$. By the supporting plane theorem (3.4.11), there exists a nonsingular supporting plane of W through X, and $E \cap W$ is a smaller extreme set than W. (3.6.4) yields the rest. □

The following proposition is an immediate consequence of (3.6.17).

(3.6.18) *Let K be an arbitrary convex set $K \subseteq R^n$. The family of sets*

$$\mathscr{J} = \{W^I \mid W \quad \text{extreme in} \quad K\}$$

is a partition of K, i.e., the members of $\mathscr{J}$ are pairwise disjoint and cover K.

The sets in $\mathscr{J}$ have important properties. For instance, they can be characterized as the maximal convex subsets of K which are relatively open in their affine hulls. Sometimes they, rather than extreme sets, are used for generalizing the concept of extreme points.

Dubins [*1*] has examined the extreme sets of the intersection of convex sets.

(3.6.19) **Theorem.** *Consider the intersection $K = K_1 \cap K_2$ of two convex sets in R^n. If W_1 and W_2 are extreme sets of K_1 and K_2, respectively, then $W_1 \cap W_2$ is an extreme set of K. Furthermore, every extreme set of K is of that kind.*

Proof. That $W_1 \cap W_2$ is an extreme set of $K = K_1 \cap K_2$ is obvious. Let then V be any extreme subset of K, and let $X \in V^I$. By (3.6.18) there exist unique extreme sets W_1 and W_2 in K_1 and K_2, respectively, so that $X \in W_1^I$ and $X \in W_2^I$. If $Y \in V$, then by (3.2.14) there exists $Z \in V$ with $X \in (Y, Z)$. Since X belongs to the extreme sets W_i, $Y \in W_i$ for $i = 1,2$. Thus $V \subseteq \hat{V} := W_1 \cap W_2$. $\hat{V}$ is again extreme in K. By (3.2.21.(i)), $\hat{V}^I = W_1^I \cap W_2^I$. Thus $X \in \hat{V}^I$. This violates the partition property (3.6.18) unless $\hat{V} = V$. □

(3.6.20) **Theorem** (Dubins [*1*]). *Consider a compact convex set $K \subseteq R^n$ and a manifold M of codimension k. Then every extreme point of $M \cap K$ is the mean of at most $k+1$ extreme points of K.*

Proof. Suppose X is an extreme point of $M \cap K$. By theorem (3.6.19), $\{X\} = M \cap W$ where W can be chosen the smallest extreme subset of K containing X. By (3.6.17), $X \in W^I$. This entails

$$\{X\} = \mathscr{M}\, W \cap M. \tag{3.6.21}$$

Indeed, if $Y \in \mathscr{M}\, W \cap M$, then by (3.2.14) there exists $Z \in W$ so that $X \in (Y, Z)$. Clearly $Z \in M$, whence $Z = X$. But $X \in (Y, X)$ is only possible if $Y = X$ (recall the convention $(X, X) = \{X\}$).

Now (3.6.21) gives $\dim W \leqslant k$, and the theorem follows from theorem (3.6.13) and Carathéodory's theorem (2.2.12). □

Dubin's theorem has been generalized by Klee [*24*].

3.7. Combinatorial Properties

In this section, we shall derive certain combinatorial properties of convex sets to be used later on. Perhaps the best known theorem of this kind is the theorem of Helly [*1*] [*2*]. It states essentially that a family $\mathscr{K}$ of convex sets in R^n has common points, if any $n+1$ members of $\mathscr{K}$ have common points. Some additional properties of the family $\mathscr{K}$ are clearly necessary to make the above statement a valid one. The family of all half lines $\{x \mid x > a\}$ in R^1, for instance, has no common points, although every pair of half lines intersects in an entire half line. The two most natural additional conditions are finiteness of the family $\mathscr{K}$ and compactness of its members. Following Rademacher and Schoenberg [*1*], we shall distinguish between the "first" and the "second" theorem of Helly.

(3.7.1) **First Theorem of Helly.** *Let K_i, $i=1,\ldots,N$ be $N \geqslant n+1$ convex sets of R^n. If any $n+1$ of them have a nonvoid intersection, then the intersection*

$$\bigcap_{i=1}^{N} K_i$$

is not empty[1].

The following elegant *proof*, due to Radon [*1*], is by induction on N. For $N=n+1$, the assertion of the theorem is trivial. Suppose that (3.7.1) holds for $N-1 \geqslant n+1$ convex sets. Then by induction hypothesis there exist N points

$$X_i \in \bigcap_{\substack{j=1\\ j\neq i}}^{N} K_j, \qquad i=1,\ldots,N.$$

Then

$$\sum_{i=1}^{N} \lambda_i X_i = 0,$$

(3.7.2)

$$\sum_{i=1}^{N} \lambda_i = 0$$

[1] A theorem which contains both the separation theorem for disjoint open convex sets and Helly's first theorem applied to open convex sets as special cases is due to Klee [*4*].

are $n+1$ linear equations for $N>n+1$ unknown numbers $\lambda_i, i=1,\ldots,N$. Hence there is a nontrivial solution

$$(3.7.3) \qquad (\lambda_1,\ldots,\lambda_N)\neq(0,\ldots,0).$$

Suppose without loss of generality

$$\lambda_1\geqslant 0,\ldots,\lambda_r\geqslant 0 \quad \textit{and} \quad \lambda_{r+1}\leqslant 0,\ldots,\lambda_N\leqslant 0.$$

Then $\sum_{i=1}^{r}\lambda_i=-\sum_{i=r+1}^{N}\lambda_i>0$ because of (3.7.3) and the point

$$Y:=\frac{\sum_{i=1}^{N}\lambda_i X_i}{\sum_{i=1}^{r}\lambda_i}$$

belongs to $\bigcap_{j=r+1}^{N} K_j$, as every X_i, $i=1,\ldots,r$, belongs to every K_j, $j=r+1,\ldots,N$. But because of (3.7.2), we have also

$$Y=\frac{\sum_{i=r+1}^{N}(-\lambda_i X_i)}{\sum_{i=r+1}^{N}(-\lambda_i)},$$

which shows that $Y\in\bigcap_{j=1}^{r} K_j$. Hence $Y\in\bigcap_{j=1}^{N} K_j\neq\emptyset$. □

The above proof by Radon is based on the observation that any finite point set can be partitioned into two nonempty parts such that the respective convex hulls have no common point. This result has been generalized by several authors, for instance De Santis [*1*], Reay [*4*].

If an infinite family of compact sets has an empty intersection, then there exists a finite subfamily whose intersection is empty. The first theorem of Helly (3.7.1) implies therefore the

(3.7.4) **Second Theorem of Helly.** *Let $\mathcal{K}=\{K_i \mid i\in I\}$ be a family of (possibly infinitely many) compact convex sets in R^n. If any $n+1$ members of $\mathcal{K}$ have a nonvoid intersection, then the intersection*

$$\bigcap_{i\in I} K_i$$

is not empty.

Helly's theorem can be reduced to the theorem of Carathéodory (2.2.11) with the help of the duality theory of cones (Sandgren [*1*], Valentine [*3*]). Elegant proofs for a variety of Helly type theorems can be obtained by this technique (consult Valentine [*4*]).

Another type of combinatorial properties does not involve the dimension of the space R^n.

(3.7.5) **Theorem.** (Berge [*3*][1]). *If* M *is a convex set and* K_i, $i=1,\dots,m$, $m \geqslant 1$, *are closed convex sets in* R^n *satisfying*

$$\text{(i)} \qquad M \cap \bigcap_{\substack{i=1 \\ i \neq j}}^{m} K_i \neq \emptyset \quad \textit{for} \quad j=1,2,\dots,m,$$

$$\text{(ii)} \qquad M \cap \bigcap_{i=1}^{m} K_i = \emptyset,$$

then M *is not contained in the union of the* K_i*:*

$$M \nsubseteq \bigcup_{i=1}^{m} K_i.$$

Proof. We may assume without loss of generality that the sets M, K_i, $i=1,\dots,m$, are compact convex sets: because of (i) there are points

$$A_j \in M \cap \bigcap_{\substack{i=1 \\ i \neq j}}^{m} K_i, \quad j=1,2,\dots,m,$$

and the sets $\hat{M} := \mathscr{H}\{A_1,\dots,A_m\} \subseteq M$, $\hat{K}_i := \hat{M} \cap K_i$ are compact convex sets satisfying (i) and (ii). Moreover,

$$\hat{M} \cap M = \hat{M} \nsubseteq \bigcup_i \hat{K}_i = \hat{M} \cap \bigcup_i K_i \;\Rightarrow\; M \nsubseteq \bigcup_i K_i.$$

The proof is now conducted by induction with respect to m. If $m=1$, then by (i) $M \neq \emptyset$ and by (ii) $M \cap K_1 = \emptyset$, giving $M \nsubseteq K_1$.

Suppose the theorem holds for $m \geqslant 1$, and assume that there are sets M, K_i, $i=1,\dots,m+1$, satisfying

$$\text{(3.7.6)} \qquad M \cap \bigcap_{\substack{i=1 \\ i \neq j}}^{m+1} K_i \neq \emptyset \quad \text{for} \quad j=1,\dots,m+1,$$

$$\text{(3.7.7)} \qquad M \cap \bigcap_{i=1}^{m+1} K_i = M \cap K_{m+1} \cap \bigcap_{i=1}^{m} K_i = \emptyset.$$

Then there is a plane E *strictly separating* the compact convex sets $M \cap K_{m+1}$ and $K := \bigcap_{i=1}^{m} K_i$. Define $K_i' := E \cap K_i$, $i=1,\dots,m$, $M' := E \cap M$.

[1] A weaker version of this theorem occurs in Levi [*1*].

Then $E \cap K = \emptyset$ implies

$$M' \cap \bigcap_{i=1}^{m} K_i' = E \cap M \cap \bigcap_{i=1}^{m} K_i = E \cap M \cap K = \emptyset,$$

that is, the sets M', K_i' satisfy condition (ii) of the theorem. Moreover, we maintain that (i) is also true:

$$M' \cap \bigcap_{\substack{i=1\\ i\neq j}}^{m} K' = M \cap E \cap \bigcap_{\substack{i=1\\ i\neq j}}^{m} K_i \neq \emptyset \quad \text{for} \quad j=1,2,\ldots,m.$$

Indeed, take an arbitrary j with $1 \leqslant j \leqslant m$. Then by (3.7.6) we have vectors X_j and Y such that

$$X_j \in M \cap \bigcap_{\substack{i=1\\ i\neq j}}^{m+1} K_i, \quad \text{hence} \quad X_j \in M \cap K_{m+1},$$

$$Y \in M \cap \bigcap_{i=1}^{m} K_i = M \cap K.$$

Moreover, E strictly separates X_j and Y. Therefore, there is a vector

$$Z_j = \lambda_j X_j + (1-\lambda_j) Y, \quad 0 < \lambda_j < 1$$

such that $Z_j \in E$. Now, M and the K_i are convex sets, hence

$$Z_j \in M \cap \bigcap_{\substack{i=1\\ i\neq j}}^{m} K_i,$$

which together with $Z_j \in E$ implies

$$Z_j \in E \cap M \cap \bigcap_{\substack{i=1\\ i\neq j}}^{m} K_i = M' \cap \bigcap_{\substack{i=1\\ i\neq j}}^{m} K_i' \neq \emptyset.$$

Therefore, the sets M', K_i', $i=1,2,\ldots,m$, satisfy the hypotheses of the theorem, and by induction hypothesis we have

$$M' = M \cap E \nsubseteq \bigcup_{i=1}^{m} K_i' = E \cap \bigcup_{i=1}^{m} K_i.$$

Because of $E \cap M \cap K_{m+1} = \emptyset$, this implies immediately

$$M \cap E \nsubseteq E \cap \bigcup_{i=1}^{m+1} K_i,$$

and, a fortiori, $M \nsubseteq \bigcup_{i=1}^{m+1} K_i$. □

Remark: The preceding proof shows that theorem (3.7.5) remains true for convex sets M and closed convex sets K_i in an arbitrary topological vector space E.

(3.7.8) **Corollary to Theorem (3.7.5).** *Let* K_i, $i=1,\ldots,m$, $m \geqslant 1$, *be closed convex sets in* R^n *whose union is convex. If the intersection of any* $m-1$ *of the sets* K_i *is nonempty, then their intersection* $\bigcap_{i=1}^{m} K_i$ *is nonempty.*

(3.7.9) **Corollary to Theorem (3.7.5).** *Let* K_i, $i=1,\ldots,m$, $m \geqslant 2$, *be closed convex sets in* R^n. *If their intersection is empty and if the intersection of any* $m-1$ *of them is not empty, then their union is not convex.*

Among the many applications of Helly's theorem to the geometry of convex sets, the theorems of Kirchberger, Santalo, and Krasnosselski stand out, each representative of an entire class of theorems. For more detailed information see Danzer, Grünbaum, and Klee [*1*], Valentine [*4*], Hadwiger [*1*].

(3.7.10) **Separation Theorem of Kirchberger** [*1*]. *Let* M_1 *and* M_2 *be finite sets of* R^n. *Then there is a plane* $E=\{X \mid A^T X = b\}$ *which separates* M_1 *and* M_2 *strictly,*

$$A^T X_1 < b < A^T X_2 \quad \textit{for} \quad X_1 \in M_1 \quad \textit{and} \quad X_2 \in M_2$$

if (and only if) for every subset $Y \subseteq M_1 \cup M_2$ *containing at most* $n+2$ *elements there exists a plane* E_Y *strictly separating* $Y \cap M_1$ *and* $Y \cap M_2$.

Proof. (Rademacher and Schoenberg [*1*]). For $X \in R^n$, denote by H_X^+ and H_X^- the following open halfspaces in R^{n+1}:

$$H_X^+ := \left\{ \begin{pmatrix} A \\ b \end{pmatrix} \,\middle|\, A^T X < b \right\}, \qquad H_X^- := \left\{ \begin{pmatrix} A \\ b \end{pmatrix} \,\middle|\, A^T X > b \right\}.$$

The union of the two families $\{H_X^+ \mid X \in M_1\}$ and $\{H_X^- \mid X \in M_2\}$ satisfies the hypothesis of Helly's first theorem (3.7.1): Any $n+2$ of them have a nonempty intersection, as the sets $Y \cap M_1$ and $Y \cap M_2$ can be strictly separated. Hence

$$\bigcap_{X \in M_1} H_X^- \cap \bigcap_{X \in M_2} H_X^+ \neq \emptyset,$$

which means that M_1 and M_2 can be strictly separated. ☐

Let $\mathcal{S}$ be a family of sets in R^2. A line which meets every member of $\mathcal{S}$, is called a

(3.7.11) *common transversal*

of $\mathcal{S}$.

(3.7.12) **Transversal Theorem of Santalo** [*1*]. *Let* $\mathcal{S}=\{S_1,\dots,S_m\}$ *be a family of parallel compact line segments* $S_i \subseteq R^2$. *Some line segments may consist of single points. If every three members of* $\mathcal{S}$ *have a common transversal, then the entire family* $\mathcal{S}$ *has a common transversal.*

Proof. Without restriction of generality, we may assume that the line segments are vertical, i. e. of the form

$$S_i = \left\{ \begin{pmatrix} x \\ y \end{pmatrix} \;\middle|\; x = x_i,\; l_i \leqslant y \leqslant u_i \right\}.$$

With each line segment S_i we associate the convex set

$$C_i = \left\{ \begin{pmatrix} a \\ b \end{pmatrix} \;\middle|\; l_i \leqslant a x_i + b \leqslant u_i \right\}.$$

By hypothesis, every three sets C_i have a nonempty intersection, and the applicability of Helly's first theorem (3.7.1) is obvious. □

A set $S \subseteq R^n$ is called

(3.7.13) *star shaped*

if there exists a point $X \in S$ so that $(X, Y) \subseteq S$ for all $Y \in S$, or in other words, if all points Y in S are "visible" via S from a single point X in S. If $T \subseteq S$, we call $X \in S$ a point of

(3.7.14) *common visibility*

of T via S, if $(X, Y) \subseteq S$ for all $Y \in T$. For a more general version of the following theorem see Valentine [*4*].

(3.7.15) **Theorem of Krasnosselski** [*1*]. *Let* S *be a compact subset of* R^n. *If every* $n+1$ *points of* S *have a point of common visibility via* S, *then* S *is star shaped.*

Proof. For each $X \in S$, define the set

$$S_X := \{ Y \in S \mid (X, Y) \subseteq S \}.$$

This is the set of all points visible from X via S. The set $\overline{\mathcal{H} S_X}$ is compact since it is a closed subset of the compact set $\mathcal{H} S$. By hypothesis, every $n+1$ of the compact convex sets $\overline{\mathcal{H} S_X}$ have nonempty intersection. Thus Helly's second theorem (3.7.4) guarantees that all sets $\overline{\mathscr{H} S_X}$ have at least one point X_0 in common.

X_0 is a point of common visibility of S. Assume to the contrary that $(X_0, Y_0) \nsubseteq S$, for some $Y_0 \in S$, and let $Z_0 \in (X_0, Y_0) \sim S$. Since S is closed there exists a ball $B(Z_0, r) = \{X \mid \|X - Z_0\| \leqslant r\}$ with $B(Z_0, r) \cap S = \emptyset$. Let

$$Z_\lambda := (1-\lambda) Z_0 + \lambda Y_0 .$$

If $B(Z_\theta, r)$ denotes the ball closest to Z_0 which just touches S, one has

$$\theta := \inf\{\lambda \geqslant 0 \mid B(Z_\lambda, r) \cap S \neq \emptyset\}.$$

Then there exists $Z \in B(Z_\theta, r) \cap S$ whereas $B(Z_\theta, r)^I \cap S = \emptyset$. The half-space $H = \{X \mid (X - Z)^T (Z_\theta - Z) \leqslant 0\}$ therefore contains all points visible from Z via S. Hence $H \supseteq \overline{\mathscr{H} S_Z}$. However, $X_0 \notin H$. To prove this, consider the projection Z^* of Z onto the line through X_0 and Z_θ, and assume $\|X_0 - Z\| < \|X_0 - Z_\theta\|$. Then $Z^* \in \mathscr{H}\{X_0, Z_\theta\}$ and $\|Z - Z^*\| \leqslant r$, which contradicts the definition of Z_θ. Thus $\|X_0 - Z\| \geqslant \|X_0 - Z_\theta\|$ and

$$(X_0 - Z)^T (Z_\theta - Z) = \tfrac{1}{2}(\|X_0 - Z\|^2 - \|X_0 - Z_\theta\|^2 + r^2) > 0.$$

This establishes $X_0 \notin H$, leading to the contradiction $X_0 \notin \overline{\mathscr{H} S_Z}$. □

3.8. Topological Properties

Compact convex sets of the same dimension are homeomorphic. We shall need this result for extending fixed point theorems which will be proved for simplices in the next section. For the proof of Kakutani's fixed point theorem, we shall need the existence of a continuous retraction of one convex set onto another. It will be verified that the euclidean "nearest point" mapping retracts the space R^n onto any convex set.

Consider any compact convex set $K \subseteq R^n$ which contains the unit sphere

$$S = \{X \in R^n \mid \|X\| = 1\}.$$

Here $\|X\|$ denotes the euclidean norm $\sqrt{X^T X}$. Then map each boundary point Y of K into the unit sphere S:

$$Y \to b(Y) := \frac{Y}{\|Y\|} \in S.$$

We call b the

(3.8.1) *boundary map*

of K into S.

(3.8.2) *The boundary map b is a homeomorphism of the boundary of K, supplied with the topology induced by R^n, and the unit sphere S.*

Proof. The boundary map b is $1-1$ onto S. Indeed, let $X \in S$, then by compactness there exists a greatest ω such that $\omega X \in K$. The point $Y := \omega X$ is a boundary point of K by lemma (3.2.9). By the accessibility lemma (3.2.11), Y is the only boundary point which is a positive multiple of X.

The euclidean norm is a continuous function by theorem (3.1.11). Since $\|Y\| \geqslant 1$ for all boundary points Y of K, b is also continuous. Finally b^{-1} is continuous since it is the inverse of a continuous mapping of a compact set onto another compact set. □

(3.8.3) **Theorem.** *Compact convex sets are homeomorphic to the unit ball of the same dimension.*

Proof. Since linear transformations from R^n into any R^m are continuous (theorem (3.1.13)), and since any linear manifold is affine equivalent to R^d of the same dimension (consequence of theorem (2.2.5)), we may consider without restriction of generality only those compact sets K which contain the unit ball $B \subseteq R^n$, where $B \leqslant \{X \mid \|X\| \leqslant 1\}$.

A homeomorphism from B onto K is then readily defined with the help of the boundary map (3.8.1):

$$f(X) := \begin{cases} \left\| b^{-1}\left(\dfrac{X}{\|X\|}\right) \right\| X & \text{if } X \neq 0, \\ 0 & \text{if } X = 0. \end{cases}$$

If $Y = f(X)$ and $X \neq 0$, then $\dfrac{Y}{\|Y\|} = \dfrac{X}{\|X\|}$. Thus

$$f^{-1}(Y) := \begin{cases} \left\| b^{-1}\left(\dfrac{Y}{\|Y\|}\right) \right\|^{-1} Y & \text{if } Y \neq 0, \\ 0 & \text{if } Y = 0. \end{cases}$$

The continuity of f and f^{-1} is obvious at points other than the origin. To prove continuity at 0, let m be an upper bound for the distances of points in K from the origin. Such a bound exists since K is compact. Then

$$\|f(X)\| \leqslant m \|X\|.$$

Since K contains the unit ball,

$$\|f^{-1}(Y)\| \leqslant \|Y\|.$$

This establishes the continuity of f and f^{-1}. □

Theorem (3.8.3) raises the interesting problem of describing the topological types of convex sets in R^n. Stoker [*1*] observed that

(3.8.4) *every closed convex set whose characteristic cone* (3.5.6) *is not a subspace is homeomorphic to a closed half space.*

It follows that there are for a given R^n only finitely many topological types which can be realized as closed convex subsets of R^n. Klee [*10*] has determined the topological types of closed convex sets in locally compact linear spaces. For any normed linear space, the topological characterization problem can be reduced to that for the unit cell of that space (Carson and Klee [*1*]). In R^n it is easily seen that every convex set of dimension n is homeomorphic to the unit ball $\{X \mid \|X\| \leqslant 1\}$ with some boundary points removed. Keller [*1*] has shown that every infinitely dimensional compact convex subset of Hilbert space is homeomorphic to the Hilbert parallelotope (result generalized by Klee [*10*]). Klee [*6*] affirmed a conjecture by Kakutani [*4*] that a Hilbert space is homeomorphic to its own unit sphere.

In R^n, the dimension of a convex set is a topological invariant, and so is the boundary. On these deep but well-known results consult P. S. Alexandroff [*1*] and Hurewicz and Wallman [*1*].

(3.8.5) **Theorem.** *Let K be a closed convex set contained in an arbitrary subset S of the R^n. Then there exists a continuous mapping g: $S \to K$ which is a retraction of S onto K,* i.e. *$g(X) = X$ for $X \in K$.*

Proof. For every point $X \in R^n$ there exists a unique point $\bar{X} \in K$ which minimizes the euclidean distance $\|X - \bar{X}\|$ between X and points of K (lemma (3.3.1)).

Consider now two points $X, Y \in R^n$, and let $\bar{X}, \bar{Y}$ be the respective closest points in K. For $0 \leqslant \varepsilon \leqslant 1$, the point $\bar{X} + \varepsilon(\bar{Y} - \bar{X})$ also belongs to K. Now

$$\|X - (\bar{X} + \varepsilon(\bar{Y} - \bar{X}))\|^2 = \|X - \bar{X}\|^2 - 2\varepsilon(X - \bar{X})^T(\bar{Y} - \bar{X}) + \varepsilon^2 \|\bar{Y} - \bar{X}\|^2 .$$

This shows that, if $(X - \bar{X})^T(\bar{Y} - \bar{X}) > 0$, then for sufficiently small $\varepsilon > 0$, $\bar{X} + \varepsilon(\bar{Y} - \bar{X})$ has a shorter distance from X than $\bar{X}$, contradicting the definition of $\bar{X}$. Therefore, and by the analogous argument for Y,

$$(X - \bar{X})^T(\bar{Y} - \bar{X}) \leqslant 0, \qquad (Y - \bar{Y})^T(\bar{X} - \bar{Y}) \leqslant 0 .$$

The identity

$$\|X - Y\|^2 - \|\bar{X} - \bar{Y}\|^2 = \|(X - \bar{X}) - (Y - \bar{Y})\|^2 - 2(X - \bar{X})^T(\bar{Y} - \bar{X}) - 2(Y - \bar{Y})^T(\bar{X} - \bar{Y})$$

then gives

$$\|X - Y\| \geqslant \|\bar{X} - \bar{Y}\| .$$

The map g: $R^n \to K$ defined by

$$g(X) := \bar{X}$$

is therefore continuous and a retraction of R^n onto K. The restriction of g to S then meets the requirements of the theorem. □

3.9. Fixed Point Theorems

Brouwer's fixed point theorem and its relatives are among the most useful and fundamental theorems in convex analysis. We present a proof of Brouwer's theorem based on a well known combinatorial lemma by Sperner [1]. The exposition partly follows Fan [7].

Let $V_0, V_1, \ldots, V_n$ be the vertices of an n-simplex (short for n-dimensional simplex)

$$S = \mathscr{H}\{V_0, V_1, \ldots, V_n\}.$$

The faces of S are again simplices. They correspond to the subsets of $\{V_0, V_1, \ldots, V_n\}$. A

(3.9.1) *simplicial subdivision*

of S is a family $\mathscr{K}$ of simplices such that (i) S is the union of the n-simplices in $\mathscr{K}$. (ii) Each pair of n-simplices intersects in a common face which

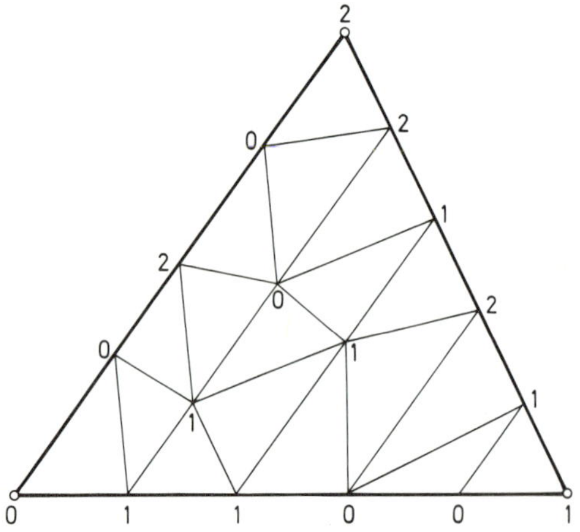

Fig. 15. Labeled subdivision of a simplex

may be empty. (iii) The faces of any simplex in $\mathscr{K}$ are again in $\mathscr{K}$. Each simplicial subdivision has a

(3.9.2) *mesh,*

which is the diameter of the largest simplex in some norm. There exist simplicial subdivisions whose mesh is arbitrarily small.

We say that the subdivision is (Fig. 15)

labeled,

if to each vertex U in $\mathscr{K}$ there is assigned one of the integers $0, 1, \dots, n$ as label $\Lambda(U)$ of U. The assignment has to be such that

$$U \in \mathscr{H}\{V_{i_0}, V_{i_1}, \dots, V_{i_k}\} \Rightarrow \Lambda(U) \in \{i_0, i_1, \dots, i_k\}. \tag{3.9.3}$$

If $U_1, \dots, U_k$ are the vertices of some simplex S_q in $\mathscr{K}$, then we consider S_q labeled by the set $\{\Lambda(U_1), \dots, \Lambda(U_k)\}$. We are now able to formulate

(3.9.4) **Sperner's Lemma.** *In a labeled subdivision $\mathscr{K}$ of a simplex S there is an odd number of n-simplices with label set $\{0, 1, \dots, n\}$.*

Proof. The lemma is trivial for $n=0$. Suppose it holds for $n-1$. Then the number $\sigma^{(n-1)}$ of $(n-1)$-simplices with label set $\{0, \dots, n-1\}$ on the facet $\mathscr{H}\{V_0, V_1, \dots, V_{n-1}\}$ of S is odd. For each n-simplex S_q in $\mathscr{K}$, let $\gamma(S_q)$ denote the number of facets of S_q with label set $\{0, 1, \dots, n-1\}$. Then

$$\gamma(S_q) = \begin{cases} 1 & \text{if } \Lambda(S_q) = \{0, 1, \dots, n\}, \\ 0 \text{ or } 2 & \text{otherwise.} \end{cases}$$

Hence

$$\sigma^{(n)} \equiv \Sigma\, \gamma(S_q) \bmod 2,$$

where $\sigma^{(n)}$ denotes the number of n-simplices in $\mathscr{K}$ with label set $\{0, 1, \dots, n\}$, and the summation runs over all n-simplices in $\mathscr{K}$. In $\Sigma\, \gamma(S_q)$ all $(n-1)$-simplices labeled $\{0, 1, \dots, n-1\}$ are counted twice, except those which lie on the boundary of S. However, all $(n-1)$-simplices labeled $\{0, 1, \dots, n-1\}$ which lie on the boundary of S must be on the facet $\mathscr{H}\{V_0, V_1, \dots, V_{n-1}\}$ by (3.9.3). Thus $\sigma^{(n-1)}$ is the number of all $(n-1)$-simplices labeled $\{0, 1, \dots, n-1\}$ which are not counted twice in $\Sigma\, \gamma(S_q)$. Therefore

$$\sigma^{(n-1)} \equiv \Sigma\, \gamma(S_q) \equiv \sigma^{(n)} \bmod 2,$$

and $\sigma^{(n)}$ is odd, since $\sigma^{(n-1)}$ is odd. □

We are now able to prove

(3.9.5) **Brouwer's Fixed Point Theorem:** *If $K \neq \phi$ is a compact convex set in R^n, then every continuous mapping $f\colon K \to K$ leaves at least one point fixed.*

Proof (Knaster, Kuratowski, Mazurkiewicz [*1*]). Since every compact convex set in a euclidean space is homeomorphic to a simplex of the same dimension (theorem (3.8.3)), it suffices to consider the case where K is the n-simplex $S = \mathscr{H}\{V_0, V_1, \dots, V_n\}$.

Each point $X \in S$ is a unique convex combination of $V_0, V_1, \dots, V_n$:

$$X = w_0 V_0 + \cdots + w_n V_n, \qquad w_i \geqslant 0, \qquad \Sigma\, w_i = 1.$$

The coefficients w_i are continuous functions $w_i(X)$ on S. The $n+1$ sets

$$F_i := \{X \in S \mid w_i(f(X)) \leqslant w_i(X)\}$$

are therefore closed. Moreover, they satisfy the following inclusion relation:

$$\mathscr{H}\{V_{i_0}, V_{i_1}, \ldots, V_{i_k}\} \subseteq F_{i_0} \cup F_{i_1} \cup \ldots \cup F_{i_k} \tag{3.9.6}$$

for every subset $\{i_0, i_1, \ldots, i_k\}$ of $\{0, 1, \ldots, n\}$. Indeed, if a point $X \in S$ is not contained in the union $F_{i_0} \cup \ldots \cup F_{i_k}$, then $w_{i_h}(f(X)) > w_{i_h}(X)$ for $0 \leqslant h \leqslant k$, and therefore

$$1 \geqslant w_{i_0}(f(X)) + \cdots + w_{i_k}(f(X)) > w_{i_0}(X) + \cdots + w_{i_k}(X).$$

This implies $X \notin \mathscr{H}\{V_{i_0}, V_{i_1}, \ldots, V_{i_k}\}$.

Now let $\mathscr{K}$ be a simplicial subdivision (3.9.1) of S, and let U be an arbitrary vertex in $\mathscr{K}$. There exists a smallest face $\mathscr{H}\{V_{i_0}, V_{i_1}, \ldots, V_{i_k}\}$ of S containing U. By (3.9.6), there exists an integer h such that $U \in F_{i_h}$. Putting $\Lambda(U) := i_h$ then defines a labeling of $\mathscr{K}$, and Sperner's lemma (3.9.4) yields the existence of an n-simplex S_q in $\mathscr{K}$ with label set $\{0, 1, \ldots, n-1\}$. The labels were defined so as to guarantee $U \in F_{\Lambda(U)}$. The simplex S_q thus meets all $n+1$ closed sets F_i.

Consider then a sequence of simplicial subdivisions $\{\mathscr{K}^{(m)}\}_{m=1,2,\ldots}$ of S such that the mesh of $\mathscr{K}^{(m)}$ tends to zero as m increases. In each $\mathscr{K}^{(m)}$ there exists an n-simplex $S_p^{(m)}$ with vertices $U_0^{(m)}, U_1^{(m)}, \ldots, U_n^{(m)}$ such that $U_i^{(m)} \in F_i$ for $i = 0, \ldots, n$. Replacing the sequence $\{\mathscr{K}^{(m)}\}_{m=1,2,\ldots}$ be a suitable subsequence, we obtain convergent sequences of vertices:

$$U_i^{(m)} \to U_i^{\infty}, \qquad i = 0, \ldots, n.$$

Since F_i is closed, we have $U_i^{(\infty)} \in F_i$. But the diameter of the simplex $S_p^{(m)} = \mathscr{H}\{U_0^{(m)}, \ldots, U_n^{(m)}\}$ tends to zero as m increases, so all vertices converge towards a single point $Z \in \bigcap F_i$.

The point Z satisfies $w_i(f(Z)) \leqslant w_i(Z)$ for each i. Since $\Sigma w_i(f(Z)) = \Sigma w_i(Z) = 1$, it follows that $w_i(f(Z)) = w_i(Z)$ for each i. Thus $Z = f(Z)$ is a fixed point. □

An important generalization of Brouwer's fixed point theorem is due to Kakutani [*3*]. This generalization consists in replacing the continuous point-to-point mapping of a convex set K into itself by an "upper semi-continuous" point-to-set mapping of K into the set of all closed convex subset of K.

Let E, F be two topological spaces, and denote by

$$\mathscr{A}(F)$$

the collection of all non-empty closed subsets of F. A point-to-set mapping $f: E \to \mathscr{A}(F)$ is said to be

(3.9.7) *upper semi-continuous*

if, for any point $X_0 \in E$ and any open set $U \subseteq F$ with $f(X_0) \subseteq U$, there exists a neighbourhood V of X_0 such that $f(X) \subseteq U$ whenever $X \in V$.

(3.9.8) **Lemma.** *Let E be an arbitrary topological space and F a regular space. Let f be an upper semi-continuous mapping from E into $\mathscr{A}(F)$. Let $X_i \to X_0$ and $Y_i \to Y_0$ for sequences in E and F, respectively. Then*

$$Y_i \in f(X_i) \quad \textit{for all} \quad i \;\Rightarrow\; Y_0 \in f(X_0).$$

Proof. Consider the collection $\mathscr{U}$ of all open sets U containing $f(X_0)$. By upper semi-continuity, there exists for each $U \in \mathscr{U}$ a neighborhood V of X_0 such that $f(X_i) \subseteq U$, whenever $X_i \in V$. Since $X_i \to X_0$, this is the case for all suitably large i. Thus $Y_i \in U$ for all suitably large i, whence $Y_0 \in \bar{U}$. By regularity, $f(X_0) = \bigcap \{\bar{U} \mid U \in \mathscr{U}\}$. Hence $Y_0 \in f(X_0)$. ▯

The euclidean space R^n is, of course, regular.

(3.9.9) **Theorem** (Kakutani [3]. *Let K be a compact convex set in R^n, and denote by $\mathscr{K}(K)$ the family of all non-empty closed sonvex subsets of K. Then for any upper semi-continuous mapping $f: K \to \mathscr{K}(K)$, there exists a point $X_0 \in K$ with $X_0 \in f(X_0)$.*

Proof. Again, we need only consider the case where K is a simplex of dimension n. Otherwise take a simplex S containing K. According to theorem (3.8.5) there exists a continuous point-to-point mapping $g: S \to K$ which is a retraction of S onto K, i.e. $g(X) = X$ for $X \in K$. Define $h: S \to \mathscr{K}(S)$ by $h(X) = f(g(X))$ for $X \in S$. Then h is upper semi-continuous. If the theorem is already proved for the simplices, there exists a point $X_0 \in S$ such that $X_0 \in h(X_0) = f(g(X_0))$. As $f(g(X_0)) \subseteq K$, we must have $X_0 \in K$ and therefore $g(X_0) = X_0$, whence $X_0 \in f(X_0)$.

Let then S be an n-simplex and let $f: S \to \mathscr{K}(S)$ be upper semi-continuous. Consider a sequence $\{\mathscr{K}^{(m)}\}_{m=1,2,\ldots}$ of simplicial subdivisions whose mesh tends to zero as m increases. For each m we define a point-to-point continuous mapping $f^{(m)}: S \to S$ by choosing for each vertex $U^{(m)}$ in $\mathscr{K}^{(m)}$ an arbitrary point in $f(U^{(m)})$ as $f^{(m)}(U)$, and extending $f^{(m)}$ linearly inside each simplex of $\mathscr{K}^{(m)}$. By Brouwer's theorem (3.9.5), $f^{(m)}$ has a fixed point $X^{(m)}$. Let $U_0^{(m)}, U_1^{(m)}, \ldots, U_n^{(m)}$ be the vertices of an n-simplex in $\mathscr{K}^{(m)}$ containing $X^{(m)}$. Then

$$X^{(m)} = w_0^{(m)} U_0^{(m)} + \cdots + w_n^{(m)} U_n^{(m)},$$

where $w_i^{(m)} \geqslant 0$ and $\Sigma w_i^{(m)} = 1$. By the definition of $f^{(m)}$ as a linear extension,

$$f^{(m)}(X^{(m)}) = w_0^{(m)} f^{(m)}(U_0^{(m)}) + \cdots + w_n^{(m)} f^{(m)}(U_n^{(m)}).$$

Abbreviating for $i = 0, \ldots, n$,

$$Z_i^{(m)} := f^{(m)}(U_i^{(m)}) \in f(U_i^{(m)}),$$

and since $f^{(m)}(X^{(m)}) = X^{(m)}$, we have

(3.9.10) $$X^{(m)} = w_0^{(m)} Z_0^{(m)} + \cdots + w_n^{(m)} Z_n^{(m)}.$$

Replacing $\{\mathscr{K}^{(m)}\}_{m=1,2,\ldots}$ by a suitable subsequence, we obtain convergent sequences of vertices and coefficients:

$$X^{(m)} \to X_0, \qquad Z_i^{(m)} \to Z_i, \qquad w_i^{(m)} \to w_i.$$

Their limits satisfy

(3.9.11) $$X_0 = w_0 Z_0 + \cdots + w_n Z_n,$$

with $\Sigma w_i = 1$ and $w_i \geqslant 0$ for $i = 0, \ldots, m$. Since $X^{(m)} \in \mathscr{H}\{U_0^{(m)}, U_1^{(m)}, \ldots, U_n^{(m)}\}$, and the mesh of $\mathscr{K}^{(m)}$ tends to zero, we have for each i:

$$U_i^{(m)} \to X_0.$$

Recall, $Z_i^{(m)} \in f(U_i^{(m)})$ and $Z_i^{(m)} \to Z_i$. Lemma (3.9.8) applies, and consequently

$$Z_i \in f(X_0).$$

Since $f(X_0)$ is convex, this gives $X_0 \in f(X_0)$ in view of (3.9.11). □

Brouwer's fixed point theorem has been extended in various degrees of generality to linear topological spaces other than R^n (Schauder [*1*], Tychonoff [*1*], Klee [*20*]). A corresponding generalization of Kakutani's fixed point theorem is due to Eilenberg and Montgomery [*1*]. Fan [*8*] uses the method of Knaster, Kuratowski, and Mazurkiewicz for proving a theorem which, while not a fixed point theorem itself, contains Tychonoff's result as a special case. Still another generalization of Schauder's theorem in Banach spaces replaces "compact" by "closed and bounded" provided the map in question is non-expansive (Browder [*4*]). The nature of fixed points is examined for instance in Bourgin [*1*], Browder [*2*], [*3*], Fenske [*1*].

A "cubical Sperner lemma" equivalent to Brouwer's fixed point theorem has been found by Kuhn [*4*]. Still another combinatorial lemma on which to base a proof of Brouwer's fixed point theorem has been given by Fan [*1*] generalizing a combinatorial lemma of Tucker [*1*].

3.10. Norms and Support Functions

We began this chapter by introducing norms (3.1.2), and we terminate it by examining the connection between norms and convex sets.

Let $p: R^n \to R$ be a norm. Then the

(3.10.1) $$\textit{norm body } B_p := \{X \in R^n \mid p(X) \leqslant 1\}$$

is a closed convex set since it is the level set of a continuous convex function (see (3.1.11)). Moreover, (3.1.10) shows that B_p is bounded on the one hand, and contains a ball around the origin on the other. Thus

(3.10.2) *the norm body B_p of an arbitrary norm in R^n is a compact convex set of dimension n which contains the origin in its interior.*

Conversely.

(3.10.3) *for every compact convex set $K \subseteq R^n$ of dimension n, which contains the origin in its interior, there is precisely one norm p such that $K = B_p$.*

Proof. Put
$$p(X) := \inf\left\{\rho \geqslant 0 \,\middle|\, \frac{1}{\rho} X \in K\right\}.$$

Clearly, $p(X) \leqslant 1$ if and only if $X \in K$. Thus $K = B_p$, provided p is a norm. To prove this, we first show that p is finite on R^n. Assume $p(X) = \infty$. Then $\frac{1}{\rho} X \notin K$ for all $\rho > 0$, which contradicts the fact that 0 lies in the proper interior of K. Trivially, $p(X) \geqslant 0$ and $p(\lambda X) = \lambda p(X)$ for $\lambda \geqslant 0$. If $p(X) = 0$, then $\{\lambda X \mid \lambda \geqslant 0\} \subseteq K$. Therefore $X = 0$, as K is bounded. Finally

$$\frac{p(X_1 + X_2)}{p(X_1) + (X_2)} = p\left(\frac{p(X_1)}{p(X_1) + p(X_2)} \cdot \frac{X_1}{p(X_1)} + \frac{p(X_2)}{p(X_1) + p(X_2)} \cdot \frac{X_2}{p(X_2)}\right) \leqslant 1,$$

since the argument of p in the middle term above belongs to K by convexity. ▯

The compact convex sets which contain the origin in their interior are therefore in $1-1$ correspondence with the norms in R^n.

For any norm p we define $q: R^n \to R$ as follows:

(3.10.4) $$q(Y) = \sup\left\{\frac{Y^T X}{p(X)} \,\middle|\, X \neq 0\right\} = \sup\{Y^T X \mid p(X) \leqslant 1\}.$$

q is again a norm, called the

$$\textit{dual norm } q = p^D$$

to p. Indeed, nonnegativity follows from:

$$q(Y) \geqslant \frac{Y^T Y}{p(Y)} \geqslant 0.$$

This inequality also yields the implication $q(Y)=0 \Rightarrow Y=0$. Homogeneity is again trivial. Finiteness is assured by the continuity of $Y^T X$ and the compactness of $\{X \mid p(X) \leqslant 1\}$. Finally

$$\begin{aligned} q(Y_1+Y_2) &= \sup\{(Y_1+Y_2)^T X \mid p(X) \leqslant 1\} \\ &\leqslant \sup\{Y_1^T X \mid p(X) \leqslant 1\} + \sup\{Y_2^T X \mid p(X) \leqslant 1\} \\ &\leqslant q(Y_1)+q(Y_2). \end{aligned}$$

The Hölder inequality

(3.10.5) $$p^D(Y)p(X) \geqslant Y^T X$$

follows immediately from the definition (3.10.4) of the dual norm.

Next we observe that

(3.10.6) *the norm body of the dual norm is the polar of the norm body of the original norm:*

$$B_q = (B_p)^\pi \quad \textit{if} \quad q = p^D.$$

Proof. Definition (3.10.4) can be also written in the form

(3.10.7) $$q(Y) := \sup\{Y^T X \mid X \in B_p\}.$$

Then (3.10.6) follows immediately from the definition (2.14.6) of the polar, since

$$q(Y) \leqslant 1 \iff \sup\{Y^T X \mid X \in B_p\} \leqslant 1 \iff Y^T X \leqslant 1 \text{ for all } X \in B_p. \quad \square$$

As $0 \in B_p$, (3.5.7) yields $(B_q)^\pi = (B_p)^{\pi\pi} = B_p$, and therefore

(3.10.8) **Theorem.** $p^{DD} = p$.

In the form (3.10.4), the definition of the dual norm q is an instance of a more general method of assigning a convex function (see chapter 4) to a set. Let S be an arbitrary nonvoid subset of R^n. Putting

$$h_S(Y) := \sup\{Y^T X \mid X \in S\}.$$

Minkowski [3] defined a function $h_S: R^n \to R \cup \{+\infty\}$, called the

(3.10.9) *support function*

of S. Support functions are positively homogeneous and sub-additive just like norms if the obvious conventions for operations involving the value $+\infty$ are adopted. Furthermore, $h_S(0)=0$. However $h_S(Y)=0$ need not imply $Y=0$ and h_S is not nonnegative in general.

(3.10.10) *If* $K=\overline{\mathscr{H}(S\cup\{0\})}$ *then* $h_K=\max\{0,h_S\}$.

Proof. For $\lambda_1,\dots,\lambda_k\geqslant 0$ with $\sum_{i=1}^{k}\lambda_i\leqslant 1$, and for arbitrary numbers $a_1,\dots,a_k\geqslant 0$, plainly

$$\sum_{i=1}^{k}\lambda_i a_i\leqslant\max\{0,a_1,\dots,a_k\}.$$

Thus

$$\sup\{Y^T X\mid X\in\mathscr{H}(S\cup\{0\})\}=\sup\left\{\sum_{i=1}^{k}\lambda_i Y^T X_i\;\middle|\; X_i\in S,\ \lambda_i\geqslant 0,\ \sum\lambda_i\leqslant 1\right\}$$
$$=\sup\{Y^T X\mid X\in S\}.$$

The transition to the closure of the convex hull is justified in view of the continuity of the scalar product. □

Proposition (3.10.6) holds for arbitrary support functions

(3.10.11) $$\{Y\mid h_S(Y)\leqslant 1\}=S^\pi.$$

The dual of a support function is defined by

(3.10.12) $$h_S^D:=h_{(S^\pi)}.$$

Recall $S^{\pi\pi}=\overline{\mathscr{H}(S\cup\{0\})}$ by (3.5.7). Thus (3.10.10) yields

(3.10.13) $$h_S^{DD}=\max\{0,h_S\}.$$

As a final observation we note without proof that

(3.10.14) *the domain of finiteness* $K(h_S):=\{Y\mid h_S(y)<\infty\}$ *is a closed convex cone, in fact, the polar of the characteristic cone of* $\mathscr{H}S$.

Any norm body in R^n is plainly compact. The normed space R^n is therefore locally compact. Conversely, if any topological linear space over R is Hausdorff and locally compact, then it must be isomorphic to R^n (for instance Schaefer [*1*]). Thus norm bodies of infinite dimensional normed linear spaces are typically not compact.

CHAPTER 4

Convex Functions

In this chapter the classical aspects of convex functions will be described. The main topic, however, will be the concept of conjugate functions introduced by Fenchel [*1,2*]. Conjugate functions will form the basis of the duality theory for convex functions as it will be developed in the subsequent chapter.

In addition, quasi- and pseudoconvex functions, as well as pseudolinear functions, are examined because of their increasing importance for optimization problems.

All considerations are again restricted to finite dimensional linear spaces over the field R of real numbers.

The reader should be familiar with the basic concepts of the theory of real functions including semi-continuity and differentiability.

4.1. Convex Functions

In this section we consider functions $f: R^n \to R \cup \{+\infty, -\infty\}$. Such a function f is called a

(4.1.1) *convex function*

if (i) $\{X \mid f(X) = -\infty\} = \emptyset$,

(ii) $\{X \mid f(X) < +\infty\} \neq \emptyset$,

(iii) $f(\lambda X + (1-\lambda) Y) \leq \lambda f(X) + (1-\lambda) f(Y)$ for $0 \leq \lambda \leq 1$.

The function f is called

strictly convex,

if strict inequality holds in (iii) for all $\lambda \neq 0, 1$ and for all pairs of distinct points $X, Y \in K(f)$. Here $K(f)$ denotes the domain of finiteness of f:

(4.1.2) $$K(f) := \{X \mid f(X) < +\infty\}.$$

Clearly, if f is convex, then the set $K(f) \subseteq R^n$ is also convex. $K(f)$ may be considered as the proper domain of definition of the convex function

f. In the next section, the concept of convex functions will be extended to include values $-\infty$. The set $K(f)$ will then include the corresponding arguments. If $-g$ is a (strictly) convex function, then g is called a

(4.1.3) *(strictly) concave function.*

Its domain of finiteness is denoted again by $K(g) := \{X \mid g(X) > -\infty\}$. Whether a function is convex or concave will usually follow from the context, and it will be clear how $K(f)$ or $K(g)$ are to be interpreted[1]. Norms p (3.1.2) are examples of convex functions; in this case $K(p) = R^n$.

(4.1.4) **Monotonicity Relations.** *If f is convex, X is in $K(f)$, and $0 < m \leqslant l$, $0 < h \leqslant k$, then*

$$\frac{f(X) - f(X - lY)}{l} \leqslant \frac{f(X) - f(X - mY)}{m} \leqslant \frac{f(X + hY) - f(X)}{h}$$

$$\leqslant \frac{f(X + kY) - f(X)}{k}.$$

Proof. Rearrange the inequalities $f(X - mY) \leqslant \frac{l-m}{l} f(X) + \frac{m}{l} f(X - lY)$, $f(X) \leqslant \frac{h}{h+m} f(X - mY) + \frac{m}{h+m} f(X + hY)$, $f(X + hY) \leqslant \frac{k-h}{k} f(X) + \frac{h}{k} f(X + kY)$. If some of the points $X - lY$, $X - mY$, $X + hY$, $X + kY$ lie outside of $K(f)$, then their function values will be infinite, and the affected relations of (4.1.4) are trivial. □

(4.1.5) **Theorem.** *At every inner point of $K(f)$, the restriction $f \mid \mathscr{M} K(f)$ of a convex (concave) function f to $\mathscr{M} K(f)$ is continuous with respect to the topology induced in $\mathscr{M} K(f)$.*

Proof. Let $X \in K(f)^I$. Then X possesses a relative neighborhood which is entirely contained in $K(f)$. Moreover, we may assume that such a neighborhood is the intersection of $\mathscr{M} K(f)$ with a closed n-cube centered at X. This neighborhood is a polyhedron P, and is by theorem (2.12.2) the convex hull of its vertices $Z_1, \ldots, Z_k$. The convexity of f implies $f(Z) = f(u_1 Z_1 + \cdots + u_k Z_k) \leqslant u_1 f(Z_1) + \cdots + u_k f(Z_k) \leqslant \max_i f(Z_i)$ for every $Z = u_1 Z_1 + \cdots + u_k Z_k$, $u_i \geqslant 0$, $\Sigma u_i = 1$. Hence f is bounded on P.

[1] In general, we shall use the letter f for convex and the letter g for concave functions.

We write $P=X+U$, where U is a neighborhood of the origin 0. Note that U is symmetric: $-U=U$. The neighborhoods $X+hU$ with $h>0$ form a neighborhood basis at X. To establish continuity, we therefore must show that for each $\varepsilon>0$ there exists a $\delta>0$ such that $|f(X+hY)-f(X)|\leqslant\varepsilon$ whenever $Y\in U$ and $h<\delta$.

Since f admits an upper bound M on P, (4.1.4) gives $f(X+hY)-f(X)\leqslant h(M-f(X))$ for $0\leqslant h\leqslant 1$. By virtue of the symmetry of U, $X+Y\in P$ implies $X-Y\in P$, and (4.1.4) gives $f(X)-f(X+hY)\leqslant h(M-f(X))$ for $0\leqslant h\leqslant 1$. The proof is now easily completed. □

Since R^n is a normed space, we may express the lower limit (=limes inferior) $\underline{f}$ of a function f as follows:

$$(4.1.6)\qquad \underline{f}(X):=\varliminf_{Z\to X} f(Z)=\lim_{\varepsilon\downarrow 0}\inf\{f(Z)\mid p(Z-X)<\varepsilon\},$$

where p is any norm. We then have $\underline{f}(X)\leqslant f(X)$ for every X. $\underline{f}(X)=f(X)$ for all $X\in R^n$ is the familiar condition for lower semicontinuity (l. s. c.) of f. Two further properties of the lower limit $\underline{f}(X)$ follow immediately from its definition. First let the sequence $\{X_i\}_{i=1,2,\dots}$ converge to X. Then the lower limit of the sequence $\{f(X_i)\}_{i=1,2,\dots}$ is bounded below by $\underline{f}(X)$:

$$\varliminf_{i\to\infty} f(X_i)=\lim_{k\to\infty}\inf\{f(X_i)\mid i>k\}\geqslant\underline{f}(X).$$

Secondly, for each $X\in K(\underline{f})$ there exists a sequence of X_i's converging to X such that $f(X_i)\to\underline{f}(X)$.

Since every inner point of $K(f)$ is a point of continuity, we have

$$f(X)=\underline{f}(X)\quad \textit{for every}\quad X\in K(f)^I.$$

The function $\underline{f}$ is called the

(4.1.7) *closure*

of the convex function f, and f is called a

(4.1.8) *closed convex function*

if $f=\underline{f}$. This important notion of closure of a function, and the theorems pertaining to it in the following section are due to Fenchel [2].

Closedness (4.1.8) is clearly equivalent to lower semicontinuity on R^n, but not necessarily to lower semicontinuity on $K(f)$, e.g.

$$f(x)=\begin{cases}+\infty & \text{if}\quad x\leqslant 0,\\ x & \text{if}\quad x>0.\end{cases}$$

Since historically the convex functions f have been considered on $K(f)$ rather than on the entire R^n, the term "closed" rather than "l.s.c." has found acceptance.

(4.1.9) *The closure $\underline{f}$ of a convex function f is again a convex function.*

Proof. Suppose $X, Y \in K(f)$. If sequences $\{X_i\}_{i=1,2,\dots}$ and $\{Y_i\}_{i=1,2,\dots}$ are chosen so that

$$\lim_{i\to\infty} X_i = X, \quad \lim_{i\to\infty} Y_i = Y, \quad \lim_{i\to\infty} f(X_i) = \underline{f}(X), \quad \lim_{i\to\infty} f(Y_i) = \underline{f}(Y),$$

then clearly

$$\underline{f}(\lambda X + \mu Y) \leqslant \lim_{i\to\infty} f(\lambda X_i + \mu Y_i) \leqslant \lim_{i\to\infty} (\lambda f(X_i) + \mu f(Y_i)) = \lambda \underline{f}(X) + \mu \underline{f}(Y)$$

for all $\lambda \geqslant 0$, $\mu \geqslant 0$, $\lambda + \mu = 1$, which was to be shown. Since (on the other hand) $\underline{\underline{f}}(X) = \underline{f}(X)$ holds for every function $f: R^n \to R \cup \{+\infty, -\infty\}$, we conclude that the closure of a convex function is always closed. □

As is readily seen, a convex function may be discontinuous. But a more surprising phenomenon does occur: The closure $\underline{f}$ of a convex function f may not be continuous on $K(f)$. Consider the function

$$f(x,y) := \begin{cases} 0 & \text{for } x = y = 0, \\ \dfrac{y^2}{x} & \text{for } x > 0, \\ +\infty & \text{otherwise.} \end{cases}$$

The identity

$$(\lambda y_1 + \mu y_2)^2 = (\lambda x_1 + \mu x_2)\left(\frac{\lambda y_1^2}{x_1} + \frac{\mu y_2^2}{x_2}\right) - \lambda \mu \frac{(x_1 y_2 - x_2 y_1)^2}{x_1 x_2}$$

may be used to verify that f is convex. Clearly $\underline{f} = f$. But f is not continuous at the origin. Note also that the domain of finiteness of f may be restricted to $\left\{\begin{pmatrix} x \\ y \end{pmatrix} \,\middle|\, x \geqslant y^4\right\}$. Then $\underline{f}$ is still discontinuos, in spite of its domain of finiteness now being closed. The function in Fig. 16 shows the same behaviour.

It is clear, however, that closed convex functions of only one variable are continuous. It follows that

(4.1.10) *the restriction of a closed convex function f to any closed, open, or half open line segment in $K(f)$ is continuous.*

An important consequence of this fact is that

(4.1.11) *a closed convex function f is uniquely determined by its values in $K(f)^I$.*

Indeed, let $U \in K(f)^I$ and $X \in K(f) \sim K(f)^I$. Then f is continuous on the line segment $\mathscr{H}\{X, U\}$. The value $f(X)$ is therefore determined

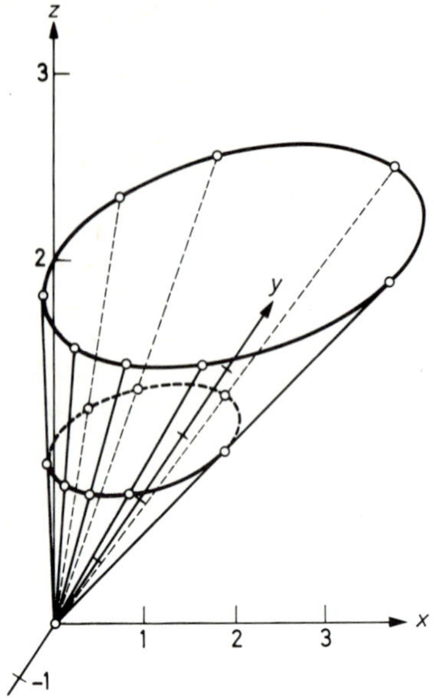

Fig. 16. $f(x,y) := \begin{cases} \dfrac{x^2+y^2}{2x} & \text{for } x>0, \\ 0 & \text{for } x=0,\ y=0 \end{cases}$

by the values of f on the open line segment (X, U). But $(X,U) \subseteq K(f)^I$ by the accessibility lemma (3.2.11). □

As a consequence of (4.1.11),

(4.1.12) $$\underline{f}(X) = \underline{\lim}_{Z \to X} f(Z) \quad \textit{where} \quad Z \in K(f)^I.$$

Returning to general convex functions, we state the following useful

(4.1.13) **Lipschitz-type Theorem.** *Suppose f is a convex (concave) function on R^n. For every compact subset K of $K(f)^I$ there exists a Lipschitz constant L such that*

We also have

$$|f(Y) - f(X)| \leqslant L \|Y - X\|$$

for all $X, Y \in K$. Here $\|Y - X\|$ denotes the euclidean distance between X and Y.

Proof. Any closed set $F \subseteq R^n$ which does not meet K, has a positive distance from the compact set K. In particular, this holds for $F = \mathscr{M} K(f) \sim K(f)^I$. Hence there is an $h > 0$ such that

$$U + hZ \in K(f)^I$$

for all $U \in K$ and all Z in the compact set

$$S := \{Z \in R^n | \, \|Z\| = 1\} \cap (\mathscr{M} K(f) - V),$$

where $V \in K(f)$. In other words, S contains all directions of euclidean length 1 which do not lead out of $\mathscr{M}\, K(f)$.

If one puts

$$Z := \frac{Y - X}{\|Y - X\|},$$

then the monotonicity relations (4.1.4) give

$$\frac{f(X) - f(X - hZ)}{h} \leqslant \frac{f(Y) - f(X)}{\|Y - X\|} \leqslant \frac{f(Y + hZ) - f(Y)}{h}.$$

Now consider the function $\phi: K \times S \to R$ defined by

$$\phi(U, Z) := \frac{f(U + hZ) - f(U)}{h}.$$

This function is continuous by theorem (4.1.5), since $U + hZ \in K(f)^I$ has been ascertained. The set $K \times S$ is compact. Therefore

$$L := \max\{\phi(U, Z) \mid U \in K, Z \in S\}$$

exists, and meets the requirements of the theorem. □

We also have

(4.1.14) **Lemma.** *The nonvoid level sets* $S(\alpha) := \{X \in R^n \,|\, f(X) \leqslant \alpha\}$ *of a closed convex function f are either all bounded or all unbounded.*

Proof. The level sets $S(\alpha)$ are closed convex sets, as f is closed and convex. Suppose that level set $S(\beta)$ is unbounded. By (3.5.1), $S(\beta)$ possesses a "direction of infinity" Z: $Z \neq 0$, $X + \theta Z \in S(\beta)$ for all $X \in S(\beta)$ and all $\theta \geqslant 0$. For $\alpha > \beta$, $S(\alpha)$ is of course unbounded, since $S(\alpha) \supseteq S(\beta)$ in this case. Let then $\alpha < \beta$ and $S(\alpha) \neq \emptyset$. Select any $X \in S(\alpha) \subseteq S(\beta)$ and $\theta \geqslant 0$. For all $\lambda, \mu > 0$, $\lambda + \mu = 1$:

$$f(X + \theta Z) \leqslant \lambda f(X) + \mu f\left(X + \frac{\theta}{\mu} Z\right) \leqslant \lambda\alpha + \mu\beta,$$

as $X + \dfrac{\theta}{\mu} Z \in S(\beta)$. Hence

$$f(X + \theta Z) \leqslant \alpha,$$

and Z is a direction of infinity of $S(\alpha)$, too. Thus all level sets are unbounded provided one of them is. □

4.2. Epigraphs

In this section, questions about convex or concave functions will be reduced to questions about convex sets. To this end we associate with any, not necessarily convex, function $f: R^n \to R \cup \{+\infty, -\infty\}$ the set

$$[f] := \left\{ \begin{pmatrix} X \\ z \end{pmatrix} \in R^{n+1} \;\middle|\; z \geqslant f(X) \right\}.$$

The set $[f]$ is a union of vertical half-lines (and some lines if f assumes the value $-\infty$), and is called the

(4.2.1) *epigraph*

of f. By $\underline{f}$ we denote the lower limit of f as in (4.1.6). Then

(4.2.2) $$[\underline{f}] = \overline{[f]}.$$

Proof. Suppose $\begin{pmatrix} X \\ z \end{pmatrix} \in \overline{[f]}$. Then there exists a sequence

$$\left\{ \begin{pmatrix} X_i \\ z_i \end{pmatrix} \right\}_{i=1,2,\dots} \to \begin{pmatrix} X \\ z \end{pmatrix} \quad \text{with} \quad \begin{pmatrix} X_i \\ z_i \end{pmatrix} \in [f].$$

$f(X_i) \leqslant z_i$ be the definition of $[f]$. Now

$$\underline{f}(X) \leqslant \varliminf_{i\to\infty} f(X_i) \leqslant \varliminf_{i\to\infty} z_i = \lim_{i\to\infty} z_i = z.$$

Hence $\begin{pmatrix} X \\ z \end{pmatrix} \in [\underline{f}]$.

Suppose $\begin{pmatrix} X \\ z \end{pmatrix} \in [\underline{f}]$. Then $\underline{f}(X) \leqslant z$. Hence there exist sequences $\{X_i\}_{i=1,2,\dots} \to X$ and $\{\bar{z}_i\}_{i=1,2,\dots} \to \bar{z} \leqslant z$ such that $f(X_i) \leqslant \bar{z}_i$.

Putting $z_i := \max\{z, \bar{z}_i\}$, one has

$$\begin{pmatrix} X_i \\ z_i \end{pmatrix} \to \begin{pmatrix} X \\ z \end{pmatrix} \quad \text{while} \quad \begin{pmatrix} X_i \\ z_i \end{pmatrix} \in [f].$$

Hence $\begin{pmatrix} X \\ z \end{pmatrix} \in \overline{[f]}$. ▯

Clearly, if f is convex, then so is $[f]$. If g is a concave function then its

(4.2.3) *hypograph*,

namely the set

$$[g] := \left\{ \begin{pmatrix} X \\ z \end{pmatrix} \in R^{n+1} \;\middle|\; z \leqslant g(X) \right\},$$

is also convex. Since in most of the following applications the functions will be understood to be either convex or concave, we will use the bracket notation "[ ]" for epigraphs of convex functions as well as for hypographs of concave functions.

As an immediate consequence of (4.2.2) we have:

(4.2.4) *A convex (concave) function is closed if and only if its corresponding epi-(hypo-)graph is closed.*

The set $\mathscr{S}$ of all epigraphs which arise from convex functions (4.1.1) almost forms a lattice under set inclusion, but not quite. If f_1 and f_2 are convex functions and if $K(f_1)\cap K(f_2)\neq\emptyset$, then $\max\{f_1,f_2\}$ is again a convex function and

$$[f_1]\cap[f_2]=[\max\{f_1,f_2\}]\in\mathscr{S}.$$

If, however, $K(f_1)\cap K(f_2)=\emptyset$, then $\max\{f_1,f_2\}\equiv+\infty$ and $[f_1]\cap[f_2]$ is the empty set, which does not belong to $\mathscr{S}$. Worse still, there exists no convex function $f\colon R\to R$ such that $[f_1]\subseteq[f]$ and $[f_2]\subseteq[f]$, where $f_1(x)=x$ and $f_2(x)=-x$.

However, the set $\mathscr{S}$ can be supplemented to a lattice by adjoining the epigraphs of functions that are convex in a "wider sense". The function $f(X):=+\infty$ for all $X\in R^n$ is such a function. Aside from this function, the generalization of convexity consists in admitting the value $-\infty$. A function $f\colon R^n\to R\cup\{-\infty,+\infty\}$ is

(4.2.5) *convex in the wider sense*

if the inequality

$$f(\lambda X+\mu Y)\leqslant\lambda f(X)+\mu f(Y),\quad \lambda,\mu\geqslant 0,\quad \lambda+\mu=1$$

holds for all X, Y for which the right hand side is not an indefinite expression of the type $\infty-\infty$. As a consequence the sets $f^{-1}(-\infty)=\{X\mid f(X)=-\infty\}$ and $K(f)=\{X\mid f(X)<+\infty\}$ must be convex. Moreover, we have the

(4.2.6) **Lemma.** *If the value $-\infty$ is assumed, then finite values cannot occur but on the (relative) boundary of $K(f)$.*

Proof. Let $f(X)$ be finite, and $f(U)=-\infty$. Putting

$$V(\theta):=X+\theta(X-U)$$

with $\theta>0$, we have $X\in(U,V(\theta))$. Assuming $f(V(\theta))<+\infty$ the values $f(V(\theta))$ and $f(U)$ can be meaningfully combined, yielding the contradiction $f(X)=-\infty$. Thus $f(V(\theta))=+\infty$ for all $\theta>0$. Since $V(\theta)\to X$ as $\theta>0$, and since $V(\theta)\in\mathscr{M}K(f)$, the point X is on the boundary of $K(f)$. □

The reader verifies readily, that

(4.2.7) *A function* $f: R^n \to R \cup \{-\infty, +\infty\}$ *is convex in the wider sense* (4.2.5) *if and only if its epigraph* $[f] := \left\{ \begin{pmatrix} X \\ z \end{pmatrix} \;\middle|\; z \geqslant f(X) \right\}$ *is convex. The set of all these epigraphs forms a lattice under set inclusion with the intersection as the cap and the convex hull as the cup operation.*

The concept of convexity in the wider sense (4.2.5) will prove to be useful in connection with the directional derivatives to be described in the next section.

We return to functions that are convex in the proper sense (4.1.1). For many considerations in chapter 5, the existence of nonvertical supporting planes of $[f]$ is essential. Indeed, the following theorem is one of the reasons for introducing epigraphs.

(4.2.8) **Theorem.** *If* f *is a convex function, then a nonsingular supporting plane* E *of* $[f]$ *through a point* $\begin{pmatrix} X_0 \\ f(X_0) \end{pmatrix}$ *with* $X_0 \in K(f)^I$ *is always nonvertical. It can be characterized as follows:*

$$E = \left\{ \begin{pmatrix} X \\ z \end{pmatrix} \;\middle|\; A^T X - z = b \right\}$$

such that $b = A^T X_0 - f(X_0)$ *and* $f(X) - f(X_0) \geqslant A^T (X - X_0)$ *for all* X.

Proof. Assume the plane E to be vertical, i.e., $E = \left\{ \begin{pmatrix} X \\ z \end{pmatrix} \;\middle|\; A^T X = b \right\}$. Then $\tilde{E} := \{X \mid A^T X = b\}$ is a supporting plane of $K(f)$ through X_0. Indeed, $[f] \subseteq \left\{ \begin{pmatrix} X \\ z \end{pmatrix} \;\middle|\; A^T X \leqslant b \right\}$ implies $K(f) \subseteq \{X \mid A^T X \leqslant b\}$. Since $\tilde{E}$ contains the inner point X_0 of $K(f)$, it must be a singular supporting plane, that is, it must contain $K(f)$ entirely (lemma (3.4.4)). But then E itself must be singular, in contradiction to the hypothesis. □

Taking into account the corollary (3.4.12) of the supporting plane theorem, we obtain the important

(4.2.9) **Corollary to Theorem (4.2.8).** *If* f *is a convex function and* $X_0 \in K(f)^I$, *then there is a vector* A *such that*

$$f(X) - f(X_0) \geqslant A^T (X - X_0)$$

holds for all X, *i.e.,* $E := \left\{ \begin{pmatrix} X \\ z \end{pmatrix} \;\middle|\; A^T X - Z = A^T X_0 - f(X_0) \right\}$ *supports* $[f]$ *at* X_0.

Theorem (4.2.8) establishes the existence of nonvertical supporting planes only in the relative interior of $K(f)$. In the general case $X_0 \in K(f)$, we have the following necessary and sufficient criterion:

(4.2.10) **Lemma.** *The epigraph* $[f]$ *of a convex function* f *in the wider sense* (4.2.5) *has a nonvertical supporting plane through* $\begin{pmatrix} X_0 \\ f(X_0) \end{pmatrix}$, $f(X_0) \in R$, *if and only if the supporting cone* $C\left([f]; \begin{pmatrix} X_0 \\ f(X_0) \end{pmatrix}\right)$ *does not contain a vertical line.*

The *proof* is an immediate consequence of the definition of supporting cones and the fact that they are the polars of normal cones (theorem (3.4.16)). Indeed if all supports are vertical, then the normal cone $N\left([f]; \begin{pmatrix} X_0 \\ f(X_0) \end{pmatrix}\right)$ (3.4.13) is contained in the plane $\left\{\begin{pmatrix} Y \\ w \end{pmatrix} \middle| w=0\right\}$. Hence $C\left([f]; \begin{pmatrix} X_0 \\ f(X_0) \end{pmatrix}\right)$ contains the vertical line $\left\{\begin{pmatrix} 0 \\ z \end{pmatrix} \middle| z \in R\right\}$. The other direction is trivial. ▯

4.3. Directional Derivatives

We shall see in this section that, except at boundary points of $K(f)$, convex functions are not only continuous but possess finite derivatives in each direction which locally does not lead out of $K(f)$. Implicitly, the proof of continuity already used this fact.

If $X \in K(f)$, then we may define the function

$$d(h) := \frac{f(X+hY) - f(X)}{h}.$$

According to the monotonicity relations (4.1.4), $d(h)$ is a monotonically increasing function of h. Therefore the directional derivative

$$f'(X; Y) := \lim_{h \downarrow 0} d(h) \tag{4.3.1}$$

exists. If $X + k_0 Y \in K(f)$ for some $k_0 > 0$, it is either finite or $-\infty$. Obviously $f'(X; 0) = 0$ for all $X \in K(f)$. Also for $h, k > 0$

$$\frac{f(X) - f(X-kY)}{k} \leqslant -f'(X; -Y) \leqslant f'(X; Y) \leqslant \frac{f(X+hY) - f(X)}{h}. \tag{4.3.2}$$

Proof. The monotonicity relations (4.1.4) give

$$\frac{f(X)-f(X-kY)}{k} \leqslant \frac{f(X+hY)-f(X)}{h}$$

and therefore

$$\sup_{k>0} \frac{f(X)-f(X-kY)}{k} = -f'(X;-Y) \leqslant f'(X;Y) = \inf_{h>0} \frac{f(X+hY)-f(X)}{h}.$$

As a consequence of (4.3.2) and of lemma (3.2.9) we have

(4.3.3) **Theorem.** *The directional derivative $f'(X;Y)$ of a convex function f is finite if $X \in K(f)^I$ and $X+Y \in \mathcal{M}K(f)$.*

Next we prove that

(4.3.4) *for every $X \in K(f)$ the directional derivative $f'(X;\cdot)$ is*
(i) *positively homogeneous, i. e.*

$$f'(X;\lambda Y) = \lambda f'(X;Y) \quad \textit{for all} \quad \lambda > 0.$$

(ii) *subadditive, i. e.*

$$f'(X;Y+Z) \leqslant f'(X;Y) + f'(X;Z) \quad \textit{if} \quad f'(X;Y)+f'(X;Z) \neq \infty - \infty.$$

Proof. (i) follows immediately from the definition (4.3.1) of $f'(X;Y)$. The inequality

$$\begin{aligned} f(X+h(Y+Z)) &= f\left(\tfrac{1}{2}(X+2hY)+\tfrac{1}{2}(X+2hZ)\right) \\ &\leqslant \tfrac{1}{2} f(X+2hY) + \tfrac{1}{2} f(X+2hZ) \end{aligned}$$

implies for $f'(X;Y), f'(X;Z) < +\infty$ that

$$\begin{aligned} f'(X;Y+Z) &\leqslant \lim_{h \downarrow 0} \frac{f(X+2hY)-f(X)}{2h} + \lim_{h \downarrow 0} \frac{f(X+2hZ)-f(X)}{2h} \\ &= f'(X;Y) + f'(X;Z). \end{aligned}$$

This proves (ii). □

As a consequence of (4.3.4), $f'(X;\cdot)$ is convex in the wider sense (4.2.5). $f'(X;\cdot)$ is properly convex in the sense (4.1.1) if $X \in K(f)^I$, since in this case $f'(X;Y) > -\infty$ for all Y according to theorem (4.3.3). The set of all arguments Y for which the values of $f'(X;Y)$ stay below $+\infty$ clearly coincides with the cone of feasible directions (3.4.15) of $K(f)$ with respect to X:

$$D(K(f);X) = \{Y \mid f'(X;Y) < +\infty\}.$$

Since $f'(X;\cdot)$ is convex in the wider sense (4.2.5), it gives rise to a convex epigraph (4.2.1).

(4.3.5) *Let f be a convex function. Then for every $X \in K(f)$ the epigraph*

$$C(f;X) := [f'(X;\cdot)] = \left\{ \begin{pmatrix} Y \\ w \end{pmatrix} \;\middle|\; w \geqslant f'(X;Y) \right\}$$

is a cone.

Proof. $C(f;X)$ contains $\begin{pmatrix} 0 \\ 0 \end{pmatrix}$. If $\begin{pmatrix} Y \\ w \end{pmatrix} \in C(f;X)$, then $\lambda \begin{pmatrix} Y \\ w \end{pmatrix} \in C(f;X)$ for all $\lambda > 0$, because of (4.3.4.(i)). If $\begin{pmatrix} Y_i \\ w_i \end{pmatrix} \in C(f;X)$, $i=1,2$, then (4.3.4.(ii)) implies

$$w_1 + w_2 \geqslant f'(X;Y_1) + f'(X;Y_2) \geqslant f'(X;Y_1+Y_2). \quad \square$$

(4.3.6) **Theorem.** *$\overline{C(f;X)}$ is the supporting cone* (3.4.14) *of $[f]$ at $\begin{pmatrix} X \\ f(X) \end{pmatrix}$:*

$$\overline{C(f;X)} = C\left([f]; \begin{pmatrix} X \\ f(X) \end{pmatrix}\right).$$

Proof. Combining the relations (i) and (ii) of theorem (3.4.16) yields $D(K;X)^{pp} = \overline{D(K;X)} = C(K;X)$. Here $D(K;X)$ denotes the cone of feasible directions (3.4.15) of a convex set K with respect to a point $X \in \bar{K}$, whereas $C(K;X)$ denotes the supporting cone of K at X. Thus,—writing simply D for the cone $D\left([f]; \begin{pmatrix} X \\ f(X) \end{pmatrix}\right)$ of feasible directions of $[f]$ with respect to $\begin{pmatrix} X \\ f(X) \end{pmatrix}$,—the theorem will be proved if we establish that

$$D \subseteq C(f;X) \subseteq \bar{D}.$$

Clearly $\begin{pmatrix} Y \\ w \end{pmatrix} \in D$ precisely if $X + \theta Y \in K(f)$ and $f(X) + \theta w \geqslant f(X+\theta Y)$ for some $\theta > 0$. The first condition is actually a consequence of the second, since $X + \theta Y \notin K(f)$ would imply $f(X+\theta Y) = +\infty$, violating the second inequality. Hence

$$\begin{pmatrix} Y \\ w \end{pmatrix} \in D \iff w \geqslant \frac{f(X+\theta Y) - f(X)}{\theta} \quad \text{for some } \theta > 0.$$

From this and (4.3.2) one concludes immediately that $D \subseteq C(f;X)$. Now suppose that $\begin{pmatrix} Y \\ w \end{pmatrix} \in C(f;X)$. If $w > f'(X;Y)$, then there exists a $\theta > 0$, such that $w \geqslant \dfrac{f(X+\theta Y) - f(X)}{\theta}$ since $f'(X;Y) = \lim\limits_{\theta \downarrow 0} \dfrac{f(X+\theta Y) - f(X)}{\theta}$.

Hence $\begin{pmatrix} Y \\ w \end{pmatrix} \in D$. If $w = f'(X; Y)$, then $\begin{pmatrix} Y \\ w \end{pmatrix}$ is the limit of points $\begin{pmatrix} Y \\ w_i \end{pmatrix}$, $w_i > w = f'(X; Y)$, and therefore $\begin{pmatrix} Y \\ w \end{pmatrix} \in \bar{D}$. □

$C(f; X)$ is not closed, for instance, if $f'(X; Y) = -\infty$ for some Y. However,

(4.3.7) *if* $X \in K(f)^I$ *then* $C(f; X)$ *is closed, and coincides therefore with the supporting cone of* $[f]$ *at* $\begin{pmatrix} X \\ f(X) \end{pmatrix}$.

Proof. If $X \in K(f)^I$, then $f'(X; Y) > -\infty$ for all Y (theorem (4.3.3)). Moreover, $D(K(f); X)$ is closed since $D(K(f); X) = \mathcal{M}(K(f) - X)$. The proof will rely on these two facts.

As a function of Y, $f'(X; Y)$ is convex in the wider sense (4.2.5). Since the value $-\infty$ has been ruled out, the function $f'(X; Y)$ must be properly convex (4.1.1), and therefore continuous in Y (theorem (4.1.5)). Hence the function $F(Y, w) := w - f'(X; Y)$ is continuous on the cartesian product $G := D(K(f); X) \times R$, and the cone $C(f; X) = \left\{ \begin{pmatrix} Y \\ w \end{pmatrix} \middle| \; F(Y, w) \geqslant 0 \right\}$ is closed in G. But G is closed in R^{n+1} since $D(K(f); X)$ is closed. Hence $C(f; X)$ is closed in R^{n+1}. □

As was mentioned in the previous section, the existence of nonvertical planes of support is of prime importance for the duality theory of convex functions. Theorem (4.2.8), in particular, yielded the existence of nonvertical supporting planes for points X_0 in the relative interior of $K(f)$. The following theorem supplements theorem (4.2.8) in that it is applicable also to boundary points of $K(f)$.

(4.3.8) **Theorem.** *If f is a convex function, then there exists a nonvertical supporting plane E of $[f]$ through a point* $\begin{pmatrix} X_0 \\ f(X_0) \end{pmatrix}$ *with* $X_0 \in K(f)$ *if and only if there exists at least one* Y_0 *with* $X_0 + Y_0 \in K(f)^I$ *and* $f'(X_0; Y_0) > -\infty$.

Proof. If $X_0 + Y_0 \in K(f)^I$ and $f'(X_0; Y_0) > -\infty$, then by (4.3.2)

$$f(X_0 + h Y_0) \geqslant f(X_0) + h f'(X_0; Y_0)$$

for all $h \in R$. Hence

$$M := \left\{ \begin{pmatrix} X \\ z \end{pmatrix} \middle| \; X = X_0 + h Y_0, \; z = f(X_0) + h f'(X_0; Y_0) \text{ for some } h \right\}$$

is a nonsingular supporting manifold of $[f]$ through $\begin{pmatrix} X_0 \\ f(X_0) \end{pmatrix}$. By

(3.4.11) there exists therefore a nonsingular supporting plane E of $[f]$ with $E \supseteq M$. Using the condition $X_0 + Y_0 \in K(f)^I$, it can be shown in much the same way as in the proof of (4.2.8) that E is nonvertical.

Conversely, if $E = \left\{ \begin{pmatrix} X \\ z \end{pmatrix} \,\middle|\, A^T X - z = b \right\}$ is a nonvertical supporting plane of $[f]$ through $\begin{pmatrix} X_0 \\ f(X_0) \end{pmatrix}$, then

$$f(X) - f(X_0) \geqslant A^T (X - X_0),$$

and therefore

$$f'(X_0; Y) \geqslant A^T Y > -\infty \qquad \text{for all} \qquad Y.$$

Hence, any Y_0 with $X_0 + Y_0 \in K(f)^I$ will do. ▯

If $Y, -Y \in D(K(f); X)$ and $f'(X; \lambda Y) = \lambda f'(X; Y)$ for all λ, then Y is called a

(4.3.9) *lineality vector*

of f at (see Bonnesen and Fenchel [*1*]). Lineality vectors are those vectors along which the "one-sided" derivatives coincide. More precisely, if Y is a lineality vector of f at X_0, then $f(X_0 + tY)$ is differentiable at $t = 0$ as a function of t.

(4.3.10) **Theorem.** *If* $Y_1, \ldots, Y_k$ *are lineality vectors of* f *at* X, *then so is any linear combination* $w_1 Y_1 + \cdots + w_k Y_k$. *Moreover, the equation.*

$$f'(X; w_1 Y_1 + \cdots + w_k Y_k) = w_1 f'(X; Y_1) + \cdots + w_k f'(X; Y_k)$$

holds for all $w_1, \ldots, w_k$.

Proof. If Y, Z are lineality vectors, then by definition (4.3.9) none of the derivatives $f'(X; \pm Y)$, $f'(X; \pm Z)$ is $+\infty$. Hence (4.3.4.(ii)) gives

$$f'(X; Y+Z) \leqslant f'(X; Y) + f'(X; Z),$$
$$f'(X; -Y-Z) \leqslant f'(X; -Y) + f'(X; -Z).$$

Since $f'(X; -Y) = -f'(X; Y)$ and $f'(X; -Z) = -f'(X; Z)$ for lineality vectors, these inequalities can be combined to give $f'(X; Y+Z) \leqslant -f'(X; -Y-Z)$. Equality then must hold in view of (4.3.2). This is the essential part of the proof. The remainder is left to the reader. ▯

In chapter 2 we introduced the lineality space (2.10.2) of a cone C as the intersection of C with $-C$. The following theorem connects the lineality space of the supporting cone with the subspace of lineality vectors.

(4.3.11) **Theorem.** *Suppose f is a convex function and $X \in K(f)^I$. Let C be the cone supporting $[f]$ at $\begin{pmatrix} X \\ f(X) \end{pmatrix}$, and L the subspace formed by all lineality vectors of f at X. Then*

$$L = \left\{ Y \;\middle|\; \begin{pmatrix} Y \\ w \end{pmatrix} \in C \cap (-C) \text{ for some } w \right\}.$$

Proof. We show that $C(f; X) \cap (-C(f; x)) = F$, where

$$F := \left\{ \begin{pmatrix} Y \\ f'(X; Y) \end{pmatrix} \;\middle|\; Y \text{ lineality vector of } f \text{ at } X \right\}.$$

In view of (4.3.7), this will prove the theorem.

If Y is a lineality vector of f at X, then

$$-\begin{pmatrix} Y \\ f'(X; Y) \end{pmatrix} = \begin{pmatrix} -Y \\ f'(X; -Y) \end{pmatrix} \in C(f; X).$$

On the other hand, if $\begin{pmatrix} Y \\ w \end{pmatrix}$, $-\begin{pmatrix} Y \\ w \end{pmatrix} \in C(f; X)$, then $w \geqslant f'(X; Y)$ and $-w \geqslant f'(X; -Y)$. Hence $f'(X; Y) \leqslant w \leqslant -f'(X; -Y)$, and $w = f'(X; Y) = -f'(X; -Y)$ by (4.3.2). Now since all directional derivatives are finite according to theorem (4.3.3), Y is a lineality vector. ▯

The examination of derivatives has been restricted so far to proper convex functions f. If the value $-\infty$ is admitted, then f need not be finitely defined in $K(f)$. If, however, $f(X)$ is finite, then all derivatives $f'(X, Y)$ exist, and may be even finite in some directions. We summarize without proofs:

(4.3.12) *If f is convex in the wider sense* (4.2.5) *and if $f(X)$ is finite, then $f'(X; Y)$ is well-defined for all Y. One has*

$$-f'(X; -Y) \leqslant f'(X; Y)$$

and

$$f'(X; \lambda Y) = \lambda f'(X; Y) \quad \text{for all} \quad \lambda > 0.$$

Again the cone of feasible directions $D(K(f); X)$ is linked to f' by

$$D(K(f); X) := \{ Y \mid f'(X; Y) < +\infty \}.$$

The epigraph

$$C(f; X) := [f'(X; \cdot)] = \left\{ \begin{pmatrix} Y \\ w \end{pmatrix} \;\middle|\; w \geqslant f'(X; Y) \right\}$$

is a cone whose closure is the supporting cone (3.4.14) *of $[f]$ at $\begin{pmatrix} X \\ f(X) \end{pmatrix}$.*

4.4. Differentiable Convex Functions

A convex function f is called

(4.4.1) *differentiable*[1]

at X_0 if the partial derivatives $\left.\frac{\partial f}{\partial x_i}\right|_{X=X_0}$ exist and are finite. The vector whose components are these derivatives, is called the

(4.4.2) *gradient* $\operatorname{grad} f(X_0)$

at the point X_0. $\operatorname{grad} f$ is a vector function, mapping points at which f is differentiable into points in R^n. Clearly, this definition of differentiability requires X_0 to be in the R^n-interior of $K(f)$, in other words, $\mathscr{M}K(f)=R^n$ and $X_0 \in K(f)^I$.

Next we note that the n coordinate vectors E_i, $i=1,\dots,n$, whose components all vanish with the exception of the i-th, are lineality vectors of f at X_0, if f is differentiable at X_0. But then *all* vectors are lineality vectors according to theorem (4.3.10), and one has

(4.4.3) $$f'(X_0;Y)=Y^T \operatorname{grad} f(X_0).$$

This relation expresses essentially the "chain rule" of differential calculus. It also shows that there exists an affine function, namely $g(X):=f(X_0)+(X-X_0)^T \operatorname{grad} f(X_0)$, such that $g(X_0)=f(X_0)$ and $g'(X_0;Y)=f'(X_0;Y)$ for all $Y \in R^n$. In geometric language: there exists a tangent plane to the graph of f at $\begin{pmatrix} X_0 \\ f(X_0) \end{pmatrix}$. The tangent plane

(4.4.4) $$T:=\left\{\begin{pmatrix} X \\ z \end{pmatrix} \;\middle|\; z=f(X_0)+(X-X_0)^T \operatorname{grad} f(X_0)\right\}$$

is a supporting plane. More precisely,

(4.4.5) *the halfspace*

$$H:=\left\{\begin{pmatrix} X \\ z \end{pmatrix} \;\middle|\; z \geqslant f(X_0)+(X-X_0)^T \operatorname{grad} f(X_0)\right\},$$

which is bounded by the tangent plane (4.4.4), *is the unique supporting halfspace of* $[f]$ *at* $\begin{pmatrix} X_0 \\ f(X_0) \end{pmatrix}$.

---

[1] This definition of differentiability differs from the most common one, which requires the partial derivatives to be also continuous (see for example Dieudonné [*1*]). Later in this section it will be seen that convexity in fact implies continuity of the derivatives, provided they exist.

Proof. Let C be the supporting cone (3.4.14) of $[f]$ at $\begin{pmatrix} X_0 \\ f(X_0) \end{pmatrix}$. According to (4.3.7)

$$C = C(f; X_0) = \left\{ \begin{pmatrix} Y \\ w \end{pmatrix} \;\middle|\; w \geqslant f'(X_0; Y) = Y^T \operatorname{grad} f(X_0) \right\} = H - \begin{pmatrix} X_0 \\ f(X_0) \end{pmatrix}.$$

This proves (4.4.5). □

In particular, (4.4.5) shows that $[f]$ admits a nonvertical supporting plane at $\begin{pmatrix} X_0 \\ f(X_0) \end{pmatrix}$, if f is differentiable at X_0. The existence of a unique supporting plane implies differentiability if this supporting plane is nonvertical. The statement in the following theorem is a trifle more general in that it requires uniqueness only for the nonvertical supporting planes.

(4.4.6) **Theorem.** *Let f be a convex function. Then f is differentiable at X_0 if and only if $X_0 \in K(f)^I$ and there is at most one nonvertical supporting plane of $[f]$ at $\begin{pmatrix} X_0 \\ f(X_0) \end{pmatrix}$.*

Proof. Let $E = \left\{ \begin{pmatrix} X \\ z \end{pmatrix} \;\middle|\; z - f(X_0) = A^T(X - X_0) \right\}$ be the only nonvertical supporting plane. Then

$$H := \left\{ \begin{pmatrix} X \\ z \end{pmatrix} \;\middle|\; z - f(X_0) \geqslant A^T(X - X_0) \right\}$$

is a supporting halfspace. In other words,

$$f(X) - f(X_0) \geqslant A^T(X - X_0)$$

holds for all X (see (4.2.9)).

We first exclude the possibility of a vertical supporting halfspace $V := \left\{ \begin{pmatrix} X \\ z \end{pmatrix} \;\middle|\; 0 \geqslant B^T(X - X_0) \right\}$. Indeed, this would imply the existence of another nonvertical supporting plane $\tilde{E} := \left\{ \begin{pmatrix} X \\ z \end{pmatrix} \;\middle|\; z - f(X_0) = (A+B)^T(X - X_0) \right\}$.

Thus E is the only supporting plane, and since E is not singular, H is the only supporting halfspace of $[f]$ at $\begin{pmatrix} X_0 \\ f(X_0) \end{pmatrix}$. We have therefore:

$$C = C(f; X_0) = H - \begin{pmatrix} X_0 \\ f(X_0) \end{pmatrix} = \left\{ \begin{pmatrix} Y \\ w \end{pmatrix} \;\middle|\; w \geqslant A^T Y \right\}.$$

Hence $C\cap(-C)=\left\{\begin{pmatrix} Y \\ A^T Y\end{pmatrix} \,\middle|\, Y \text{ arbitrary}\right\}$. Every Y is therefore a lineality vector by theorem (4.3.11). This proves that f is differentiable at X_0. □

For arbitrary functions, relation (4.4.3), which shows $f'(X;Y)$ to be linear in Y, is a consequence of continuous differentiability.

(4.4.7) **Theorem.** *If a convex function f is differentiable in an open region U of the R^n, then the derivative $f'(X;Y)$ is continuous in $U\times R^n$ as a function of both X and Y.*

Proof. It is obviously no restriction of generality to assume that the region U contains the origin 0, that $f(0)=0$, and to prove continuity only at 0. Relation (4.4.3) shows that we may further assume without restriction of generality that

$$f'(0;Y)=0 \quad \textit{for all} \quad Y\in R^n. \tag{4.4.8}$$

Indeed, if necessary, we can always replace $f(X)$ by $f(X)-X^T \operatorname{grad} f(0)$.

Under these circumstances it suffices to prove that $|f'(X;Y)|\leqslant \varepsilon p(Y)$ holds for all points X in a suitable neighborhood of 0, p being a suitable norm on R^n.

Let $p_0(X):=\max|x_i|$ denote the maximum norm, and let Z_i, $i=1,\dots,2^n$, be the vertices of the unit n-cube $Q:=\{X\mid p_0(X)\leqslant 1\}$. There exists $k>0$ such that $kQ\subseteq U$. For every X with $p_0(X)\leqslant\frac{1}{2}h$ and any $Y\neq 0$, there exists $\lambda>0$ such that $p_0(X+\lambda Y)=h$ (continuity of norms (3.1.11)). Putting $Z:=X+\lambda Y$, we then have by (4.3.2):

$$f(Z)-f(X)\geqslant \lambda f'(X;Y), \qquad f(X)=f(X)-f(0)\geqslant f'(0;X)=0,$$

and therefore

$$f(Z)\geqslant \lambda f'(X;Y).$$

On the other hand, $f(Z)\leqslant \max_i f(hZ_i)\leqslant \frac{h}{k}\max_i f(kZ_i)$ if $0<h<k$ (convexity, and monotonicity relations (4.1.4)). Furthermore

$$h=p_0(Z)<p_0(X)+\lambda p_0(Y)\leqslant \tfrac{1}{2}h+\lambda p_0(Y),$$

and consequently $2\lambda p_0(Y)\geqslant h$. Hence

$$f'(X;Y)\leqslant 2p_0(Y)\max_i \frac{f(kZ_i)}{k} \quad \text{if} \quad k\geqslant 2p_0(X) \quad \text{and} \quad kQ\subseteq U.$$

Applied to $-Y$ this inequality gives

$$-f'(X;Y)=f'(X;-Y)\leqslant 2p_0(Y)\max_i \frac{f(kZ_i)}{k},$$

hence

$$|f'(X;Y)| \leqslant 2p_0(Y)\max_i \frac{f(kZ_i)}{k} \quad \text{if} \quad k \geqslant 2p_0(X) \quad \text{and} \quad kQ \subseteq U.$$

Now $\lim_{k\downarrow 0} \frac{f(kZ_i)}{k} = f'(0;Z_i) = 0$ by (4.4.8). For every $\varepsilon > 0$, there exists therefore $k > 0$ such that $kQ \subseteq U$ and $2\max_i \frac{f(kZ_i)}{k} \leqslant \varepsilon$. Then $|f'(X;Y)| \leqslant \varepsilon p_0(Y)$ for all X with $p_0(X) \leqslant k/2$. The theorem is thus proved. □

Let us now turn to the second derivatives of a convex function f. If f depends only on one variable, then f' is monotonically increasing by (4.3.2), and f'', wherever it exists, must be nonnegative. In R^n, the condition that f'' be nonnegative has to be replaced by the condition that the

(4.4.9) *Hessian matrix* $$H(X) = \begin{vmatrix} \frac{\partial^2 f}{\partial x_1 \partial x_1}, \ldots, \frac{\partial^2 f}{\partial x_1 \partial x_n} \\ \vdots \qquad\qquad \vdots \\ \frac{\partial^2 f}{\partial x_n \partial x_1}, \ldots, \frac{\partial^2 f}{\partial x_n \partial x_n} \end{vmatrix}$$

be positive semidefinite. This presupposes that $H(X)$ is a symmetric matrix, which is indeed the case provided the second partial derivatives of f are continuous at X. That this is not necessarily the case is shown by the convex function

$$f(x,y) := \begin{cases} 0 & \text{if} \quad x = y = 0, \\ xy\dfrac{x^2 - y^2}{x^2 + y^2} + 13x^2 + 13y^2 & \text{otherwise,} \end{cases}$$

whose mixed derivatives differ at the origin.

(4.4.10) **Theorem.** *Let U be an open convex domain in R^n and N a subset of U consisting of isolated points. Suppose that a differentiable function $f|U \to R$ has continuous second partial derivatives in the open set $U \sim N$. Then*

(i) *f is convex in U if and only if its Hessian $H(X)$ is positive semidefinite in $U \sim N$.*

(ii) *f is strictly convex in U if its Hessian $H(X)$ is positive definite in $U \sim N$.*

Proof. For $X, Y \in U$, $X \neq Y$, the function $g(t) := f(X + t(Y - X))$ has a continuous second derivative $g''(t) = (Y-X)^T H(X + t(Y-X))(Y-X)$ for all but isolated values t in $[0,1]$. If H is positive semidefinite

(definite), then $g''(t) \geqslant 0$ ($g''(t) > 0$) with only isolated exceptions, and g' is consequently piecewise (strictly) monotonic. Hence the one-sided limits

$$\lim_{t \uparrow t_0} g'(t), \quad \lim_{t \downarrow t_0} g'(t) \in R \cup \{+\infty\}$$

exist for $t \in (0,1)$. By a well-known property of derivatives—they assume all values between $g'(a)$ and $g'(b)$ in the interval $[a,b]$, and consequently cannot have discontinuities of the first kind (e. g. Aumann [*1*], Rudin [*1*]),—this implies continuity of g'. Hence g' is (strictly) monotonic, and g is therefore (strictly) convex.

Assume $D^T H(X_0) D < 0$ for some $X_0 \in U \sim N$. For sufficiently small $\varepsilon > 0$, $X := X_0 - \varepsilon D \in U \sim N$, $Y := X_0 + \varepsilon D \in U \sim N$, and $(X, Y) \subseteq U \sim N$. Then $g''(\tfrac{1}{2}) = 4\varepsilon^2 D^T H(X_0) D < 0$, contradicting convexity. □

The result of Anderson and Klee [*1*] concerning points of non-smoothness on a convex surface gives immediately for convex functions that the set of arguments for which a convex function $f\colon R^n \to R$ is not differentiable is the union of a countable number of compact sets of dimension lower than n. Convex functions are thus almost everywhere differentiable. They are also almost everywhere twice differentiable (A. D. Alexandroff [*1*]).

4.5. A Regularity Condition

In the previous section, differentiability of a convex function f was defined only at such points of $K(f)$ that are interior points of $K(f)$ with respect to the topology in R^n. Some examinations in chapters 5 and 6 will require a class of convex functions f which are not only differentiable everywhere in $K(f)^I$, but which satisfy the following stronger

(4.5.1) **Regularity Condition.** *Through every boundary point of the epigraph* $[f]$ *there passes at most one nonvertical supporting plane* (Fig. 17).

By theorem (4.4.6), the regularity condition implies differentiability at every inner point of $K(f)$, but not conversely. For example, consider the following functions:

$$f_1(x) := \begin{cases} -\sqrt{x} & \text{for} \quad 0 \leqslant x, \\ +\infty & \text{for} \quad x < 0, \end{cases} \qquad f_2(x) := \begin{cases} x & \text{for} \quad x < 1, \\ +\infty & \text{for} \quad x \geqslant 1. \end{cases}$$

The function f_1 satisfies the regularity condition, whereas f_2 does not. Indeed the point $\begin{pmatrix} 1 \\ 1 \end{pmatrix}$ is a boundary point of $[f_2]$ with several non-

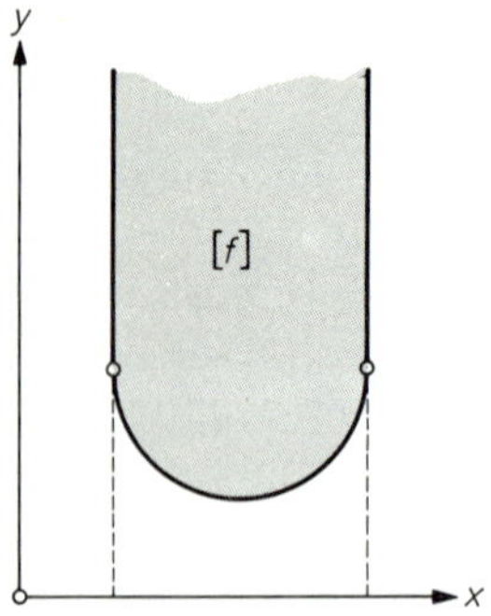

Fig. 17. Epigraph of function satisfying regularity condition

vertical supporting planes. The function

$$f(x,y) := \begin{cases} -\sqrt[4]{xy} & \text{for} \quad x \geqslant 0, \quad y \geqslant 0, \\ +\infty & \text{otherwise} \end{cases}$$

provides an example of a function with nonunique vertical supporting planes satisfying the regularity condition.

As part of the proof of theorem (4.4.6) is has been shown, that if a vertical and a nonvertical supporting plane pass through the same point $\begin{pmatrix} X \\ z \end{pmatrix} \in \overline{[f]}$, then there exist additional nonvertical supporting planes through $\begin{pmatrix} X \\ z \end{pmatrix}$. This gives rise to another formulation of the regularity condition:

(4.5.2) *A convex function f satisfies the regularity condition if and only if f is differentiable everywhere in $K(f)^I$ and if all supporting planes of $[f]$ at points other than $\begin{pmatrix} X \\ f(X) \end{pmatrix}$ with $X \in K(f)^I$ are vertical.*

For closed convex functions, the regularity condition can be expressed in terms of directional derivatives:

(4.5.3) **Theorem.** *For a closed convex function f the following statements are equivalent:*

(i) *f satisfies the regularity condition* (4.5.1),

(ii) *f is differentiable everywhere in $K(f)^I$, and for each point $X \in K(f)$, $X \notin K(f)^I$ there exists a vector Y such that $X + Y \in K(f)^I$ and $f'(X; Y) = -\infty$.*

In particular, if $K(f) = K(f)^I$, and if f is differentiable everywhere in $K(f)$, then f satisfies the regularity condition.

Proof. Suppose that f satisfies the regularity condition. We have already seen that in this case f is differentiable everywhere in $K(f)^I$, and that $K(f)^I$ is open relative to the entire R^n.

Let then $X \in K(f)$, but $X \notin K(f)^I$, and choose Y such that $X+Y \in K(f)^I$. This is always possible since $K(f)$ is nonvoid, and possesses therefore inner points (lemma (3.2.8)). Clearly $Y \neq 0$. Since $f(X+hY) < +\infty$, we have

$$f'(X;Y) = \inf_{h>0} \frac{f(X+hY)-f(X)}{h} < +\infty.$$

Assume that $f'(X;Y) > -\infty$. Then theorem (4.3.8) shows that there is a nonvertical plane through $\begin{pmatrix} X \\ f(X) \end{pmatrix}$ supporting $[f]$, in contradiction to (4.5.2). Therefore, $f'(X;Y) = -\infty$.

Conversely, suppose that the closed convex function f satisfies (ii). Since it is easily verified that any supporting plane through a boundary point $\begin{pmatrix} X \\ z \end{pmatrix}$ of $[f]$ with $z > f(X)$ must be vertical, it is sufficient to show that supporting planes through $\begin{pmatrix} X \\ f(X) \end{pmatrix}$ where $X \in K(f)$ but $X \notin K(f)^I$ are necessarily vertical. This is the case if (and only if) the supporting cone C of $[f]$ at $\begin{pmatrix} X \\ f(X) \end{pmatrix}$ contains $\begin{pmatrix} 0 \\ -1 \end{pmatrix}$. Indeed, by the definition of the supporting cone (3.4.14), if $\begin{pmatrix} 0 \\ -1 \end{pmatrix} \in C$, then the vertical line $V := \left\{ \begin{pmatrix} X \\ z \end{pmatrix} \,\middle|\, z \in R \right\}$ lies in the intersection of all supporting halfspaces of $[f]$ at $\begin{pmatrix} X \\ f(X) \end{pmatrix}$. Thus V is contained in all supporting planes through $\begin{pmatrix} X \\ f(X) \end{pmatrix}$. All these planes must therefore be vertical.

Now C is the closure of the cone $C(f;X) = \left\{ \begin{pmatrix} Y \\ w \end{pmatrix} \,\middle|\, w \geqslant f'(X;Y) \right\}$ by theorem (4.3.6). We know that $f'(X;Y) = -\infty$ for some Y. Hence $\begin{pmatrix} Y \\ w \end{pmatrix} \in C(f;X)$ for this Y and all $w \in R$, and $\dfrac{-1}{w}\begin{pmatrix} Y \\ w \end{pmatrix} \in C(f;X)$ for all $w < 0$. Now $\dfrac{-1}{w}\begin{pmatrix} Y \\ w \end{pmatrix} \to \begin{pmatrix} 0 \\ -1 \end{pmatrix}$ as $w \to -\infty$. Thus $\begin{pmatrix} 0 \\ -1 \end{pmatrix} \in \overline{C(f;X)} = C$, which remained to be shown. □

The last theorem is used to prove

(4.5.4) **Theorem.** *Let f_1 and f_2 be closed convex functions with $K(f_1) \subseteq K(f_2)$. If f_1 and f_2 are both differentiable for all $X \in K(f_1)^I$, and f_1 satisfies the regularity condition* (4.5.1), *then the convex function $f(X) := f_1(X) + f_2(X)$ with $K(f) = K(f_1)$ also satisfies the regularity condition.*

Proof. As is easily seen, the function $f := f_1 + f_2$ is a closed convex function, which is differentiable for all $X \in K(f)^I$. Moreover, if $X \in K(f)$, $X \notin K(f)^I$ and Y is chosen such that

$$X + Y \in K(f)^I = K(f_1)^I \subseteq K(f_2)^I,$$

then $f_2'(X; Y) < +\infty$ and $f_1'(X; Y) = -\infty$, by the proof of the last theorem. Therefore,

$$f'(X; Y) = f_1'(X; Y) + f_2'(X; Y) = -\infty.$$

Thus, theorem (4.5.3) can be applied, and f satisfies the regularity condition (4.5.1). □

4.6. Conjugate Functions

Consider any function $f: R^n \to R \cup \{+\infty, -\infty\}$. The function $f^c: R^n \to R \cup \{+\infty, -\infty\}$ defined by

(4.6.1) $$f^c(Y) := \sup\{Y^T X - f(X) \mid X \in R^n\} = \sup_X (Y^T X - f(X))$$

is called the

convex conjugate function

of f. The function g^c defined by

(4.6.2) $$g^c(Y) := \inf\{Y^T X - g(X) \mid X \in R^n\} = \inf_X (Y^T X - g(X))$$

is called the

concave conjugate function

of g. Typically, one considers the convex conjugate of a convex function and the concave conjugate of a concave function. Since usually it will be clear from the context whether a function is considered convex or concave, we speak simply of the

(4.6.3) *conjugate function* f^c

of a function f. A similar convention was introduced for the symbol $[f]$: Depending on whether f was convex or concave, $[f]$ was understood to be the epigraph or the hypograph, respectively. Note, that if f is a convex function, then the conjugate g^c of the concave function $g(X) := -f(X)$ is

$$g^c(Y) = \inf\{Y^T X - g(X)\} = \inf\{Y^T X + f(X)\} = -f^c(-Y)$$

and not simply $-f^c(Y)$.

It is left to the reader to verify that the convex conjugate function f^c of any function f is convex in the wider sense (4.2.5). Moreover,

(4.6.4) *if f is a convex function in the proper sense* (4.1.1), *then its conjugate f^c is also convex in this sense.*

Proof. Since $K(f) \neq \emptyset$, we obtain $f^c(Y) \geq Y^T X_0 - f(X_0)$ for all $X_0 \in K(f)$ and all $Y \in R^n$. Therefore $\{Y \mid f^c(Y) = -\infty\} = \emptyset$. Further, the set $K(f^c) = \{Y \mid f^c(Y) < \infty\}$ is not empty, since for every $X_0 \in K(f)^I$ there exists a Y_0 such that

$$f^c(Y_0) = Y_0^T X_0 - f(X_0) \geq Y_0^T X - f(X)$$

holds for all X (Corollary (4.2.9)). $\square$

Parallel to conjugate functions we define conjugate sets. To each set $M \subseteq R^{n+1}$ we assign the

(4.6.5) *upper conjugate set*

$$M^c := \left\{ \begin{pmatrix} Y \\ w \end{pmatrix} \;\middle|\; Y^T X - z \leq w \quad \text{for all} \begin{pmatrix} X \\ z \end{pmatrix} \in M \right\}.$$

The set M^c is an epigraph. The

(4.6.6) *lower conjugate set*

$$M^c := \left\{ \begin{pmatrix} Y \\ w \end{pmatrix} \;\middle|\; Y^T X - z \geq w \quad \text{for all} \begin{pmatrix} X \\ z \end{pmatrix} \in M \right\}$$

of M is a hypograph. One works usually with the upper conjugate set of an epigraph, and the lower conjugate set of a hypograph. It will follow from the context how M^c is to be interpreted.

Conjugate functions and conjugate sets are, of course, closely connected. For the epigraph $[f]$ of any function f,

(4.6.7) $$[f]^c = [f^c].$$

In words, the upper conjugate set of the epigraph, and the epigraph of convex conjugate function (4.6.1) of a function $f: R^n \to R \cup \{+\infty, -\infty\}$ coincide.

The geometric significance of conjugate functions is best illustrated in connection with their related epigraphs. By definition (4.6.5), $\binom{Y}{w} \in M^c$ is equivalent to

$$M \subseteq H(Y, w) := \left\{ \binom{X}{z} \,\middle|\, Y^T X - z \leqslant w \right\}.$$

It follows immediately from definition (4.6.1) that if $M = [f]$ and $w_0 = f^c(Y) \neq \pm\infty$, then w_0 is the smallest w such that $M \subseteq H(Y, w)$. This fact is expressed by the following theorem:

(4.6.8) **Theorem.** *If* $f^c(Y)$ *is finite, then the nonvertical plane* $E := \left\{ \binom{X}{z} \,\middle|\, Y^T X - z = f^c(Y) \right\}$ *is a nonsingular supporting plane of* $[f]$. *If* $\binom{X_0}{z_0}$ *is the point of support of* E, *then* $z_0 = f(X_0)$.

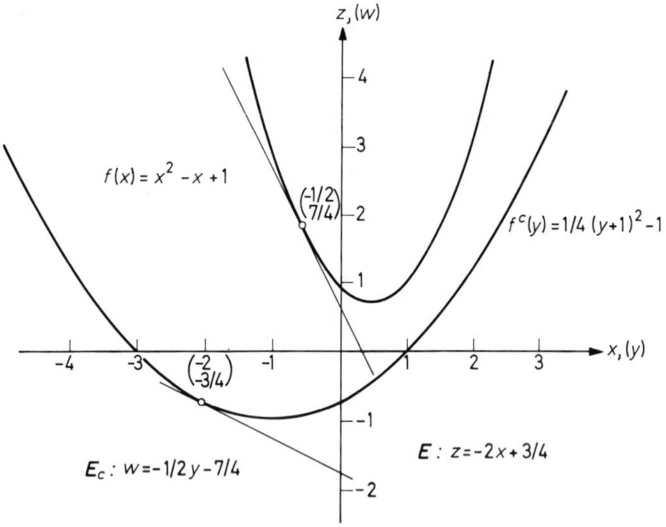

Fig. 18. Illustration of the symmetry theorem (4.6.9)

The equation of the above supporting plane corresponds to the point of support of a supporting plane to $[f^c]$. Indeed, one has the

(4.6.9) **Symmetry Theorem.** *If f is convex, and if* $\begin{pmatrix} X_0 \\ z_0 \end{pmatrix}$ *is a point of support of a supporting plane* $E := \left\{ \begin{pmatrix} X \\ z \end{pmatrix} \middle| Y_0^T X - z = w_0 \right\}$ *of* $[f]$, *then* $E_c := \left\{ \begin{pmatrix} Y \\ w \end{pmatrix} \middle| X_0^T Y - w = z_0 \right\}$ *is a supporting plane of* $[f^c]$, *a point of support being* $\begin{pmatrix} Y_0 \\ w_0 \end{pmatrix}$ (Fig. 18).

Proof. By definition (4.6.5) and relation (4.6.7), and since $\begin{pmatrix} X_0 \\ z_0 \end{pmatrix} \in [f]$:

$$[f^c] = [f]^c = \left\{ \begin{pmatrix} Y \\ w \end{pmatrix} \middle| \; Y^T X - z \leqslant w \quad \text{for all} \begin{pmatrix} X \\ z \end{pmatrix} \in [f] \right\}$$

$$\subseteq \left\{ \begin{pmatrix} Y \\ w \end{pmatrix} \middle| \; Y^T X_0 - z_0 \leqslant w \right\}.$$

Thus $[f^c]$ lies above E_c. Now $\begin{pmatrix} X_0 \\ z_0 \end{pmatrix} \in E$. Hence $Y_0^T X_0 - z_0 = w_0$. This shows that $\begin{pmatrix} Y_0 \\ w_0 \end{pmatrix} \in E_c$. As was seen above, E is a supporting plane if and only if $w_0 = f^c(Y_0)$. Thus $\begin{pmatrix} Y_0 \\ w_0 \end{pmatrix} \in [f^c]$. □

4.7. Strongly Closed Convex Functions

A conjugate set M^c is the intersection of *nonvertical* halfspaces. In general, we call sets which are intersections of nonvertical halfspaces

(4.7.1) *strongly closed.*

Strongly closed sets are convex and topologically closed in R^{n+1}. On the other hand, not every closed convex epigraph is strongly closed. If such an epigraph contains a full vertical line, then it is not strongly closed unless it coincides with the entire R^{n+1}. In most other cases, however, closed convex epigraphs are automatically strongly closed. This is the content of the following important

(4.7.2) **Lemma.** *Every closed convex epigraph M is strongly closed if $M^c \neq \emptyset$.*

Proof. Since M is a closed convex set, it is an intersection of half-spaces (theorem (3.3.7)), some of them possibly vertical. The point is to show that we may ignore those vertical halfspaces. We pick any $\begin{pmatrix} Y \\ w \end{pmatrix} \in M^c$. Then the nonvertical halfspace

$$H(Y,w) := \left\{ \begin{pmatrix} X \\ z \end{pmatrix} \;\middle|\; Y^T X - z \leqslant w \right\}$$

contains M. Hence M is an intersection of nonvertical halfspaces and of sets of the form $H(Y,w) \cap H$, where H is a vertical halfspace containing M. All we have to show therefore is that every $H(Y,w) \cap H$ is an intersection of nonvertical halfspaces.

Now, if $H = \left\{ \begin{pmatrix} X \\ z \end{pmatrix} \;\middle|\; A^T X \leqslant b \right\}$, then

$$H(Y,w) \cap H = \bigcap_{\theta \geqslant 0} H_\theta, \quad \textit{where} \quad H_\theta := \left\{ \begin{pmatrix} X \\ z \end{pmatrix} \;\middle|\; (Y+\theta A)^T X - z \leqslant w + \theta b \right\}. \tag{4.7.3}$$

Indeed, $\begin{pmatrix} X \\ z \end{pmatrix} \in \bigcap_{\theta \geqslant 0} H_\theta$ implies $A^T X - b \leqslant \frac{1}{\theta}(-Y^T X + z + w)$ for all $\theta > 0$, and therefore $A^T X - b \leqslant 0$. Hence $\bigcap_{\theta \geqslant 0} H_\theta \subseteq H$. Since $H(Y,w) = H_\theta$ for $\theta = 0$, we have $\bigcap_{\theta \geqslant 0} H_\theta \subseteq H(Y,w) \cap H$. On the other hand, $H_\theta \supseteq H(Y,w) \cap H$ holds for all $\theta > 0$. This establishes (4.7.3), completing the proof of the lemma. □

The conjugates of sets have properties analogous to those of orthogonal complements S (2.6.1) or polar sets S^p (2.7.1). The reader verifies readily

(4.7.4) (i) $M \subseteq N$ *implies* $M^c \supseteq N^c$.
(ii) $M^{cc} = M$ *if and only if* M *is a strongly closed epigraph.*
(iii) $M^{ccc} = M^c$.

The strongly closed epigraphs form a lattice under set inclusion with the intersection as a cap and the strong closure of the convex hull as a cup product. Forming the conjugate set is an anti-automorphism of this lattice.

We call f a

(4.7.5) *strongly closed function,*

if $[f]$ is strongly closed. Then (4.7.4) translates into:

(4.7.6) (i) $f_1 \geqslant f_2$ *implies* $f_1^c \leqslant f_2^c$,
(ii) $f^{cc} = f$ *if and only if* f *is strongly closed,*
(iii) $f^{ccc} = f^c$, *that is,* f^c *is strongly closed.*

There are only two strongly closed functions that are convex in the wider sense (4.2.5) but not in the proper sense (4.1.1): $f\equiv+\infty$ and $f\equiv-\infty$. Indeed if $f(X)=-\infty$, for some X, then $f^c\equiv+\infty$ and $f^{cc}\equiv-\infty$. On the other hand, if $f(X)>-\infty$ for all X and $f(Y)<+\infty$ for some Y, then f is convex in the sense (4.1.1). For functions f which are properly convex there is no difference between "closed" and "strongly closed". Indeed, f^c is again properly convex by (4.6.4), therefore $[f^c]=[f]^c\neq\emptyset$, and lemma (4.7.2) applies. As a consequence we have the

(4.7.7) **Theorem.** *If f is convex* (4.1.1) *then*

$$f^{cc}=\underline{f}.$$

It will be useful to keep in mind that

(4.7.8) *A conjugate function f^c is properly convex provided it assumes at least one finite value.*

Indeed, f^c is strongly closed by (4.7.6.(ii)) and (4.7.6.(iii)), and it was shown above that $f\equiv\infty$ and $f\equiv-\infty$ are the only strongly closed functions which are not properly convex.

4.8. Examples of Conjugate Functions

(1) Consider a linear function

$$f(X):=A^T X+b.$$

Then

$$f^c(Y)=\sup_X\,(Y^T X-A^T X-b)=\begin{cases}+\infty & \text{if}\quad Y\neq A,\\ -b & \text{if}\quad Y=A.\end{cases}$$

By theorem (4.7.7), we have $f^{cc}=f$. Hence the conjugate of a function with just one finite value is a linear function.

(2) Consider a piecewise linear function

$$f(X):=\max\{A_1^T X+b_1,\, A_2^T X+b_2\},\qquad A_1\neq A_2.$$

We show first that $f^c(Y)=+\infty$ for all $Y\notin\mathscr{H}\{A_1,A_2\}$. Plainly, there exists a halfspace $H:=\{U\,|\,U^T Z\leqslant d\}$ which contains A_1 and A_2 but not Y. In other words,

$$Y^T Z>d\geqslant\max\{A_1^T Z, A_2^T Z\}$$

holds for a suitable Z. Restricting the arguments X in the definition (4.6.1) of the conjugate function to arguments of the form θZ, $\theta>0$, we derive:

$$f^c(Y)\geqslant\theta\left(Y^T Z-\max\left\{A_1^T Z+\frac{b_1}{\theta},\ A_2^T Z+\frac{b_2}{\theta}\right\}\right).$$

The expression on the right hand side is clearly not bounded.

Next we assume $Y\in\mathscr{H}\{A_1,A_2\}$. Then there exist $\lambda,\mu\geqslant 0$ such that $Y=\lambda A_1+\mu A_2$ and $\lambda+\mu=1$. We have

$$f^c(Y)=\sup_X\,(\lambda A_1^T X+\mu A_2^T X-\max\{A_1^T X+b_1,\ A_2^T X+b_2\})\leqslant\lambda s_1+\mu s_2,$$

where

$$s_i:=\sup_X\,(A_i^T X-\max\{A_1^T X+b_1,\ A_2^T X+b_2\}),\qquad i=1,2.$$

It follows that $s_i\leqslant -b_i$. Hence $f^c(Y)\leqslant-\lambda b_1-\mu b_2$. Since $A_1\neq A_2$, then there exists an X such that $A_1^T X+b_1=A_2^T X+b_2$. For this X one finds at once that $\lambda A_1^T X+\mu A_2^T X-f(X)=-\lambda b_1-\mu b_2$. Thus

$$f^c(Y)=\begin{cases}+\infty & \text{if}\quad Y\notin\mathscr{H}\{A_1,A_2\},\\ -\lambda b_1-\mu b_2 & \text{if}\quad Y=\lambda A_1+\mu A_2,\quad \lambda,\mu\geqslant 0,\quad \lambda+\mu=1.\end{cases}$$

We observe that $[f^c]$ is the convex hull of the two vertical half-lines $\left\{\begin{pmatrix}A_i\\ z\end{pmatrix}\,\middle|\, z\geqslant -b_i\right\}$, $i=1,2$. The latter are the epigraphs $[f_i^c]$, where $f_i(X):=A_i^T X+b_i$ for $i=1,2$. This result could have been also gained by using the fact that forming the conjugate set is an anti-automorphism of the lattice of all strongly closed epigraphs. Indeed $[f]=[f_1]\cap[f_2]$ and this implies that $[f^c]$ is the strong closure of the convex hull $\mathscr{H}\{[f_1^c]\cup[f_2^c]\}$. Since this convex hull is strongly closed in the above example, it coincides with $[f^c]$.

(3) Consider the function

$$f(X):=\begin{cases} b & \text{if}\quad A^T X\leqslant a,\\ +\infty & \text{otherwise}\end{cases}\qquad\text{with}\quad A\neq 0.$$

Let $A^T Z=a$ and $A^T D=0$. Then $X=Z+\theta D$ satisfies $A^T X\leqslant a$ for all θ, and $Y^T X=Y^T Z+\theta Y^T D$ goes to $+\infty$ if $\theta\,\mathrm{sign}(Y^T D)\to\infty$. Finite values of $f^c(Y)$ can be expected therefore only if $Y^T D=0$ whenever $A^T D=0$, in other words, only if $Y=\rho A$ for some ρ. Now

$$f^c(\rho A)=\sup_X\,(\rho A^T X-f(X))=\sup\{\rho A^T X-b\mid A^T X\leqslant a\}.$$

It follows that

$$f^c(Y) = \begin{cases} \rho a - b & \text{if} \quad Y = \rho A \quad \text{and} \quad \rho \geqslant 0, \\ +\infty & \text{otherwise.} \end{cases}$$

(4) All three examples and their conjugates are representatives of a class of functions which we call

(4.8.1) *polyhedral functions.*

These functions are strongly closed convex functions whose epigraphs are polyhedra. The functions $f \equiv +\infty$ and $f \equiv -\infty$ are admitted so that the polyhedral functions form a lattice with respect to the partial ordering $f \leqslant g$ if $f(X) \leqslant g(X)$ for all X. It will be seen below that the operation $f \to f^c$ is an anti-automorphism of this lattice.

A finite linear inequality system which describes a strongly closed epigraph M other than R^{n+1} or $\emptyset$ must be necessarily equivalent to a system of the form (compare (1.1.9))

$$\begin{aligned} A_i^T X - z &\leqslant a_i, \qquad i = 1, \ldots, k, \\ B_j^T X \quad &\leqslant b_j, \qquad j = 1, \ldots, l, \end{aligned}$$

with $k > 0$ and $\begin{pmatrix} 0 \\ -1 \end{pmatrix} \notin \mathscr{C}\left\{\begin{pmatrix} B_j \\ b_j \end{pmatrix} \middle| \, j = 1, \ldots, l\right\}$. The corresponding polyhedral function is given by

(4.8.2)

$$f(X) := \begin{cases} \max\{A_i^T X - a_i \mid i = 1, \ldots, k\} & \text{if} \quad B_j^T X \leqslant b_j \quad \text{for} \quad 1 \leqslant j \leqslant l, \\ +\infty & \text{otherwise.} \end{cases}$$

On the other hand, any set

$$\mathscr{H} G + \mathscr{C}\left\{H, \begin{pmatrix} 0 \\ 1 \end{pmatrix}\right\}$$

where

$$G = \left\{\begin{pmatrix} A_i \\ a_i \end{pmatrix} \middle| \, i = 1, \ldots, k\right\} \quad \text{and} \quad H = \left\{\begin{pmatrix} B_j \\ b_j \end{pmatrix} \middle| \, j = 1, \ldots, l\right\},$$

with

$$\begin{pmatrix} 0 \\ -1 \end{pmatrix} \notin \mathscr{C} H \quad \text{and} \quad k > 0,$$

is a strongly closed epigraph $M' \neq R^{n+1}, \phi$, which is also a polyhedron by theorem (2.5.8).

The corresponding polyhedral function is then given by

(4.8.3)

$$h(Y)=\inf\{\Sigma\lambda_i a_i+\Sigma\mu_j b_j \mid Y=\Sigma\lambda_i A_i+\Sigma\mu_j B_j,\ \Sigma\lambda_i=1,\ \lambda_i,\ \mu_j\geqslant 0\}.$$

By the converse (2.11.4) of the finite basis theorem, an expression of the type (4.8.3) defines always a polyhedral function. The infimum in (4.8.3) is attained for all $X\in\mathscr{H}\bar{G}+\mathscr{C}\bar{H}$, where $\bar{G}:=\{A_i \mid i=1,\dots,k\}$ and $\bar{H}:=\{B_j \mid j=1,\dots,l\}$.

Every polyhedral function with finite values can be written in the form (4.8.2) as well as in the form (4.8.3). Both forms are conjugates of each other:

(4.8.4) *If f is defined by* (4.8.2) *and h by* (4.8.3), *then* $f^c=h$ *and* $h^c=f$.

We have proved (4.8.4) for special cases at the beginning of this section. It is left to the reader to extend the proofs given there.

(5) Consider differentiable (4.4.1) convex functions f. If $X_0\in K(f)^I$, then the tangent plane

$$T=\left\{\begin{pmatrix}X\\ z\end{pmatrix}\ \middle|\ z=f(X_0)+(X-X_0)^T\operatorname{grad} f(X_0)\right\}$$

is the only supporting plane of $[f]$ through the point $\begin{pmatrix}X_0\\ z_0\end{pmatrix}$, $z_0=f(X_0)$. By the symmetry theorem (4.6.9) this implies

$$f^c(Y_0)=X_0^T\operatorname{grad} f(X_0)-f(X_0)$$

for

$$Y_0:=\operatorname{grad} f(X_0)\in K(f^c).$$

Examples show that not every $Y\in K(f^c)$ can be represented as a gradient, even if f is differentiable for all $X\in K(f)^I$. This will cause difficulties in establishing duality theorems for differentiable functions. However, one can cope with these difficulties if f satisfies the regularity condition (4.5.1). In this case, the gradients of f fill almost entirely the domain of finiteness $K(f^c)$ of the conjugate function f^c:

(4.8.5) **Lemma.** *If a closed convex function f fulfills the regularity condition* (4.5.1), *then for every $Y\in K(f^c)^I$ there exists an $X\in K(f)^I$ such that $Y=\operatorname{grad} f(X)$ and*

$$f^c(Y)=f^c(\operatorname{grad} f(X))=X^T\operatorname{grad} f(X)-f(X)$$

holds.

Proof. Since f is closed, we have $[f]=[f^c]^c$ by (4.6.7) and by theorem (4.7.7). We apply the symmetry theorem (4.6.9) to f^c. If $Y_0\in K(f^c)^I$, then there exists a nonvertical supporting plane

$$\left\{\begin{pmatrix} Y \\ w \end{pmatrix} \;\middle|\; X_0^T Y - w = z_0\right\}$$

of $[f^c]$ through the point $\begin{pmatrix} Y_0 \\ w_0 \end{pmatrix}$, $w_0 = f^c(Y_0)$, by corollary (4.2.9). By theorem (4.6.8), $z_0 = f(X_0)$, and by the symmetry theorem

$$\left\{\begin{pmatrix} X \\ z \end{pmatrix} \;\middle|\; Y_0^T X - z = w_0\right\}$$

is a nonvertical supporting plane of $[f]$ through $\begin{pmatrix} X_0 \\ z_0 \end{pmatrix}$. We recall (4.5.2) which states that if the regularity condition is satisfied, then nonvertical supporting planes have their arguments of support in $K(f)^I$. Thus $X_0 \in K(f)^I$, and as regularity implies differentiability in $K(f)^I$, $Y_0 = \operatorname{grad} f(X_0)$ and $f^c(Y_0) = X_0^T \operatorname{grad} f(X_0) - f(X_0)$. □

Note, that the conjugate $f^c(Y)$ of a convex function f is convex. However, if $\operatorname{grad} f(X)$ is substituted for Y, the function

$$\phi(X) := f^c(\operatorname{grad} f(X)) = X^T \operatorname{grad} f(X) - f(X)$$

need not be convex on $K(f)^I$.

(6) Consider a convex function h which is positively homogeneous,

$$h(\lambda X) = \lambda h(X) \qquad \text{for all} \qquad \lambda \geqslant 0.$$

The set C of all possible values $Y^T X - h(X)$ is a cone in R^1. Indeed if $Y^T X - h(X) \in C$, then $\lambda(Y^T X - h(X)) = Y^T(\lambda X) - h(\lambda X) \in C$ for all $\lambda \geqslant 0$. It follows that

$$h^c(Y) = \sup_X \{Y^T X - h(X)\}$$

is capable of only two values, namely $+\infty$ and 0. The latter will be assumed precisely for all points Y with $Y^T X - h(X) \leqslant 0$ for all X.

(4.8.6) *If h is a positively homogeneous convex or concave function, then its conjugate h^c vanishes on $K(h^c)$.*

In particular, if h is a norm $p: R^n \to R$, then $K(p^c)$ is the norm body of the dual norm (3.10.4)

$$(4.8.7) \qquad p^c(Y) = \begin{cases} 0 & \text{if} \quad p^D(Y) \leqslant 1, \\ \infty & \text{otherwise.} \end{cases}$$

(7) Consider the support function (3.10.9) of a set $S \subseteq R^n$:

$$h_s(Y) := \sup\{Y^T X \mid X \in S\}.$$

This function is obviously the convex conjugate of the function $\gamma: R^n \to R \cup \{+\infty\}$ defined by

$$\gamma(X) := \begin{cases} 0 & \text{if} \quad X \in S, \\ +\infty & \text{otherwise.} \end{cases}$$

Since h_s is positively homogeneous one has by (4.8.6)

$$\gamma^{cc}(X) := \begin{cases} 0 & \text{if} \quad X \in \overline{\mathscr{H} S}, \\ +\infty & \text{otherwise.} \end{cases}$$

Indeed, $X \in K(h_S^c)$ holds if and only if

$$\begin{aligned} & Y^T X \leqslant h_S(Y) \quad \text{for all} \quad Y \\ \Leftrightarrow\ & Y^T X \leqslant \sup\{Y^T Z \mid Z \in S\} \quad \text{for all} \quad Y \\ \Leftrightarrow\ & X \in \overline{\mathscr{H} S}. \end{aligned}$$

(8) Consider two functions $f_1: R^n \to R \cup \{+\infty\}$ and $f_2: R^n \to R \cup \{+\infty\}$ not necessarily convex or concave. The function h defined by

$$h(Z) := \inf\{f_1(X_1) + f_2(X_2) \mid X_1 + X_2 = Z\} = \inf_X \{f_1(X) + f_2(Z - X)\}$$

is called the

(4.8.8) *infimum convolution* $h = f_1 \square f_2$

of f_1 and f_2. It plays an important role in duality theory.

(4.8.9) *If f_1 and f_2 are convex functions, then h is again convex, possibly in the wider sense* (4.2.5).

Proof. Consider any two numbers $\lambda, \mu \geqslant 0$ with $\lambda + \mu = 1$. Then by definition (4.8.8)

$$h(\lambda Z_1 + \mu Z_2) = \inf_X (f_1(X) + f_2(\lambda Z_1 + \mu Z_2 - X)).$$

One can replace any infimum that ranges over the points X of an open convex set K by an infimum that ranges over the pairs $(X_1, X_2) \in K \times K$:

$$\begin{aligned} h(\lambda Z_1 + \mu Z_2) &= \inf_{X_1, X_2} (f_1(\lambda X_1 + \mu X_2) + f_2(\lambda(Z_1 - X_1) + \mu(Z_2 - X_2))) \\ &\leqslant \inf_{X_1, X_2} (\lambda(f_1(X_1) + f_2(Z_1 - X_1)) + \mu(f_1(X_2) + f_2(Z_2 - X_2))) \\ &= \lambda \inf_{X_1}(f_1(X_1) + f_2(Z_1 - X_1)) + \mu \inf_{X_2}(f_1(X_2) + f_2(Z_2 - X_2)) \end{aligned}$$

provided the sum of the two infima is defined. Thus

$$h(\lambda Z_1+\mu Z_2)\leqslant\lambda h(Z_1)+\mu h(Z_2)$$

whenever the right hand side is not the indefinite expression $\infty-\infty$. □

(4.8.10) *If both* $f_1: R^n\to R\cup\{+\infty\}$ *and* $f_2: R^n\to R\cup\{+\infty\}$ *assume finite values at all, then*

$$(f\,\Box\,g)^c=f^c+g^c.$$

Proof. $-h^c(Y)=\inf\limits_Z(h(Z)-Y^TZ)=\inf\limits_{Z,X}(f_1(X)+f_2(Z-X)-Y^TZ)$

$$=\inf_{Z,X}(f_1(X)-Y^TX+f_2(Z-X)-Y^T(Z-X))$$

$$=\inf_X((f_1(X)-Y^TX)+\inf_Z(f_2(Z-X)-Y^T(Z-X)))$$

$$=-f_1^c(Y)-f_2^c(Y),$$

provided the sum of the infima is defined. This, however, is assured by the hypothesis of (4.8.10). □

(9) We proceed to examine the effect of a linear transformation on the conjugate of a convex function. Let

$$M:=\{AU-B\mid U\in R^m\}$$

be a linear manifold given in "slope-intercept" form: A is an $n\times m$-matrix and B is a vector of R^n. For an arbitrary function $f: R^n\to R\cup\{+\infty\}$ we define

$$h(V):=\inf\{f(Y)+Y^TB\mid V=A^TY\}.$$

Then

$$h^c(U)=f^c(AU-B). \tag{4.8.11}$$

Proof. $h^c(U)=\sup\limits_V\{V^TU-h(V)\}$

$$=\sup_V\{\sup\{V^TU-Y^TB-f(Y)\mid V=A^TY\}\}$$

$$=\sup_Y\{Y^T(AU-B)-f(Y)\}.\quad □$$

(10) The following formula is trivial but it involves an unexpected change of sign:

$$(-f)^c(Y)=-f^c(-Y). \tag{4.8.12}$$

Here f^c is understood to mean the *convex* conjugate of f, whereas $(-f)^c$ stands for the *concave* conjugate of $-f$.

(11) Consider real functions on the cartesian product $R^k\times R^l$. Since $R^k\times R^l$ can be canonically identified with $R^{k+l}=R^n$, concepts

like convexity and continuity carry over. A function f on $R^k \times R^l$ is called

(4.8.13) *separable*

if for each $(X_1, X_2) \in R^k \times R^l$

$$f(X_1, X_2) = f_1(X_1) + f_2(X_2),$$

where f_1 and f_2 are functions on R^k and R^l, respectively. One of the most useful properties of forming conjugates is that this operation preserves separability. Indeed

$$f^c(Y_1, Y_2) = f_1^c(Y_1) + f_2(Y_2), \tag{4.8.14}$$

where f^c is again defined on $R^k \times R^l$.

Proof.
$$\begin{aligned} f^c(Y_1, Y_2) &= \sup_{X_1, X_2} (Y_1^T X_1 + Y_2^T X_2 - f(X_1, X_2)) \\ &= \sup_{X_1, X_2} (Y_1^T X_1 + Y_2^T X_2 - f_1(X_1) - f_2(X_2)) \\ &= \sup_{X_1} (Y_1^T X_1 - f_1(X_1)) + \sup_{X_2} (Y_2^T X_2 - f_2(X_2)) \\ &= f_1^c(Y_1) + f_2^c(Y_2). \end{aligned}$$

(12) A symmetric $n \times n$-matrix $D = D^T$ is called

(4.8.15) *positive semidefinite*

if $X^T D X \geqslant 0$ for all $X \in R^n$. D is called

(4.8.16) *positive definite*

if, additionally, $X^T D X = 0$ implies $X = 0$, in other words, if D admits an inverse D^{-1}. Consider the nullspace of D,

$$L_D := \{X \mid D X = 0\}.$$

Then by (2.6.12.(ii))

$$L_D^{\perp} = \{D X \mid X \in R^n\}. \tag{4.8.17}$$

Now consider the quadratic function

$$f(X) := \tfrac{1}{2} X^T D X + P^T X$$

where $P \in R^n$. We assume D to be positive semidefinite since this is necessary and sufficient for f to be convex. The conjugate f^c of f is now readily determined. One has

$$K(f^c) = L_D^{\perp} + P = \{D X + P \mid X \in R^n\} \tag{4.8.18}$$

and

$$f^c(D X + P) = \tfrac{1}{2} X^T D X. \tag{4.8.19}$$

Proof. Suppose $Y \in K(f^c)$. Then for any $\bar{X}$ with $D\bar{X}=0$

$$+\infty > f^c(Y) \geq \sup_{\lambda} (Y^T(\lambda \bar{X}) - f(\lambda \bar{X})) = \sup_{\lambda} \lambda (Y-P)^T \bar{X}$$

must hold, and that requires $(Y-P)^T \bar{X}=0$. Thus $Y \in L_D^{\perp}+P$ for all $Y \in K(f^c)$, or

$$K(f^c) \subseteq L_D^{\perp}+P=\{DX+P \mid X \in R^n\}$$

in view of (4.8.17).

In order to prove the converse inclusion and, at the same time, prove (4.8.19), suppose $Y=D\bar{X}+P$ for some $\bar{X} \in R^n$. Then

$$\begin{aligned} f^c(Y)=f^c(D\bar{X}+P) &= \sup_X ((D\bar{X}+P)^T X - \tfrac{1}{2} X^T D X - P^T X) \\ &= \sup_X (\bar{X}^T D X - \tfrac{1}{2} X^T D X) \\ &= \tfrac{1}{2} \bar{X}^T D \bar{X} + \sup_X (-\tfrac{1}{2}(X-\bar{X})^T D (X-\bar{X})) = \tfrac{1}{2} \bar{X}^T D \bar{X}, \end{aligned}$$

since D is positive semidefinite. □

If, furthermore, D is positive definite, then clearly

(4.8.20) $$f^c(Y)=\tfrac{1}{2}(Y-P)^T D^{-1}(Y-P)$$

and $K(f^c)=R^n$.

4.9. Generalization of Convexity

In this section we consider real functions $f: K \to R$ on a convex set $K \subseteq R^n$. We are interested in various ways of weakening convexity while retaining some of the important properties of convex functions (compare Arrow and Enthoven [*1*], Mangasarian [*2*]). One such property with geometric implications is the convexity of the level sets

$$L_\alpha := \{X \in K \mid f(X) \leq \alpha\},$$

where $\alpha \in R$. Functions with this property are called

(4.9.1) *quasiconvex*

on K.

In connection with minimization methods it is important to have some kind of local optimality criterion. For convex functions we know that a "local" minimum, that is, a point X_0 with $f(X_0)=\min f(U \cap K)$ for some neighborhood U of X_0, is also a "global" minimum. Convexity can now be relaxed while preserving this property. A function $f: K \to R$ is called

(4.9.2) *pseudoconvex*

on K if $f(Z)<f(\bar{X})$, whenever $\underline{X}$, $\bar{X}\in K$, $f(\underline{X})<f(\bar{X})$, and Z is in the open segment $(\underline{X},\bar{X})$. The reader verifies easily that every local minimum of a pseudoconvex function is indeed a global one. Note that this is not true for general quasiconvex functions. Indeed, every monotonic function on the real line is quasiconvex. Consider then a monotonic function which increases, remains constant, and increases again. The points in the interior of the constant interval are local minima but not global ones.

Quasiconvexity is essentially more general than pseudoconvexity. However, there are some rather pathological functions which are pseudoconvex, but not quasiconvex on K, e.g.

$$K=R \quad \text{and} \quad f(x):=\begin{cases}1 & \text{if} \quad x=0,\\ 0 & \text{otherwise}\end{cases}$$

or

$$K=\{X\in R^2 \mid |x_1|+|x_2|\leqslant 2\} \quad \text{and} \quad f(X):=\begin{cases}1 & \text{if} \quad |x_1|=|x_2|=1,\\ 0 & \text{otherwise.}\end{cases}$$

We call functions which are both quasi- and pseudoconvex

(4.9.3) *strongly quasiconvex.*

A quasiconvex function is

(4.9.4) *strictly quasiconvex*

if it has at most one local minimum. A strictly quasiconvex function is automatically strongly quasiconvex. A strictly convex function is strictly quasiconvex in its domain of finiteness. We say that a (pseudo-, quasi-) convex function $f: K\to R$, $K\subseteq R^n$ is

(4.9.5) *closed (pseudo-, quasi-) convex,*

if all level sets $L_\alpha=\{X\in K \mid f(X)\leqslant\alpha\}$, $\alpha\in R$ are closed sets in R^n.

A function f is

(4.9.6) *(closed) (strictly, strongly) (pseudo-, quasi-) concave*

if $-f$ is (closed) (strictly, strongly) (pseudo-, quasi-) convex.

Contrary to closed convex functions (compare (4.1.14)), closed quasiconvex functions may have both bounded and unbounded level sets at the same time. In this case, however, there exists always a "smallest" unbounded level set. More precisely:

(4.9.7) **Lemma:** *If a closed quasiconvex function $f: K\to R$ has nonempty bounded as well as nonempty unbounded level sets* $L_\alpha:=\{X\in K \mid f(X)\leqslant\alpha\}$, *then there exists an* $\bar{\alpha}\in R$ *such that* L_α *is unbounded for* $\alpha\geqslant\bar{\alpha}$ *and bounded otherwise.*

Proof: Suppose that L_{α_2} is unbounded and $L_{\alpha_1}\neq\emptyset$ is bounded and define

$$\bar{\alpha}:=\inf\{\alpha\mid L_\alpha \text{ is unbounded}\}.$$

Since $\alpha\leqslant\beta$ implies $L_\alpha\subseteq L_\beta$, it follows that $\alpha_1<\alpha_2$, $\alpha_1\leqslant\bar{\alpha}\leqslant\alpha_2$, L_α is unbounded for all $\alpha>\bar{\alpha}$ and L_α is bounded for all $\alpha<\bar{\alpha}$. We only have to show that $L_{\bar{\alpha}}$ itself is unbounded. As every L_α, $\alpha>\bar{\alpha}$ is an unbounded convex set, its characteristic cone C_α (3.5.6) is a closed nontrivial cone: $C_\alpha=\bar{C}_\alpha\neq\{0\}$ and the sets $S_\alpha:=\{Z\in C_\alpha\mid \|Z\|=1\}$ of all directions of infinity of L_α of euclidean length 1 are closed subsets of the compact unit sphere of R^n for $\alpha>\bar{\alpha}$. Since $L_\alpha\supseteq L_\beta$ for $\alpha\geqslant\beta$ and therefore $S_\alpha\supseteq S_\beta$ we have that any finite intersection $S_{\alpha_1}\cap\cdots\cap S_{\alpha_k}$ is nonempty, if $\alpha_i>\bar{\alpha}$ for $i=1,\ldots,k$. Therefore by a known theorem on compact sets, $S:=\bigcap_{\alpha>\bar{\alpha}} S_\alpha\neq\emptyset$. Hence, there is a $Z\in S$ which is a direction of infinity for all L_α, $\alpha>\bar{\alpha}$. As $L_\alpha\supseteq L_{\bar{\alpha}}\supseteq L_{\alpha_1}\neq\emptyset$ for all $\alpha>\bar{\alpha}$, there is an $X_0\in L_\alpha$ for all $\alpha\geq\bar{\alpha}$. Since each L_α, $\alpha>\bar{\alpha}$, is a closed convex set by the quasiconvexity of f, we have by (3.5.1) $X_0+\lambda Z\in L_\alpha$ for all $\alpha>\bar{\alpha}$ and all $\lambda\geqslant 0$, that is $f(X_0+\lambda Z)\leqslant\alpha$ for all $\alpha>\bar{\alpha}$ and $\lambda\geqslant 0$. But this implies immediately, $f(X_0+\lambda Z)\leqslant\bar{\alpha}$ for all $\lambda\geqslant 0$, showing that $L_{\bar{\alpha}}$ is unbounded. □

There are topological conditions, which enforce quasiconvexity when applied to pseudoconvex functions.

(4.9.8) **Theorem.** *A pseudoconvex function* $f\colon K\to R$ *which is lower semicontinuous (l.s.c.) on* K *is quasiconvex, and therefore strongly quasiconvex, on* K. *In particular, a closed pseudoconvex function is strongly quasiconvex.*

Proof. If f is not quasiconvex, then there exist a number α and points X_1, X_2, X_3 with $X_3\in(X_1,X_2)$ such that

$$f(X_1)\leqslant\alpha,\quad f(X_2)\leqslant\alpha,\quad f(X_3)>\alpha.$$

Clearly $X_1\neq X_2$. Assume that $f(X_1)\neq f(X_2)$, say $f(X_1)<f(X_2)$, then the pseudoconvexity of f would give $f(X_3)<f(X_2)\leqslant\alpha$, contradicting the relation above. Hence $f(X_1)=f(X_2)<f(X_3)$. Since f is l.s.c. at X_3, there is an $X_4\in(X_1,X_3)$ with $f(X_1)<f(X_4)$. Since $X_3\in(X_4,X_2)$ and $f(X_2)=f(X_1)<f(X_4)$, it follows, again by pseudoconvexity, that

$$f(X_3)<f(X_4).$$

On the other hand, $X_4\in(X_1,X_3)$ and the same argument yields the contradiction

$$f(X_4)<f(X_3).\quad □$$

We proceed to characterize (strongly) quasiconvex functions. To this end we observe that

(4.9.9) *A function* $f: K \to R$ *is quasi-(pseudo-) convex on* K *if and only if it is quasi- (pseudo-) convex when restricted to any line segment on* K.

Thus it will suffice to characterize (strongly) quasiconvex functions of a single variable only. Let I be an open, half open, or closed interval of the real line, and consider a sequence of mutually exclusive intervals $\{I_1, \ldots, I_k\}$, each of which may be open, half open, closed, or empty. The intervals are moreover ordered in such a way that for each $l<k$, I_l lies to the left of I_{l+1}. We say that such a sequence of intervals

(4.9.10) *sequentially partitions* I

if $I = I_1 \cup \cdots \cup I_k$.

(4.9.11) **Theorem.** *A function* $f: I \to R$ *is quasiconvex if and only if there exists a sequential partition of* I *into two possibly empty intervals* $I_1 = I^{\geqslant}$ *and* $I_2 = I^{\leqslant}$ *such that* f *does not increase on* $I^{\geqslant}$ *and not decrease on* $I^{\leqslant}$. *A function* $f: K \to R$ *is quasiconvex if and only if it is quasiconvex, when restricted to any line segment in* K.

Proof. We prove only the "only if" direction of the first part. The proof of the other direction can be left to the reader. The second part of the theorem had been stated in (4.9.9). Define the sets

$$I^{\geqslant} := \{x \in I \mid y \in I \text{ exists with } x<y \text{ and } f(x)>f(y)\},$$
$$I^{\leqslant} := I \sim I^{\geqslant}.$$

We first show that

(4.9.12) *if* $x_0 \in I^{\geqslant}$, *then* $f(z) \geqslant f(x_0)$ *for all* $z \in I$ *with* $z<x_0$; *if* $x_0 \in I^{\leqslant}$, *then* $f(z) \geqslant f(x_0)$ *for all* $z \in I$ *with* $z>x_0$.

Assume that, $z<x_0$, $f(z)<f(x_0)$, and $x_0 \in I^{\geqslant}$. Then there exists $y>x_0$ such that $f(y)<f(x_0)$, and the level set $\{x \in I \mid f(x) \leqslant \max\{f(z), f(y)\}\}$ is clearly not convex. Assume then that $z>x_0$ and $f(z)<f(x_0)$. This gives $x_0 \in I^{\geqslant}$ by definition of $I^{\geqslant}$. Hence x_0 cannot be in $I^{\leqslant}$.

It remains to be shown that $I^{\geqslant}$ and $I^{\leqslant}$ are intervals. We prove this for $I^{\geqslant}$ by observing that if $x_0 \in I^{\geqslant}$, then $I^{\geqslant}$ contains all points of I to the left of x_0. Indeed, if $z<x_0$, then $f(z) \geqslant f(x_0)$ by (4.9.11), and there exists $y>x_0>z$ such that $f(y)<f(x_0) \leqslant f(z)$. Since $I^{\geqslant}$ is the union of intervals whose complements are again intervals, the complement $I^{\leqslant} = I \sim I^{\geqslant}$ is the intersection of intervals, and therefore an interval itself. ☐

(4.9.13) *a function* $f: I \to R$ *is strongly quasiconvex if and only if there exists a sequential partition of* I *into three intervals* $I_1 = I^{>}$, $I_2 = I^{=}$, $I_3 = I^{<}$

such that f decreases on $I^{>}$, *is constant (and minimum) on* $I^{=}$, *and increases on* $I^{<}$. *A function* $f\colon K\to R$ *is strongly quasiconvex if and only if it is strongly quasiconvex when restricted to any line segment in* K.

Proof. Again, we prove only the "only if" direction. Suppose that f is strongly quasiconvex. Put

$$\beta := \inf\{f(x) \mid x\in I\},$$

and define

$$I^{=} := \{x\in I \mid f(x)\leqslant\beta\},$$
$$I^{>} := I^{\geqslant}\sim I^{=}, \qquad I^{<} := I^{\leqslant}\sim I^{=},$$

where $I^{\geqslant}$ and $I^{\leqslant}$ are as in (4.9.11). By quasiconvexity, $I^{=}$ is convex and therefore an interval. To show that $I^{>}$ is an interval, suppose $x,y\in I^{>}$ and $x<y$. Then the open interval (x,y) is contained in $I^{\geqslant}$ since $x,y\in I^{\geqslant}$ and $I^{\geqslant}$ is known to be an interval. $f(y)>\beta$ since $y\notin I^{=}$. Moreover, $f(z)\geqslant f(y)$ for all $z<y$. Hence $(x,y)\cap I^{=}=\emptyset$, and therefore $(x,y)\subseteq I^{>}$, which was to be shown. The convexity of $I^{<}$ is established analogously.

The function f does not increase on $I^{>}$ since $I^{>}\subseteq I^{\geqslant}$. It cannot stay constant at any level other than the minimum, because this would imply the existence of local minima, which are not global, thus violating pseudoconvexity. □

4.10. Pseudolinear Functions

The notion of a linear function can be weakened in a manner which preserves most of the properties of linear minimization (Charnes and Cooper [*1*], Martos [*1*], [*2*]). A function $f\colon K\to R$ on a convex set $K\subseteq R^n$ which is both pseudoconvex (4.9.2) and pseudoconcave (4.9.6) is called

(4.10.1) *pseudolinear.*

In other words, f is pseudolinear if and only if the image $f((X, Y))$ of each open segment (X, Y) with $X, Y\in K$ is contained in the open segment $(f(X), f(Y))$. Such a function is either constant or strictly monotonic on each line segment in K. By (4.9.3),

(4.10.2) *a pseudolinear function* f *is (strongly) quasiconvex and quasiconcave.*

It follows that the level sets of f, as well as those of $-f$, are convex.

(4.10.3) *A pseudolinear function assumes its minimum (maximum) on an extreme subset, which may be empty, of* K.

Proof. We denote by

$$S_{\min}$$

the set of all minima. It is a level set and therefore convex. Let X_1 and X_2 be points in K such that the open segment (X_1, X_2) meets $S_{\min}$ in some point X_0. Then for $0 \leqslant \theta \leqslant 1$,

$$h(\theta) := f(X_1 + \theta(X_2 - X_1))$$

cannot strictly increase since this would imply $f(X_1) < f(X_0)$ contradicting $X_0 \in S_{\min}$. The same argument rules out a strict decrease of h. Since f is pseudolinear, h must be constant, whence $X_1, X_2 \in S_{\min}$, and the characteristic property of extreme sets has been established. ▯

The set $S_{\min}$ need not be exposed. Let K be the union of the upper half of the unit circle $\left\{\begin{pmatrix} x \\ y \end{pmatrix} \in R^2 \;\middle|\; y \geqslant 0,\ x^2 + y^2 \leqslant 1\right\}$ and triangle $\mathscr{H}\left\{\begin{pmatrix} -1 \\ 0 \end{pmatrix}, \begin{pmatrix} 1 \\ 0 \end{pmatrix}, \begin{pmatrix} 1 \\ -1 \end{pmatrix}\right\}$. The point $\begin{pmatrix} 1 \\ 0 \end{pmatrix}$ is extreme but not exposed. In order to construct a continuous pseudolinear function f which has $\left\{\begin{pmatrix} 1 \\ 0 \end{pmatrix}\right\}$ as minimum set, take any homeomorphism g of the circular arc $A := \left\{\begin{pmatrix} x \\ y \end{pmatrix} \;\middle|\; x^2 + y^2 = 1, y \geqslant 0\right\}$ onto the line segment $S := \mathscr{H}\left\{\begin{pmatrix} 1 \\ 0 \end{pmatrix}, \begin{pmatrix} 1 \\ -1 \end{pmatrix}\right\}$ such that $\begin{pmatrix} -1 \\ 0 \end{pmatrix} \in A$ corresponds to $\begin{pmatrix} 1 \\ -1 \end{pmatrix} \in S$ and $\begin{pmatrix} 1 \\ 0 \end{pmatrix} \in A$ to $\begin{pmatrix} 1 \\ 0 \end{pmatrix} \in S$. Consider line segments $\mathscr{H}\left\{\begin{pmatrix} x_1 \\ y_1 \end{pmatrix}, \begin{pmatrix} x_2 \\ y_2 \end{pmatrix}\right\}$ with $\begin{pmatrix} x_1 \\ y_1 \end{pmatrix} \in A$, $\begin{pmatrix} x_2 \\ y_2 \end{pmatrix} \in S$, and $\begin{pmatrix} x_2 \\ y_2 \end{pmatrix} = g\begin{pmatrix} x_1 \\ y_1 \end{pmatrix}$. Then define f on each segment to be constant and of value $-y_2$.

The following theorem is stated for polyhedra since in this case exposed sets and extreme sets coincide (theorem (2.4.12)).

(4.10.4) **Theorem.** *A continuous function f is pseudolinear on a polyhedron P in R^n if and only if f is either constant on P, or each contour set*

$$S_z := \{X \in P \mid f(X) = z\}, \qquad z \in R,$$

is the intersection $S_z = E_z \cap P$ of a plane E_z with P.

Proof. Suppose that f is pseudolinear and not constant on P:

$$\underline{z} := \inf\{f(X) \mid X \in P\} < \bar{z} := \sup\{f(X) \mid X \in P\}.$$

If $z=\underline{z}$ is a minimum, then S_z is a face of P by (4.10.3). Since every face is an exposed set, it is the intersection of P with a supporting plane. The same holds for $z=\bar{z}$.

We can assume therefore that

$$\underline{z}<z<\bar{z}.$$

The sets

$$S_z^< := \{X \in P \mid f(X)<z\}, \qquad S_z^> := \{X \in P \mid f(X)>z\}$$

are both convex by virtue of (4.10.2). Moreover, since $\underline{z}<z<\bar{z}$, they are nonvoid and satisfy

$$S_z^< \cap S_z^> = \emptyset.$$

By the separation theorem (3.3.9) there exists therefore a plane E_z separating $S_z^<$ and $S_z^>$ such that

(4.10.5) $$E_z \not\supseteq S_z^< \cup S_z^>.$$

All we have to show is that $S_z = E_z \cap P$. Since f is continuous, and E_z is closed and separates $S_z^<$ and $S_z^>$, we have

$$E_z \supseteq \overline{S_z^<} \cap \overline{S_z^>} = S_z,$$

whence

$$E_z \cap P \supseteq S_z \cap P = S_z.$$

In order to prove the converse inclusion $E_z \cap P \subseteq S_z$, it is sufficient to show that

$$E_z \cap S_z^< = E_z \cap S_z^> = \emptyset,$$

in view of $P = S_z^< \cup S_z^> \cup S_z$. Assume that there is an $X_1 \in S_z^< \cap E_z$. Then choose an arbitrary $X_2 \in S_z^>$. Since by hypothesis

$$f(X_1)<z, \qquad f(X_2)>z,$$

f is continuous, and $E_z \supseteq S_z$, there is an X with

$$X = \lambda X_1 + (1-\lambda) X_2 \in S_z \subseteq E_z, \qquad 0<\lambda<1.$$

Since $X_1 \in E_z$, $X \in E_z$ and E_z is a linear manifold it follows that $X_2 \in E_z$, and therefore

$$E_z \supseteq S_z^>,$$

as $X_2 \in S_z^>$ was arbitrary. This in turn implies by the same argument

$$E_z \supseteq S_z^<,$$

hence

$$E_z \supseteq S_z^< \cup S_z^>,$$

in contradiction to (4.10.5). Therefore $E_z \cap P = S_z$.

Proving the remaining direction of the theorem is left to the reader.

CHAPTER 5

Duality Theorems

This chapter discusses topics which concern the problem of minimizing a convex function on a convex set, in other words, the convex programming problem. Conceptual rather than algorithmic aspects of convex programming are considered.

Convex programming problems frequently come in pairs composed of a minimization and a maximization problem, which are called dual to each other. Typically, if one problem has an optimal solution, then so has the other, and the extreme values of both objective functions are equal. We shall examine the interplay of dual problems and study the structure of duality statements.

The concept of the conjugate of a convex function due to Fenchel [*1*] is fundamental to these studies. It leads to Fenchel's duality theorem, which will be generalized and used as a foundation for the duality theory of convex programs. Rockafellar [*1*] introduced the concept of stability for convex functions in order to obtain a generalization of Fenchel's theorem. He proved this generalization in an algebraic manner. In this chapter, different, more geometrical proofs are given, which permit a somewhat wider generality.

Special cases of the generalized Fenchel duality theorem include the duality theorem of linear programming by Gale, Kuhn, and Tucker [*1*] described in section 1.7, nonlinear duality theorems developed by Dennis [*1*], [*2*], Dorn [*1*], [*2*], Eisenberg [*1*], and Cottle [*1*], as well as natural generalizations of these theorems.

5.1. The Duality Theorem of Fenchel

Let $M=[f]$ be any epigraph corresponding to a convex function f. We start out examining the effect of a vertical translation of M on the conjugate set $M^c=[f^c]$. We observe that if M moves parallel to the z-axis, then M^c moves in the opposite direction. More precisely,

$$[f-\mu]^c=[f^c+\mu] \tag{5.1.1}$$

for every constant μ. In particular, the domain of finiteness $K(f^c)$ is not changed by a vertical translation of f.

In addition, we consider a concave function g and its associated hypograph $N:=[g]$ with $N^c=[g^c]$. We assume that $K(f^c)\cap K(g^c)$ is nonvoid. Hence by a suitable translation, M^c can be made to overlap N^c. Every element $\begin{pmatrix} Y \\ w \end{pmatrix}$ of $M^c\cap N^c$ then has a simple geometric meaning: it determines a nonvertical plane $E:=\left\{\begin{pmatrix} X \\ z \end{pmatrix} \,\middle|\, Y^T X - z = w\right\}$ which separates M and N. Indeed, the epigraph M lies above E, and the hypograph N lies below E. If also $K(f)\cap K(g)\neq\emptyset$, then we may repeat the same argument, showing that M can be translated to intersect N, and this amounts to M^c and N^c being separable by a nonvertical plane.

This leads to the conjecture that, roughly speaking, M and N start to separate at the same moment at which M^c and N^c start to overlap. In fact, this is essentially the content of the duality theorem of Fenchel [*1*]. In order to cast Fenchel's theorem into a form which is customary for duality theorems, we formulate a "primal" and "dual" program.

(5.1.2) *Given a convex function f, a concave function g, and their conjugate functions f^c and g^c, respectively, we consider*

program I (primal): *Find* $\inf_X (f(X)-g(X))$,

program II (dual): *Find* $\sup_Y (g^c(Y)-f^c(Y))$.

By definitions (4.1.1) and (4.1.3), convex and concave functions are mappings into the extended real line $R\cup\{+\infty,-\infty\}$. With the conventions $\sup M=-\infty$ and $\inf M=+\infty$ if $M=\emptyset$, the above programs can be written equivalently as follows:

program I: *Find* $\inf(f(X)-g(X))$
subject to $X\in K(f)\cap K(g)$,

program II: *Find* $\sup(g^c(Y)-f^c(Y))$,
subject to $Y\in K(f^c)\cap K(g^c)$.

Consequently, a point X_0 is called a feasible solution of program I, if $X_0\in K(f)\cap K(g)$, and X_0 is said to be an optimal solution of program I, if X_0 is feasible and if $\inf(f(X)-g(X))=f(X_0)-g(X_0)$. Feasible and optimal solutions of program II are defined analogously. Note that the above programs are not entirely symmetric, since f^c and g^c are necessarily closed functions, which need not be the case for f and g.

As a first step in the direction of the announced duality theorem, we state the

(5.1.3) **Lemma.** *If* $K(f)\cap K(g)\neq\emptyset$ *and* $K(f^c)\cap K(g^c)\neq\emptyset$, *then*

$$-\infty<\sup_Y(g^c(Y)-f^c(Y))\leqslant\inf_X(f(X)-g(X))<+\infty.$$

Moreover, $f(X_0)-g(X_0)=g^c(Y_0)-f^c(Y_0)$ *holds for a pair of feasible solutions* $X_0\in K(f)\cap K(g)$ *and* $Y_0\in K(f^c)\cap K(g^c)$ *of programs I and II, respectively, if and only if*

$$f^c(Y_0)=Y_0^T X_0-f(X_0)$$

and

$$g^c(Y_0)=Y_0^T X_0-g(X_0).$$

Proof. For an arbitrary $Y\in K(f^c)\cap K(g^c)$ we have by the definition (4.6.3) of conjugate functions

$$f^c(Y)\geqslant Y^T X-f(X)\quad\text{for all}\quad X,$$
$$g^c(Y)\leqslant Y^T X-g(X)\quad\text{for all}\quad X.$$

Hence

$$g^c(Y)-f^c(Y)\leqslant f(X)-g(X)\quad\text{for all}\quad X,$$

and equality can hold only if equality holds in the two relations above.

This proves the lemma, since $K(f)\cap K(g)$ and $K(f^c)\cap K(g^c)$ are nonvoid by hypothesis. □

If the sets $K(f)$ and $K(g)$ overlap properly, that is, if $K(f)^I\cap K(g)^I\neq\emptyset$, then we are able to establish equality in (5.1.3), and thereby the

(5.1.4) **Duality Theorem of Fenchel.** *Consider the programs I and II* (5.1.2). *If* $K(f)^I\cap K(g)^I$ *is nonvoid and* $\inf_X(f(X)-g(X))$ *is finite, then program II has an optimal solution* Y_0, *and*

$$\inf_X(f(X)-g(X))=g^c(Y_0)-f^c(Y_0)=\max_Y(g^c(Y)-f^c(Y)).$$

Proof. $\mu:=\inf_X(f(X)-g(X))$ is finite by hypothesis. Consider the translated function $\tilde f(X):=f(X)-\mu$ and its conjugate function $\tilde f^c(Y)=f^c(Y)+\mu$. Then

$$\tilde f(X)-g(X)\geqslant 0. \tag{5.1.5}$$

The epigraph $[\tilde f]$ and the hypograph $[g]$ have no inner points in common. Indeed, suppose $\binom{X}{z}\in[\tilde f]\cap[g]$. Then $X\in K(\tilde f)\cap K(g)$

$=K(f)\cap K(g)$ and $\tilde{f}(X)\leqslant z\leqslant g(X)$. By (5.1.5), $\tilde{f}(X)=z=g(X)$. Hence $\begin{pmatrix}X\\z\end{pmatrix}$ is a boundary point of both $[\tilde{f}]$ and $[g]$.

Since $[\tilde{f}]$ and $[g]$ have no inner points in common, they can be separated by a plane E which does not contain both sets (separation theorem (3.3.9)). Assume E is vertical. Then the plane $\tilde{E}:=\left\{X \,\middle|\, \begin{pmatrix}X\\z\end{pmatrix}\in E\right\}$ would separate $K(\tilde{f})=K(f)$ and $K(g)$ without containing both. But this contradicts the fact that $K(f)$ and $K(g)$ have inner points in common (separation theorem (3.3.9)).

Thus the separating plane E is nonvertical and can be written in the form

$$E:=\left\{\begin{pmatrix}X\\z\end{pmatrix} \,\middle|\, Y_0^T X - z = w_0\right\}.$$

Then $\begin{pmatrix}Y_0\\w_0\end{pmatrix}$ is a point of the intersection $[\tilde{f}^c]\cap[g^c]$. This implies $\tilde{f}^c(Y_0)\leqslant w_0\leqslant g^c(Y_0)$, since $[\tilde{f}^c]$ is an epigraph, and $[g^c]$ a hypograph. Hence

$$g^c(Y_0)-\tilde{f}^c(Y_0)=g^c(Y_0)-\mu-f^c(Y_0)\geqslant 0$$

or

$$g^c(Y_0)-f^c(Y_0)\geqslant\mu=\inf_X(f(X)-g(X)).$$

Together with lemma (5.1.3), this proves the theorem. □

The above proof shows that Fenchel's duality theorem (5.1.4) is essentially a separation theorem for the convex sets $[f]$ and $[g]$.

(5.1.6) **Corollary to Theorem (5.1.4).** *Let f be a convex function and g a concave function. If $[f]^I\cap[g]^I=\emptyset$ and $K(f)^I\cap K(g)^I\neq\emptyset$ then there exists a nonvertical plane* $E=\left\{\begin{pmatrix}X\\z\end{pmatrix} \,\middle|\, Y^T X - z = w\right\}$ *which separates $[f]$ and $[g]$.*

The point is, of course, that the separating plane is nonvertical. We remark further that by (5.1.3) the hypothesis

$$\text{“}\inf_X(f(X)-g(X))=\text{finite”}$$

of theorem (5.1.4) may be replaced by

$$\text{“}K(f^c)\cap K(g^c)\neq\emptyset\text{”.}$$

Therefore, and since Fenchel's theorem may be applied to f^c and g^c as well as to f and g, we obtain the following important

(5.1.7) **Corollary to Theorem (5.1.4).** *If the functions f and g of the programs* (5.1.2) *are closed, and if $K(f)^I \cap K(g)^I \neq \emptyset$ and $K(f^c)^I \cap K(g^c)^I \neq \emptyset$, then both programs I and II have optimal solutions X_0 and Y_0, respectively. For these solutions one has:*

$$\min_X (f(X)-g(X)) = f(X_0)-g(X_0) = g^c(Y_0)-f^c(Y_0) = \max_Y (g^c(Y)-f^c(Y)).$$

Fenchel's duality theorem can be used to determine the conjugate of the sum of two convex functions. If we write $f(X):=f_1(X)-Y_0^T X$, $g(X):=-f_2(X)$, where f_1 and f_2 are convex functions, then

$$f^c(Y)=f_1^c(Y+Y_0),$$
$$g^c(Y)=-f_2^c(-Y).$$

Programs I and II become

$$\textit{Find} \quad \inf_X (f_1(X)+f_2(X)-Y_0^T X) = -(f_1+f_2)^c(Y_0).$$

$$\textit{Find} \quad \sup_Y (-f_2^c(-Y)-f_1^c(Y+Y_0)) = -\inf_Y (f_1^c(Y_0-Y)+f_2^c(Y)),$$

and the duality theorem of Fenchel (5.1.4) yields

(5.1.8) **Theorem.** *If f_1 and f_2 are convex functions satisfying $K(f_1)^I \cap K(f_2)^I \neq \emptyset$, then the conjugate of the convex function $h(X):=f_1(X)+f_2(X)$ is*

$$h^c(Y_0)=\min_Y (f_1^c(Y_0-Y)+f_2^c(Y))$$

or formulated somewhat weaker in terms of the infimum convolution (4.8.8)

$$(f_1+f_2)^c = f_1^c \,\square\, f_2^c.$$

The counterpart (4.8.10) of this relation holds under much more general conditions

5.2. Duality Gaps

In this section let us examine the hypotheses of Fenchel's duality theorem. It has been maintained (see Fenchel [*1*] and Karlin [*3*]) that for general closed functions f and g the hypotheses $K(f) \cap K(g) \neq \emptyset$, $\inf_X (f(X)-g(X)) > -\infty$ together suffice to establish equality in (5.1.3). The following example shows that this is not so, that

(5.2.1) *duality gaps*

may occur.

Take as convex sets in R^2

$$K(f) := \left\{ \begin{pmatrix} x \\ y \end{pmatrix} \in R^2 \;\middle|\; x=0,\; y \geqslant 0 \right\},$$

$$K(g) := \left\{ \begin{pmatrix} x \\ y \end{pmatrix} \in R^2 \;\middle|\; x \geqslant 0,\; y \geqslant 0 \right\},$$

and the functions defined by

$$f(x,y) := \begin{cases} 0 & \text{for all} \quad \begin{pmatrix} x \\ y \end{pmatrix} \in K(f), \\ +\infty & \text{for all} \quad \begin{pmatrix} x \\ y \end{pmatrix} \notin K(f), \end{cases}$$

$$g(x,y) := \begin{cases} 1 & \text{for all} \quad \begin{pmatrix} x \\ y \end{pmatrix} \in K(g) \quad \text{with} \quad xy \geqslant 1, \\ \sqrt{xy} & \text{for all} \quad \begin{pmatrix} x \\ y \end{pmatrix} \in K(g) \quad \text{with} \quad xy \leqslant 1, \\ -\infty & \text{otherwise.} \end{cases}$$

Clearly f is convex. The identity

$$(\lambda x_1 + \mu x_2)(\lambda y_1 + \mu y_2) = (\lambda \sqrt{x_1 y_1} + \mu \sqrt{x_2 y_2})^2 + \lambda \mu (\sqrt{x_1 y_2} - \sqrt{x_2 y_1})^2$$

shows that the function $\sqrt{xy}$ is concave in the positive orthant of the R^2. Hence g is a concave function. Both functions are closed. Clearly,

$$K(f) \cap K(g) = K(f) \neq \emptyset,$$

and $f(x,y)=0$, $g(x,y)=0$ for all $\begin{pmatrix} x \\ y \end{pmatrix} \in K(f) \cap K(g)$.

We proceed to study the conjugate functions f^c and g^c of f and g, respectively. $K(f^c)$ is the set of all points $\begin{pmatrix} \xi \\ \eta \end{pmatrix}$ for which $\sup\{\xi x + \eta y \mid x=0, y \geqslant 0\} < \infty$, that is, $K(f^c) = \left\{ \begin{pmatrix} \xi \\ \eta \end{pmatrix} \;\middle|\; \eta \leqslant 0 \right\}$. Similarly we get $K(g^c) = \left\{ \begin{pmatrix} \xi \\ \eta \end{pmatrix} \;\middle|\; \xi \geqslant 0, \eta \geqslant 0 \right\}$ and therefore

$$K(f^c) \cap K(g^c) = \left\{ \begin{pmatrix} \xi \\ \eta \end{pmatrix} \;\middle|\; \xi \geqslant 0,\; \eta = 0 \right\} \neq \emptyset.$$

Further, we have

$$\inf\{\xi x - g(x,y) \mid x \geqslant 0,\ y \geqslant 0\} = -1 \quad \text{for} \quad \xi \geqslant 0.$$

Indeed, $\xi x - g(x,y) \geqslant -1$ for all $x \geqslant 0$, $y \geqslant 0$. On the other hand, if $\xi > 0$, choose x and y such that $\varepsilon/\xi \geqslant x > 0$ and $y > 1/x$ in order to achieve $\varepsilon - 1 \geqslant \xi x - g(x,y) \geqslant -1$. If $\xi = 0$, choose x and y such that $xy \geqslant 1$. Therefore $g^c(\xi,\eta) = -1$ for all $\begin{pmatrix}\xi\\ \eta\end{pmatrix} \in K(f^c) \cap K(g^c)$. In the same way, one may show $f^c(\xi,\eta) = 0$ for all $\begin{pmatrix}\xi\\ \eta\end{pmatrix} \in K(f^c) \cap K(g^c)$. This proves

$$\inf_{x,y} (f(x,y) - g(x,y)) = 0 > -1 = \sup_{\xi,\eta} (g^c(\xi,\eta) - f^c(\xi,\eta)).$$

The same example shows that the duality theorem (5.1.4) is false if the condition $K(f)^I \cap K(g)^I \neq \emptyset$ is relaxed to $K(f)^I \cap K(g) \neq \emptyset$. Note that $K(f^c) \cap K(g^c)$ can be void even if

$$\inf_X (f(X) - g(X))$$

is finite. Indeed, choose the same sets $K(f)$ and $K(g)$ as in the preceding example. Set

$$f(x,y) := \begin{cases} 0 & \text{for all} \quad \begin{pmatrix}x\\ y\end{pmatrix} \in K(f), \\ +\infty & \text{otherwise,} \end{cases}$$

$$g(x,y) := \begin{cases} \sqrt{xy} & \text{for all} \quad \begin{pmatrix}x\\ y\end{pmatrix} \in K(g), \\ -\infty & \text{otherwise.} \end{cases}$$

Then $f(x,y) = 0$ and $g(x,y) = 0$ for all $\begin{pmatrix}x\\ y\end{pmatrix} \in K(f) \cap K(g)$. Hence

$$\inf_{x,y} (f(x,y) - g(x,y)) = 0.$$

Further

$$K(f^c) = \left\{ \begin{pmatrix}\xi\\ \eta\end{pmatrix} \;\middle|\; \eta \leqslant 0 \right\},$$

$$K(g^c) \subseteq \left\{ \begin{pmatrix}\xi\\ \eta\end{pmatrix} \;\middle|\; \xi \geqslant 0,\ \eta \geqslant 0 \right\},$$

and therefore

$$K(f^c)\cap K(g^c)\subseteq\left\{\begin{pmatrix}\xi\\ \eta\end{pmatrix}\;\middle|\;\xi\geqslant 0,\ \eta=0\right\}.$$

But for $\xi\geqslant 0$ we have

$$\inf\left\{\xi x-g(x,y)\;\middle|\;\begin{pmatrix}x\\ y\end{pmatrix}\in K(g)\right\}=\inf_{x,y}\{\xi x-\sqrt{xy}\mid x\geqslant 0,\ y\geqslant 0\}=-\infty.$$

Hence $K(f^c)\cap K(g^c)$ must be empty.

The following necessary and sufficient conditions for the absence of duality gaps are due to Rockafellar [1]. His analysis is based on the convolution function (compare (4.8.8)):

$$h(Z):=\inf_X(f(X)-g(X-Z))=f\,\square\,(-\tilde g)(Z)\qquad\text{with}\qquad \tilde g(X):=g(-X).$$

which represents a perturbation of Fenchel's first problem (5.1.2).

(5.2.2) **Theorem.** *Let f be a convex function, g a concave function on R^n. Then with $h:=f\,\square\,(-\tilde g)$ and $\tilde g(X):=g(-X)$,*

(i) $$\inf_X(f(X)-g(X))=\max_Y(g^c(Y)-f^c(Y))$$

holds if and only if there is a Y_0 such that $h(Z)-h(0)\geqslant Y_0^TZ$ for all Z.

(ii) $$\inf_X(f(X)-g(X))=\sup_Y(g^c(Y)-f^c(Y))$$

holds if and only if $h^{cc}(0)=h(0)$.

Proof. Since h is the infimum convolution (4.8.8) of f and $-\tilde g$, both of which are convex, h is convex at least in the wider sense (4.2.5). Moreover, (4.8.10) and (4.8.12) give $h^c=f^c-g^c$. By lemma (5.1.3),

$$\inf_X(f(X)-g(X))=\max_Y(g^c(Y)-f^c(Y))$$

is equivalent to

$$h(0)=\inf_X(f(X)-g(X))\leqslant\max_Y(g^c(Y)-f^c(Y))=\max_Y(-h^c(Y)),$$

that is, to the existence of a Y_0 such that

$$h(0)\leqslant -h^c(Y_0)=\inf_Z(h(Z)-Y_0^TZ).$$

This proves (i). Next,

$$\sup_Y(g^c(Y)-f^c(Y))=\sup_Y(-h^c(Y))=\sup_Y(0^TY-h^c(Y))=h^{cc}(0),$$

which yields (ii). □

Since f and $-g$ are proper convex functions, so are f^c, $-g^c$ according to (4.6.4). As $h^c = f^c - g^c$, $h^c \not\equiv -\infty$ and therefore $h \not\equiv +\infty$. Thus h is either properly convex or it assumes the value $-\infty$. In the first case,

$$h^{cc}(0) = \underline{h}(0) = \varliminf_{Z \to 0} h(Z)$$

by theorem (4.7.7). In the second case, $h^c \equiv +\infty$ and $h^{cc} \equiv -\infty$. Since, in this case, we have also

$$h(Z) = -\infty \qquad \text{for all} \qquad Z \in K(h)^I$$

by (4.2.6), we find that $h^{cc}(0) \neq h(0)$ if and only if $h^{cc}(0) = -\infty$ and $\underline{h}(0) = +\infty$. This and (5.2.2), (ii) proves

(5.2.3) **Corollary to Theorem (5.2.2).**

$$\inf_X (f(X) - g(X)) = \sup_Y (g^c(Y) - f^c(Y))$$

holds if and only if

$$h(0) = \underline{h}(0),$$

except in the trivial case, where

$$\sup_Y (g^c(Y) - f^c(Y)) = -\infty,$$
$$\underline{h}(0) = +\infty.$$

The necessary and sufficient condition for (5.2.2.(i)) can be reformulated in terms of the directional derivative $h'(0; Z)$ of h:

(5.2.4) **Theorem.** *Suppose that*

$$h(0) = \inf_X (f(X) - g(X))$$

is finite. Then

$$\inf_X (f(X) - g(X)) = \max_Y (g^c(Y) - f^c(Y))$$

holds if and only if the directional derivative of h satisfies

$$h'(0; Z) := \lim_{\lambda \downarrow 0} \frac{h(\lambda Z) - h(0)}{\lambda} > -\infty$$

for all Z.

Proof. As $h(0)$ is finite, $h'(0; Z)$ is well defined and is a convex function in the wider sense (see (4.3.4)). Now, if $\inf (f(X) - g(X)) = \max_Y (g^c(Y) - f^c(Y))$, then it follows at once from (5.2.2.(i)), that

$$h'(0; Z) \geqslant Y_0^T Z > -\infty$$

for all Z.

To prove the converse, suppose that

$$h'(0;Z) > -\infty \tag{5.2.5}$$

for all Z. Then h must be in fact properly convex; otherwise $h(Z) = -\infty$ for all $Z \in K(h)^I$ by (4.2.6), and since for any $Z \in K(h)^I$ the open segment $(0, Z)$ lies in $K(h)^I$,

$$h'(0;Z) = \lim_{\lambda \downarrow 0} \frac{h(\lambda Z) - h(0)}{\lambda} = -\infty,$$

contradicting (5.2.5).

Now by (5.2.5) and theorem (4.3.8), applied to the convex function h and the point $X_0 = 0$, there is a nonvertical supporting plane

$$E = \left\{ \begin{pmatrix} Z \\ z \end{pmatrix} \;\middle|\; Y_0^T Z - z = -h(0) \right\}$$

of $[h]$ through $\begin{pmatrix} 0 \\ h(0) \end{pmatrix}$:

$$Y_0^T Z - h(Z) \leqslant -h(0) \qquad \text{for all} \qquad Z.$$

Hence $\inf_X (f(X) - g(X)) = \max_Y (g^c(Y) - f^c(Y))$ by (5.2.2.(i)). □

The conditions in (5.2.3) and (5.2.4) illustrate that duality gaps are only possible if the functions $h(Z) = \inf_X (f(X) - g(X - Z))$ changes abruptly in the neighborhood of $Z = 0$, or in other words, if the program

$$(I_Z) \qquad \textit{Find} \quad \inf_X (f(X) - g(X - Z))$$

is very sensitive against small changes of Z at the origin.

Unfortunately, the conditions of theorem (5.2.2) are difficult to verify. In the following two sections, we shall give conditions for Fenchel's duality theorem, which are more useful in practice, though they are only sufficient conditions. We shall characterize a large class of functions, for which the hypotheses of Fenchel's theorem can indeed be weakened.

5.3. Generalization of Fenchel's Duality Theorem

The corollary (5.1.6) of Fenchel's duality theorem (5.1.4) indicates the close connection between duality and separation statements. A scrutiny of these separation properties will yield more refined duality theorems.

To begin with, let $M \subseteq R^n$ be a linear manifold in R^n. We say that a set $N \subseteq R^{n+1}$

(5.3.1) *covers* M

if $\pi(N) = M$, where $\pi: R^{n+1} \to R^n$ is the projection of R^{n+1} into R^n in the direction of the z-axis:

$$\pi\begin{pmatrix} X \\ z \end{pmatrix} := X \in R^n \quad \text{for} \quad \begin{pmatrix} X \\ z \end{pmatrix} \in R^{n+1}. \tag{5.3.2}$$

By

$$N_M$$

we denote a *nonvertical* linear manifold in R^{n+1} covering the linear manifold M in R^n.

If f is a convex function on R^n and N is a linear manifold in R^{n+1} covering M, then we say that N

(5.3.3) *lies below*

the epigraph $[f]$ if

$$M \cap K(f) \neq \emptyset$$

and

$$f(X) \geqslant z \quad \text{for all} \quad \begin{pmatrix} X \\ z \end{pmatrix} \in N.$$

As is easily seen, every linear manifold N covering M and lying below $[f]$ is nonvertical. Moreover, we say that N_M

(5.3.4) *lies properly below*

the epigraph $[f]$ if (i) N_M lies below $[f]$, and (ii) there exists an $\varepsilon > 0$ such that the translated linear manifold

$$N_{M,\varepsilon} := \left\{ \begin{pmatrix} X \\ z+\varepsilon \end{pmatrix} \;\middle|\; \begin{pmatrix} X \\ z \end{pmatrix} \in N_M \right\}$$

still lies below $[f]$. For concave functions g, lying (properly) above the hypograph $[g]$, is defined analogously.

Now, given a convex function f on R^n, we are going to classify the points $X_0 \in K(f)$ in terms of separation properties. We call the point $X_0 \in K(f)$

(5.3.5) *stable*

if every linear (nonvertical) manifold N in R^{n+1} lying below $[f]$ and satisfying $X_0 \in \pi(N)$ can be separated from $[f]$ by a *nonvertical* plane $E := \left\{ \begin{pmatrix} X \\ z \end{pmatrix} \;\middle|\; Y^T X - z = b \right\}$ in R^{n+1}:

$$Y^T X_1 - z_1 \leqslant b \leqslant Y^T X_2 - z_2$$

for all $\begin{pmatrix} X_1 \\ z_1 \end{pmatrix} \in [f]$, $\begin{pmatrix} X_2 \\ z_2 \end{pmatrix} \in N$.

The point $X_0 \in K(f)$ is said to be

(5.3.6) *weakly stable*

if every linear manifold N of R^{n+1} lying *properly* below $[f]$ and satisfying $X_0 \in \pi(N)$ can be separated from $[f]$ by a nonvertical plane E of R^{n+1}.

Clearly, if $X_0 \in K(f)$ is (weakly) stable, then, by a vertical translation of E, we can always achieve that the plane E separating N from $[f]$ passes through N:

$$Y^T X_1 - z_1 \leqslant b = Y^T X_2 - z_2$$

for all $\begin{pmatrix} X_1 \\ z_1 \end{pmatrix} \in [f]$, $\begin{pmatrix} X_2 \\ z_2 \end{pmatrix} \in N$.

If $X_0 \in K(f)$ is (weakly) stable, then we call f

(5.3.7) *(weakly) X_0-stable.*

If f is (weakly) X_0-stable for all $X_0 \in K(f)$ then it is said to be

(5.3.8) *(weakly) stable.*

Let f be any convex function. Clearly

(5.3.9) *every stable point $X_0 \in K(f)$ is also weakly stable.*

Moreover,

(5.3.10) *every inner point of $K(f)$ is stable. As a consequence, f is stable if $K(f)^I = K(f)$.*

Proof. Let N be an arbitrary linear manifold in R^{n+1} lying below $[f]$ and satisfying $X_0 \in \pi(N) \cap K(f)^I$. Then, as mentioned above, N must be nonvertical, and therefore

$$g_N(X) := \begin{cases} z & \text{if } \begin{pmatrix} X \\ z \end{pmatrix} \in N, \\ -\infty & \text{otherwise,} \end{cases}$$

is a well defined concave function on R^n. g_N is a linear function on the projection $\pi(N)$ of N into R^n. We call g_N therefore

(5.3.11) *partially linear.*

Partially linear functions are useful proof-theoretical tools, and they will be used again in this capacity at the end of this section.

For the proof at hand, we note that $X_0 \in K(g_N)^I \cap K(f)^I \neq \emptyset$ and $[f]^I \cap [g_N]^I = \emptyset$, since N lies below $[f]$. Then the corollary (5.1.6) of Fenchel's duality theorem yields the existence of a nonvertical plane E which separates $[f]$ and $[g_N] \supset N$.

Thus instability is restricted to boundary points of $K(f)$. The following examples illustrate the various degrees of stability:

1. The partially linear functions (5.3.11) are examples of stable functions.

2. The point $x_0 := 0 \in K(f) = \{x \geqslant 0\}$ of the convex function

$$f(x) := \begin{cases} -\sqrt{x} & \text{for all} \quad x \geqslant 0, \\ +\infty & \text{otherwise,} \end{cases}$$

on R^1 is weakly stable. But there is no nonvertical line

$$E = \left\{ \begin{pmatrix} x \\ z \end{pmatrix} \,\middle|\, ax - z = b \right\}$$

in R^2 which separates the linear manifold

$$N := \left\{ \begin{pmatrix} 0 \\ 0 \end{pmatrix} \right\},$$

which lies below $[f]$, from $[f]$, whence x_0 is not stable.

3. Finally, the function

$$g(x,y) := \begin{cases} \sqrt{xy} & \text{for all} \quad \begin{pmatrix} x \\ y \end{pmatrix} \in K(g) := \{x \geqslant 0, y \geqslant 0\}, \\ -\infty & \text{otherwise,} \end{cases}$$

which appeared in the counterexample at the end of the preceding section, is not even weakly X_0-stable at the point

$$X_0 := \begin{pmatrix} 0 \\ 0 \end{pmatrix} \in K(g).$$

The linear manifold

$$N := \left\{ \begin{pmatrix} x \\ y \\ z \end{pmatrix} \,\middle|\, x = 0, z = 1 \right\}$$

lies properly above $[g]$ and satisfies $\pi(N) \ni \begin{pmatrix} 0 \\ 0 \end{pmatrix}$ but cannot be separated from $[g]$ by a nonvertical plane.

For stable and weakly stable functions we are now able to relax the hypotheses of Fenchel's duality theorem (5.1.4), as we promised in section 5.1. The extended Fenchel theorem reads as follows

(5.3.12) **Duality Theorem.** *Let f be a convex function, g a concave function on R^n. Consider the dual programs*

program I: *Find* $\inf_X (f(X)-g(X))$,

program II: *Find* $\sup_Y (g^c(Y)-f^c(Y))$.

Assume further that $\inf_X (f(X)-g(X))$ *is finite. Then:*

(i) *If there are points $X_f, X_g \in K(f)\cap K(g)$ such that f is X_f-stable and g is X_g-stable, then program II has an optimal solution Y_0 which satisfies*

$$\inf_X (f(X)-g(X)) = g^c(Y_0)-f^c(Y_0) = \max_Y (g^c(Y)-f^c(Y)).$$

(ii) *If there are points $X_f, X_g \in K(f)\cap K(g)$ such that f is weakly X_f-stable and g is weakly X_g-stable, then*

$$\inf_X (f(X)-g(X)) = \sup_Y (g^c(Y)-f^c(Y)).$$

By (5.3.10), this duality theorem (Stoer [*2*]) is a generalization of Fenchel's duality theorem (5.1.4) It also extends a theorem of Rockafellar [*1*], who proved an analogue of (i) under the stronger hypothesis that f and g are stable functions. A proof to theorem (5.3.12) will be given in the following section. We conclude this section by showing that theorem (5.3.12) is essentially of the greatest possible generality.

More precisely, we shall show that the (weakly) X_0-stable functions constitute the broadest class of convex and concave functions for which the duality statements remain generally true. To this end, we recall the partially linear functions (5.3.11), which were seen to be stable. Our assertion will then follow immediately from the

(5.3.13) **Theorem.** *Let f be a convex function on R^n and let X_0 be a point of $K(f)$. If the duality relation*

(i) $$\inf_X (f(X)-g_N(X)) = \sup_Y (g_N^c(Y)-f^c(Y))$$

is true for every partially linear concave function g_N with $X_0 \in K(g_N)$, then f is weakly X_0-stable. Moreover, if

(ii) $$\inf_X (f(X)-g_N(X)) = \max_Y (g_N^c(Y)-f^c(Y))$$

holds for every partially linear concave function g_N with $X_0 \in K(g_N)$, then f is X_0-stable.

Proof. In order to show that f is weakly X_0-stable, if (i) holds, let N be a linear manifold which lies properly below $[f]$. Then there is an $\varepsilon > 0$ such that the translated manifold

$$N_\varepsilon := \left\{ \begin{pmatrix} X \\ z+\varepsilon \end{pmatrix} \middle| \begin{pmatrix} X \\ z \end{pmatrix} \in N \right\}$$

also lies below $[f]$. For the partially linear function g_{N_ε}, (i) implies

$$0 \leqslant \mu := \inf_X (f(X) - g_{N_\varepsilon}(X)) = \sup_Y (g^c_{N_\varepsilon}(Y) - f^c(Y)).$$

Hence, for every η with $\varepsilon \geqslant \eta > 0$ there is a vector $Y_\eta \in K(g^c_{N_\varepsilon}) \cap K(f^c)$ such that

$$\mu - \eta \leqslant g^c_{N_\varepsilon}(Y_\eta) - f^c(Y_\eta),$$

which implies

$$\begin{aligned} Y_\eta^T X - f(X) \leqslant f^c(Y_\eta) &\leqslant g^c_{N_\varepsilon}(Y_\eta) - \mu + \eta \\ &\leqslant Y_\eta^T X - g_{N_\varepsilon}(X) + \eta \\ &\leqslant Y_\eta^T X - g_N(X) - \varepsilon + \eta \\ &\leqslant Y_\eta^T X - g_N(X). \end{aligned}$$

It follows that the nonvertical plane $N_\eta := \left\{ \begin{pmatrix} X \\ z \end{pmatrix} \middle| \; Y_\eta^T X - z = f^c(Y_\eta) \right\}$ separates $[f]$ from N. Hence, f is weakly X_0-stable.

As to the second part of this theorem, the preceding proof carries over with obvious modifications. □

5.4. Proof of the Generalized Fenchel Theorem

The proof of theorem (5.3.12) will be based on a simple and intuitive lemma. This lemma will enable us to infer separation by nonvertical planes from separation by vertical planes.

For an illustration, consider two nonvertical lines E_1 and E_2, and one vertical line E_v in the plane R^2, all three lines intersecting in one point. Suppose that E_v separates two arbitrary sets K_1 and K_2, and that E_1 passes below K_1, whereas E_2 passes above K_2. Then at least one of the two nonvertical lines $E_i, i=1,2,$ separates K_1 and K_2 (Fig. 19). In its general form the lemma reads as follows:

(5.4.1) **Lemma.** *Let K_1 and K_2 be arbitrary sets of R^{n+1}. Suppose that there are a vertical plane*

$$E_v := \left\{ \begin{pmatrix} X \\ z \end{pmatrix} \middle| (Y_v^T X = b_v \right\},$$

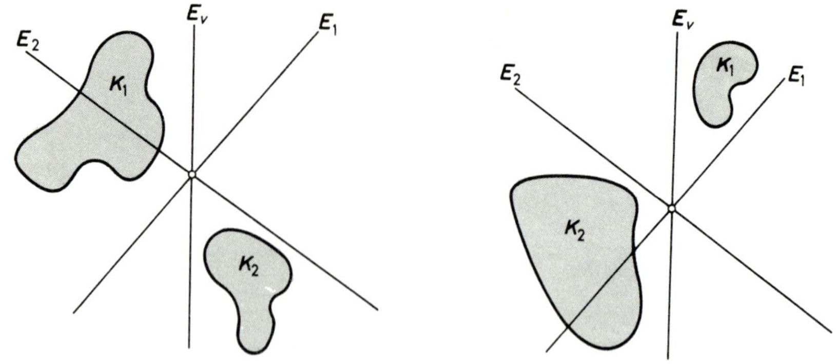

Fig. 19. Illustration of separation lemma (5.4.1)

two nonvertical planes

$$E_i := \left\{ \begin{pmatrix} X \\ z \end{pmatrix} \middle| \, Y_i^T X - z = b_i \right\}, \qquad i = 1,2,$$

and a vertical linear manifold

$$\hat{M} := \left\{ \begin{pmatrix} X \\ z \end{pmatrix} \middle| \, C^T X = D \right\}$$

in R^{n+1} with the following properties:

(i) E_v *separates* K_1 *and* K_2:

$$Y_v^T X_1 \leqslant b_v \leqslant Y_v^T X_2 \quad \textit{for all} \quad \begin{pmatrix} X_i \\ z_i \end{pmatrix} \in K_i, \qquad i = 1,2;$$

(ii) E_1 *lies below* K_1:

$$Y_1^T X_1 - z_1 \leqslant b_1 \quad \textit{for all} \quad \begin{pmatrix} X_1 \\ z_1 \end{pmatrix} \in K_1;$$

(iii) E_2 *lies above* K_2:

$$Y_2^T X_2 - z_2 \geqslant b_2 \quad \textit{for all} \quad \begin{pmatrix} X_2 \\ z_2 \end{pmatrix} \in K_2;$$

(iv) $E_1 \cap E_v \cap \hat{M} = E_2 \cap E_v \cap \hat{M} \neq \emptyset$.

Then at least one of the nonvertical planes E_1 or E_2 separates $K_1 \cap \hat{M}$ and $K_2 \cap \hat{M}$.

Proof. Condition (iv) can be stated as follows: The linear system

$$\begin{aligned} C^T X &= D, \\ Y_v^T X &= b_v, \\ Y_1^T X - z &= b_1, \\ Y_2^T X - z &= b_2, \end{aligned}$$

is solvable, and each of the last two equations is a consequence of the other equations. Such superfluous equations are linear combinations of the remaining equations of the system (theorem (1.4.4)). Thus

(5.4.2)
$$\begin{aligned} Y_2 &= C U + \lambda Y_v + Y_1, \\ b_2 &= D^T U + \lambda b_v + b_1. \end{aligned}$$

That the coefficient of Y_1 in this combination takes the value 1 is due to the variable z, which occurs only in the two bottom equations.

If $\lambda \geqslant 0$ in (5.4.2), then

$$Y_2^T X_1 - z_1 = U^T C^T X_1 + \lambda Y_v^T X_1 + Y_1^T X_1 - z_1 \leqslant U^T D + \lambda b_v + b_1 = b_2$$

for all $\begin{pmatrix} X_1 \\ z_1 \end{pmatrix} \in K_1 \cap \hat{M}$. Together with (iii) this shows that E_2 separates $K_1 \cap \hat{M}$ and K_2. If $\lambda < 0$ in (5.4.2), then (5.4.2) is rearranged to

$$\begin{aligned} Y_1 &= -C U - \lambda Y_v + Y_2, \\ b_1 &= -D^T U - \lambda b_v + b_2, \end{aligned}$$

and the same argument as above shows that E_1 separates K_1 and $K_2 \cap \hat{M}$. □

Now let us turn to the proof of the duality theorem (5.3.12). We first prove the statement (5.3.12.(i)). As in the proof of Fenchel's duality theorem (5.1.4), we assume without loss of generality that

(5.4.3)
$$\inf_X (f(X) - g(X)) = 0,$$

and then try to separate $[f]$ and $[g]$ by a nonvertical plane

(5.4.4)
$$E = \left\{ \begin{pmatrix} X \\ z \end{pmatrix} \,\middle|\, Y_0^T X - z = b_0 \right\}.$$

For such a plane $f^c(Y_0) \leqslant b_0 \leqslant g^c(Y_0)$, and therefore

$$g^c(Y_0) - f^c(Y_0) \geqslant 0 = \inf_X (f(X) - g(X)).$$

Lemma (5.1.3) then yields the relation

$$\inf_X (f(X)-g(X)) = g^c(Y_0)-f^c(Y_0) = \max_Y (g^c(Y)-f^c(Y)),$$

which was asserted by (5.3.12.(i)).

The problem thus is to construct a nonvertical separating plane (5.4.4). Starting with $M_n := R^n$, we shall first define a decreasing sequence of manifolds in R^n

(5.4.5) $$R^n := M_n \supset M_{n-1} \supset \cdots \supset M_k$$

together with a (possibly empty) sequence of planes in R^n

$$E_{n-1}, \ldots, E_k$$

such that $M_j = E_j \cap M_{j+1}$ for $j=n-1,\ldots,k$. Then starting with k and working our way upwards, we shall use lemma (5.4.1) and the stability properties of f and g for constructing a sequence of nonvertical linear manifolds N_{M_j} in R^{n+1} which cover M_j and satisfy

(5.4.6) $$f(X) \geqslant z \geqslant g(X) \quad \text{for all} \quad \binom{X}{z} \in N_{M_j}.$$

Then N_{M_j} lies above $[g]$ and below $[f]$. Since in particular $M_n = R^n$, the manifold N_{M_n} will be the desired plane (5.4.4).

For the construction of the sequence (5.4.5), we assume that M_j and E_j have been defined down to some $i \leqslant n$ such that for $i \leqslant j \leqslant n$

(5.4.7) (i) *E_j separates $K(f) \cap M_{j+1}$ and $K(g) \cap M_{j+1}$,*
(ii) $M_j = E_j \cap M_{j+1} \neq M_{j+1}$,
(iii) $M_j \supseteq K(f) \cap K(g)$.

We then consider the restrictions of f and g to the set M_i:

$$f_{M_i}(X) := \begin{cases} f(X) & \text{for} \quad X \in M_i, \\ +\infty & \text{otherwise}, \end{cases}$$

$$g_{M_i}(X) := \begin{cases} g(X) & \text{for} \quad X \in M_i, \\ -\infty & \text{otherwise}. \end{cases}$$

Obviously

$$K(f_{M_i}) = K(f) \cap M_i, \qquad K(g_{M_i}) = K(g) \cap M_i.$$

If $K(f_{M_i})^I \cap K(g_{M_i})^I \neq \emptyset$, then we put $k := i$, and the construction of the sequence (5.4.5) is completed. If $K(f_{M_i})^I \cap K(g_{M_i})^I = \emptyset$, then the general separation theorem (3.3.9) yields the existence of a plane E_{i-1} in R^n which separates $K(f_{M_i})$ from $K(g_{M_i})$ without containing both sets, thus satisfying (5.4.7.(i)) for $j = i-1$. We then define

$$M_{i-1} := E_{i-1} \cap M_i$$

and proceed to verify (5.4.7.(ii)) and (5.4.7.(iii)). By their definition, $K(f_{M_i})$ and $K(g_{M_i})$ are both subsets of M_i. If E_{i-1} contains M_i, then it contains both $K(f_{M_i})$ and $K(g_{M_i})$, which was ruled out above. Hence $M_{i-1} \neq M_i$. In order to prove (5.4.7.(iii)), we note that E_{i-1} must contain the intersection of $K(f_{M_i})$ and $K(g_{M_i})$ since it separates these sets. Knowing that M_i satisfies (5.4.7.(iii)), we obtain therefore

$$E_{i-1} \supseteq K(f_{M_i}) \cap K(g_{M_i}) = K(f) \cap K(g) \cap M_i = K(f) \cap K(g),$$

and

$$M_{i-1} = E_{i-1} \cap M_i \supseteq K(f) \cap K(g).$$

Clearly the planes E_i thus constructed are linearly independent, and the construction process must therefore terminate with some $i = k \geqslant 0$, for which one has

$$K(f_{M_k})^I \cap K(g_{M_k})^I \neq \emptyset. \tag{5.4.8}$$

Now we have to construct the nonvertical linear manifolds N_{M_i} which cover M_i and satisfy (5.4.6). We start with the construction of N_{M_k}. Clearly $[f_{M_k}] = [f] \cap \hat{M}_k$ with $\hat{M}_k = \left\{ \begin{pmatrix} X \\ z \end{pmatrix} \middle| X \in M_k \right\} = \pi^{-1}(M_k)$.

Since $f(X) \geqslant g(X)$ and therefore $f_{M_k}(X) \geqslant g_{M_k}(X)$ for all X, we have

$$[f_{M_k}]^I \cap [g_{M_k}]^I = \emptyset.$$

This and (5.4.8) are precisely the hypotheses of the corollary (5.1.6) of Fenchel's duality theorem. By this corollary there exists a nonvertical plane E in R^{n+1} separating $[f_{M_k}]$ and $[g_{M_k}]$. We thus define

$$N_{M_k} := E \cap \hat{M}_k \quad \text{where} \quad \hat{M}_k := \pi^{-1}(M_k).$$

The manifold N_{M_k} is nonvertical, covers M_k, and satisfies (5.4.6) for $j = k$.

Assume that manifolds N_{M_i} have been defined up to some $i < n$, each N_{M_j} covering M_j and satisfying (5.4.6) for $j \leqslant i$. Our goal is to construct $N_{M_{i+1}}$. By the hypothesis of theorem (5.3.12), the set $K(f) \cap K(g)$ contains the points X_f and X_g. Since $M_i \supseteq K(f) \cap K(g)$ by (5.4.7.(iii)), we have $X_f, X_g \in M_i$. By the X_f-stability of f and the X_g-stability of g, there exist in R^{n+1} nonvertical planes E_f and E_g through N_{M_i}, which lie below $[f]$ and above $[g]$, respectively. With

$$\hat{M}_i := \pi^{-1}(M_i),$$

we have

$$E_f \cap \hat{M}_i = E_g \cap \hat{M}_i = N_{M_i}.$$

Indeed $E_f \cap M_i \supseteq N_{M_i}$ by the construction of E_f. But this implies $E_f \cap \hat{M}_i = N_{M_i}$, since E_f is nonvertical and therefore no two points of E_f have the same projection into R^n. From (5.4.7.(ii)),

$$E_f \cap E_v \cap \hat{M}_{i+1} = E_g \cap E_v \cap \hat{M}_{i+1} = N_{M_i} \neq \emptyset$$

where

$$E_v := \pi^{-1}(E_i).$$

The plane E_i separates $K(f) \cap M_{i+1}$ and $K(g) \cap M_{i+1}$ (5.4.7.(i)). Hence E_v is a vertical plane which separates $[f] \cap \hat{M}_{i+1}$ and $[g] \cap \hat{M}_{i+1}$. The manifold $\hat{M}_{i+1}$, the planes E_f, E_g, E_v, and the sets $[f] \cap \hat{M}_{i+1}$, $[g] \cap \hat{M}_{i+1}$ satisfy the hypotheses of the lemma (5.4.1). Thus one of the two nonvertical planes E_f and E_g separates $[f] \cap \hat{M}_{i+1}$ and $[g] \cap \hat{M}_{i+1}$. The intersection of this plane with $\hat{M}_{i+1}$ then is a nonvertical manifold $N_{M_{i+1}}$ which covers M_{i+1}, lies below $[f]$, and above $[g]$, and thereby satisfies (5.4.6) for $j = i+1$. This completes the proof of (5.3.12.(i)).

The proof of (5.3.12.(ii)) proceeds along the same lines as the proof of (5.3.12.(i)). Again we assume

$$\inf_X (f(X) - g(X)) = 0,$$

without loss of generality. Then we choose an arbitrary $\varepsilon > 0$ and show the existence of a nonvertical plane

$$E_\varepsilon = \left\{ \begin{pmatrix} X \\ z \end{pmatrix} \middle| \, Y_\varepsilon^T X - z = b_\varepsilon \right\}$$

separating the translated sets $[f + n\varepsilon]$ and $[g - n\varepsilon]$. Clearly, if we have found such a plane E_ε, then

$$Y_\varepsilon^T X_1 - f(X_1) - n\varepsilon \leqslant b_\varepsilon \leqslant Y_\varepsilon^T X_2 - g(X_2) + n\varepsilon$$

holds for all X_1, X_2, which shows

$$f^c(Y_\varepsilon) \leqslant b_\varepsilon + n\varepsilon \quad \text{and} \quad g^c(Y_\varepsilon) \geqslant b_\varepsilon - n\varepsilon.$$

Hence $g^c(Y_\varepsilon) - f^c(Y_\varepsilon) \geqslant -2n\varepsilon$, and therefore

$$\sup_Y (g^c(Y) - f^c(Y)) \geqslant 0,$$

since $\varepsilon > 0$ was arbitrary. Lemma (5.1.3) then establishes

$$\inf_X (f(X) - g(X)) = \sup_Y (g^c(Y) - f^c(Y)).$$

The construction of the nonvertical plane E_ε which separates $[f + n\varepsilon]$ and $[g - n\varepsilon]$ uses again the sequences of manifolds M_i and planes E_i, which satisfy (5.4.7). Also the nonvertical manifold N_{M_k}, which lies below $[f]$ and above $[g]$, is defined as before. A fortiori, N_{M_k} lies properly below $[f+k\varepsilon]$ and properly above $[g-k\varepsilon]$. Assume then inductively that N_{M_i} is a nonvertical linear manifold which covers M_i, lies properly below $[f+i\varepsilon]$, and properly above $[g-i\varepsilon]$. By the weak X_f-stability of f and the weak X_g-stability of g, there is a nonvertical plane E_f lying below $[f+i\varepsilon]$ and a nonvertical plane E_g lying above $[g-i\varepsilon]$ such that

$$E_f \cap \hat{M}_i = E_g \cap \hat{M}_i = N_{M_i},$$

where $\hat{M}_i := \pi^{-1}(M_i)$, and the argument continues precisely as in the proof of (5.3.12.(i)). □

5.5. Alternative Characterizations of Stability

The extended duality theorem (5.3.12) is useful only if one knows manageable characterizations of stable functions, and function operations which preserve stability. The results of the following two sections pertain to such questions.

(5.5.1) **Lemma.** *The point $X_0 \in K(f)$ of the convex function f on R^n is weakly stable if and only if for every $\varepsilon > 0$ the vertical translate*

$$N_{-\varepsilon} := \left\{ \begin{pmatrix} X \\ z-\varepsilon \end{pmatrix} \middle| \begin{pmatrix} X \\ z \end{pmatrix} \in N \right\}$$

of every nonvertical linear manifold N with $X_0 \in \pi(N)$ and which lies below $[f]$ is not a supporting manifold (3.4.9) *of* $[f]$.

This property, which characterizes weak X_0-stability, could also be expressed by requiring that no vertical translate N_ε, $\varepsilon \neq 0$, of a linear supporting manifold N lying below $[f]$ remains a supporting manifold of $[f]$.

To *prove* lemma (5.5.1), we first assume that f is weakly X_0-stable. Let ε be a positive number and N an arbitrary nonvertical linear manifold lying below $[f]$ and satisfying $X_0 \in \pi(N)$. Then for every η with $0 < \eta < \varepsilon$ the vertical translate

$$N_{-\eta} := \left\{ \begin{pmatrix} X \\ z-\eta \end{pmatrix} \middle| \begin{pmatrix} X \\ z \end{pmatrix} \in N \right\}$$

of N lies properly below $[f]$. By the weak stability of X_0 (5.3.6), there exists a nonvertical plane E which separates $N_{-\eta}$ from $[f]$. A fortiori, E separates $N_{-\varepsilon}$ from $[f]$. Since $\varepsilon > \eta$, moreover, the distance between $N_{-\varepsilon}$ and E is positive. Hence $N_{-\varepsilon}$ cannot be a supporting manifold (3.4.9) of $[f]$.

Conversely, suppose that a linear nonvertical manifold N lies properly below $[f]$ and $X_0 \in \pi(N)$. By the definition (5.3.4) of lying properly below, one can move N up a little bit and still stay below $[f]$. Then, by hypothesis, N is not a supporting manifold of $[f]$. In other words, the distance d between N and $[f]$ is positive. For some $\eta > 0$ with $\eta < d$, we consider the "tube"

$$T_\eta := \left\{ \begin{pmatrix} X \\ z \end{pmatrix} \,\middle|\, \inf_{\binom{Y}{w} \in N} \left\| \begin{pmatrix} X - Y \\ z - w \end{pmatrix} \right\| \leqslant \eta \right\},$$

which is the set of all points in R^{n+1} whose distance from N does not exceed η. Clearly, T_η is convex with N in its interior. Since T_η was defined so as to be disjoint from $[f]$, there exists a plane

$$E := \left\{ \begin{pmatrix} X \\ z \end{pmatrix} \,\middle|\, Y^T X - a z = w \right\}$$

which separates T_η, and therefore N, from $[f]$ (separation theorem (3.3.9)):

$$Y^T X_1 - a z_1 \leqslant w \leqslant Y^T X_2 - a z_2 \tag{5.5.2}$$

for all $\begin{pmatrix} X_1 \\ z_1 \end{pmatrix} \in [f]$, $\begin{pmatrix} X_2 \\ z_2 \end{pmatrix} \in T_\eta$. We have to show that E is indeed nonvertical, in other words, that $a \neq 0$.

To this end we note that E does not meet the interior of T_η. If it did, it would be a singular supporting plane of T_η, and would then contain T_η by lemma (3.4.4), contradicting $\dim T_\eta = n+1$. Thus $E \cap N = \emptyset$, and (5.5.2) becomes, for $a = 0$:

$$Y^T X_1 \leqslant w < Y^T X_2,$$

whenever $X_1 \in K(f)$, $X_2 \in \pi(N)$. But this contradicts $X_0 \in K(f) \cap \pi(N)$. □

We proceed to characterize X_0-stability algebraically. Every nonvertical linear manifold N_M covering $M = \{AU - B \mid U \in R^m\}$ is the intersection

$$N_M = E \cap \hat{M}$$

of a nonvertical plane $E = \left\{ \begin{pmatrix} X \\ z \end{pmatrix} \,\middle|\, Y^T X - z = w \right\}$ in R^{n+1} with the vertical manifold $\hat{M} := \pi^{-1}(M)$. Or, equivalently, every such N_M has the form

$$\begin{aligned} N_M &= \left\{ \begin{pmatrix} X \\ z \end{pmatrix} \,\middle|\, Y^T X - z = w \right\} \cap \left\{ \begin{pmatrix} AU - B \\ z \end{pmatrix} \,\middle|\, U \in R^m \right\} \\ &= \left\{ \begin{pmatrix} AU - B \\ z \end{pmatrix} \,\middle|\, V^T U - z = w',\ U \in R^m \right\} \end{aligned} \tag{5.5.3}$$

with

(5.5.4) $$V := A^T Y \quad \text{and} \quad w' := w + Y^T B.$$

Moreover, N_M lies below $[f]$ and covers M, if and only if $\infty > w' \geqslant \sup_U (V^T U - f(AU - B))$. With the abbreviation

$$\hat{f}(U) := f(AU - B),$$

we thus can state that

(5.5.5) *the nonvertical manifold N_M covering M lies (properly) below $[f]$ if and only if $\hat{f}^c(V) \leqslant w'$ $(\hat{f}^c(V) < w')$.*

Now consider a manifold N_M which lies properly below $[f]$ and which is determined by V and w'. Each vector Y and constant w satisfying (5.5.4) determine a nonvertical plane E such that $N_M = E \cap \pi^{-1}(M)$. If $X_0 \in M$, and if the function f is weakly X_0-stable, then, by definition (5.3.7), there exists a plane E of the above kind which lies below $[f]$. In other words, there exist a vector Y_0 and a constant w_0 which satisfy (5.5.4) as well as the inequality

$$w_0 \geqslant f^c(Y_0).$$

This necessary condition for weak X_0-stability can be reformulated in view of (5.5.5): If f is weakly X_0-stable, then $\hat{f}^c(V) < w'$ implies the existence of Y_0, w_0 such that $V = A^T Y_0$, $w' = w_0 + Y_0^T B$, and $w_0 \geqslant f^c(Y_0)$. Hence

$$\hat{f}^c(V) \geqslant \inf\{f^c(Y) + Y^T B \mid V = A^T Y\}.$$

This relation is trivial for $V \notin K(\hat{f}^c)$. The function

$$h(V) := \inf\{f^c(Y) + Y^T B \mid V = A^T Y\}$$

has been examined in section 4.8. By (4.8.11), $h^c(U) = f^{cc}(AU - B)$. Since $f^{cc} = \underline{f}$, we have

$$f^{cc}(AU - B) \leqslant \varliminf_{W \to U} f(AW - B) = \varliminf_{W \to U} \hat{f}(W) = \hat{f}^{cc}(U),$$

whence $h^c \leqslant \hat{f}^{cc}$ and $\hat{f}^c \leqslant h^{cc} \leqslant h$. Consequently

(5.5.6) $$\hat{f}^c(V) = \inf\{f^c(Y) + Y^T B \mid V = A^T Y\},$$

or equivalently,

$$h = h^{cc},$$

is necessary for f to be weakly X_0-stable.

The sufficiency of (5.5.6) is straightforward. Assume that f is not even weakly X_0-stable. Then there exists a manifold N_M covering M and lying properly below $[f]$ so that none of the nonvertical planes E with $N_M = E \cap \pi^{-1}(M)$ lies below $[f]$. In other words, there exist

V and w' with $\hat{f}^c(V) < w'$ such that for all Y, w with $V = A^T Y$ and $w' = w + Y^T B$,

$$w < f^c(Y),$$

and therefore

$$\hat{f}^c(V) < w' < f^c(Y) + Y^T B.$$

This implies

$$\hat{f}^c(V) < w' \leqslant \inf\{f^c(Y) + Y^T B \mid V = A^T Y\},$$

contradicting (5.5.6). Thus we have proved the

(5.5.7) **Lemma.** *A convex function f on R^n is weakly X_0-stable for $X_0 \in K(f)$, if and only if for every $n \times m$-matrix A and for every vector $B \in R^n$ with $X_0 = AU_0 - B$ for some $U_0 \in R^m$, the relation*

$$\hat{f}^c(V) = \inf\{f^c(Y) + Y^T B \mid V = A^T Y\}$$

holds for all $V \in R^m$. Here $\hat{f}(U) := f(AU - B)$.

An analogous algebraic characterization exists for proper X_0-stability. Here "min" replaces "inf":

(5.5.8) **Lemma.** *Let $X_0 \in K(f)$. A convex function f on R^n is X_0-stable if and only if for every $n \times m$-matrix A and for every vector $B \in R^n$ with $X_0 = AU_0 - B$ for some $U_0 \in R^m$ the relation*

$$\hat{f}^c(V) = \min\{f^c(Y) + Y^T B \mid V = A^T Y\}$$

holds for all $V \in R^m$. Here $\hat{f}(U) := f(AU - B)$.

Proof. Suppose f is X_0-stable. Then

$$\hat{f}^c(V) = h(V) = \inf\{f^c(Y) + Y^T B \mid V = A^T Y\}$$

by lemma (5.5.7) and (5.3.9). We proceed to show that the infimum is assumed for all $V \in K(f^c)$. For any such V, the manifold

$$N_M = \left\{ \begin{pmatrix} AU - B \\ z \end{pmatrix} \;\middle|\; V^T U - z = \hat{f}^c(V) \right\}$$

covering M lies below $[f]$ by (5.5.5). Hence there exists by X_0-stability a nonvertical plane $E = \left\{ \begin{pmatrix} X \\ z \end{pmatrix} \;\middle|\; Y_0^T X - z = w_0 \right\}$ which contains N_M and lies below $[f]$. In other words, $V = A^T Y_0$, $\hat{f}^c(V) = w_0 + Y_0^T B$, and $w_0 \geqslant f^c(Y_0)$ for some Y_0, w_0. Then $\hat{f}^c(V) \geqslant f^c(Y_0) + Y_0^T B$, whence "inf" = "min".

The sufficiency of $\hat{f}^c(V) = \min\{f^c(Y) + Y^T B \mid V = A^T Y\}$ for X_0-stability can be shown in almost precisely the same fashion as the analogous statement in lemma (5.5.7). The remainder of the proof is therefore left to the reader. ▯

As an application of the algebraic characterization (5.5.8) of X_0-stability, we derive the following theorem, which states that polyhedral convex functions (4.8.1) are well-behaved functions from the point of view of duality theory:

(5.5.9) **Theorem.** *Every properly convex polyhedral function f is stable.*

Proof. With the intention of applying lemma (5.5.8), we examine the functions

$$\hat{f}(U) := f(AU - B),$$
$$h(V) := \inf\{f^c(Y) + Y^T B \mid V = A^T Y\}.$$

The first observation is that $\hat{f}$ is a properly convex function, i.e. $\hat{f} \not\equiv +\infty$ and $\hat{f}(U) > -\infty$ for all $U \in R^m$, provided there is a U with $AU - B \in K(f)$. Hence by (4.8.11),

$$h(V) \geqslant h^{cc}(V) = \hat{f}^c(V) > -\infty.$$

The second observation is that $t = h(V)$ is the solution of the following program:

$$\textit{Find } \inf(t) \textit{ for all } Y \textit{ and } t \textit{ with}$$
$$t \geqslant f^c(Y) + Y^T B,$$
$$V = A^T Y.$$

Since $f^c(Y)$ is again a polyhedral function by (4.8.4), the above program is actually a linear program, and theorem (2.5.9) establishes

$$h(V) = \min\{f^c(Y) + Y^T B \mid V = A^T Y\}$$

since $h(V) > -\infty$ for all $V \in R^m$, and in view of the convention $\min \emptyset = +\infty$.

The set

$$P := \left\{ \begin{pmatrix} V \\ t \end{pmatrix} \;\middle|\; t \geqslant f^c(Y) + Y^T B \text{ for some } Y \text{ with } V = A^T Y \right\}$$

is a polyhedron since f^c is a polyhedral function. But

$$\begin{pmatrix} V \\ t \end{pmatrix} \in P \Leftrightarrow t \geqslant \min\{f^c(Y) + Y^T B \mid V = A^T Y\} = h(V) \Leftrightarrow \begin{pmatrix} V \\ t \end{pmatrix} \in [h],$$

whence $[h] = P$. Since $h(V) > -\infty$ for all $V \in R^m$ and $h \not\equiv +\infty$, this implies that h is a polyhedral function. Consequently, $h = h^{cc}$. The equation $\hat{f}^c = h$ then follows from (4.8.11). Thus lemma (5.5.8) applies, and the theorem is proved. ▯

5.6. Generation of Stable Functions

The following lemmas deal with the question of how to generate (weakly) X_0-stable functions from functions which are known to be (weakly) X_0-stable.

(5.6.1) **Lemma.** *If the convex function f on R^n is (weakly) X_0-stable, then so is the convex function*

$$h(X) := f(X) + Y_0^T X.$$

Moreover, if A is an $n \times m$-matrix, $B \in R^n$, and $U_0 \in R^m$ is a vector such that $X_0 = A U_0 - B$, then the function

$$\hat{f}(U) := f(AU - B)$$

is (weakly) U_0-stable.

Proof. We first assume that f is X_0-stable. In order to show that $h(X) := f(X) + Y_0^T X$ is X_0-stable, it is sufficient to remark that E is a nonvertical plane lying below $[f]$ and passing through a nonvertical linear manifold $\bar{E}$ which lies below $[f]$ and satisfies $X_0 \in \pi(\bar{E})$, if and only if the nonvertical plane

$$E' := \left\{ \begin{pmatrix} X \\ (z + Y_0^T X) \end{pmatrix} \;\middle|\; \begin{pmatrix} X \\ z \end{pmatrix} \in E \right\}$$

lies below $[h]$ and contains the linear manifold

$$\bar{E}' := \left\{ \begin{pmatrix} X \\ (z + Y_0^T X) \end{pmatrix} \;\middle|\; \begin{pmatrix} X \\ z \end{pmatrix} \in \bar{E} \right\}$$

lying below $[h]$ and satisfying $X_0 \in \pi(\bar{E}')$.

To prove that the function $\hat{f}(U) := f(AU - B)$ is U_0-stable, if f is X_0-stable, we use the algebraical characterization (5.5.8) of U_0-stability. We must show that for all $m \times p$-matrices C and all vectors $D \in R^m$ which satisfy

$$C W_0 - D = U_0$$

for some $W_0 \in R^p$, the relation

$$\hat{\hat{f}}^c(Q) = \min \{ \hat{f}^c(V) + V^T D \mid Q = C^T V \}$$

holds for the function $\hat{\hat{f}}(W) := \hat{f}(CW - D)$. But because of

$$\hat{\hat{f}}(W) = f(ACW - AD - B)$$

and

$$ACW_0 - AD - B = AU_0 - B = X_0,$$

the X_0-stability of f yields

$$\hat{\hat{f}}^c(Q) = \min\{f^c(Y) + Y^T(AD+B) \mid Q = C^T A^T Y\}.$$

Setting $V := A^T Y$, we may continue

$$\begin{aligned}\hat{\hat{f}}^c(Q) &= \min_{V:Q=C^T V}\left(V^T D + \min_{Y:V=A^T Y}(f^c(Y)+Y^T B)\right)\\ &= \min\{V^T D + \hat{f}^c(V) \mid Q = C^T V\},\end{aligned}$$

again by the X_0-stability of f. Therefore, $\hat{f}$ is U_0-stable.

The case of a weakly X_0-stable function f is treated similarly. This completes the proof of the lemma. ▯

Now, we easily obtain a sharpened version of theorem (5.1.8):

(5.6.2) **Theorem** *Let f_1 and f_2 be convex functions in R^n. If f_1 is X_1-stable and f_2 is X_2-stable for some points X_1, $X_2 \in K(f_1) \cap K(f_2)$, then the conjugate function of $h(X) := f_1(X) + f_2(X)$ is*

$$h^c(Y_0) = \min_Y (f_1^c(Y_0 - Y) + f_2^c(Y)).$$

If f_1 is weakly X_1-stable and f_2 is weakly X_2-stable for some X_1, $X_2 \in K(f_1) \cap K(f_2)$, then the conjugate function of $h(X) := f_1(X) + f_2(X)$ is the infimum convolution

$$h^c(Y_0) = \inf_Y (f_1^c(Y_0 - Y) + f_2^c(Y)) = f_1^c \,\square\, f_2^c.$$

The *proof* is a mere paraphrase of the proof of the corresponding theorem (5.1.8). We need only observe that by lemma (5.6.1) $f_1(X) - Y_0^T X$ is also (weakly) X_1-stable, if so is f_1. Then an application of the sharper duality theorem (5.3.12) instead of Fenchel's theorem (5.1.4) completes the proof. ▯

The preceding theorem can be used to show that the class of (weakly) X_0-stable functions is closed under addition:

(5.6.3) **Theorem.** *If the convex functions f_1, f_2 on R^n are both (weakly) X_0-stable, then so is the convex function $f(X) := f_1(X) + f_2(X)$.*

Proof. We show that for every $n \times m$-matrix A and every $B \in R^n$ with $X_0 = AU_0 - B$ for some $U_0 \in R^m$, the equation

$$\hat{f}^c(V) = \min\{f^c(Y) + Y^T B \mid V = A^T Y\}$$

holds for all $V \in R^m$, where $f := f_1 + f_2$, $\hat{f}(U) := f(AU - B)$, if f_1 and f_2 are X_0-stable. By setting

$$\hat{f}_i(U) := f_i(AU - B) \quad \text{for} \quad i = 1, 2,$$

we obtain $\hat{f}=\hat{f}_1+\hat{f}_2$. Now, f_1, f_2 are X_0-stable. Hence by (5.5.8)

$$(5.6.4)\qquad \hat{f}_i^c(V)=\min\{f_i^c(Y)+Y^T B \mid V=A^T Y\},\qquad i=1,2.$$

For the same reason, theorem (5.6.2) yields

$$f^c(Y)=\min_{\bar{Y}}(f_1^c(Y-\bar{Y})+f_2^c(\bar{Y})).$$

Further, by lemma (5.6.1), $\hat{f}_1$ and $\hat{f}_2$ are U_0-stable. Hence, again by theorem (5.6.2) and by (5.6.4)

$$\begin{aligned}
\hat{f}^c(V) &= \min_W(\hat{f}_1^c(V-W)+\hat{f}_2^c(W))\\
&= \min_W\Big(\min_{Y:V-W=A^TY}(f_1^c(Y)+Y^T B)+\min_{\bar{Y}:W=A^T\bar{Y}}(f_2^c(\bar{Y})+\bar{Y}^T B)\Big)\\
&= \min_{\bar{Y}}\Big(\min_{Y:V=A^T(Y+\bar{Y})}(f_1^c(Y)+f_2^c(\bar{Y})+(Y+\bar{Y})^T B)\Big)\\
&= \min_{\bar{Y}}\Big(\min_{Y:V=A^TY}(f_1^c(Y-\bar{Y})+f_2^c(\bar{Y})+Y^T B)\Big)\\
&= \min_{Y:V=A^TY}\Big(Y^T B+\min_{\bar{Y}}(f_1^c(Y-\bar{Y})+f_2^c(\bar{Y}))\Big)\\
&= \min_{Y:V=A^TY}(Y^T B+f^c(Y))
\end{aligned}$$

which was to be shown. Obviously, this proof carries over to the weakly X_0-stable case. □

5.7. Rockafellar's Duality Theorem

In the following sections, we shall derive most duality theorems in linear and nonlinear programming from the general duality theorem (5.3.12). To this end, it will be advantageous to recast theorem (5.3.12) in a form described by Rockafellar [*1*].

Let f be a convex function in R^n, g a concave function in R^m, A a $m\times n$-matrix, $B\in R^m$, $C\in R^n$. Consider the following dual programs

(5.7.1) *program I:* *Find* $\inf_{X\in R^n}(f(X)-g(AX-B)+C^T X)$,

(5.7.2) *program II:* *Find* $\sup_{Y\in R^m}(g^c(Y)-f^c(A^T Y-C)+B^T Y)$.

Every point $X_0\in K(f)\cap\{X\mid AX-B\in K(g)\}$ is called a feasible solution of program I (5.7.1). Moreover, X_0 is an optimal solution if X_0 is feasible, and

$$\min_X(f(X)-g(AX-B)+C^T X)=f(X_0)-g(AX_0-B)+C^T X_0.$$

Feasible and optimal solutions of program II (5.7.2) are defined similarly.

Parallel to lemma (5.1.3), we have

(5.7.3) **Lemma.** *If X and Y are feasible solutions of programs* (5.7.1) *and* (5.7.2), *respectively, then the inequality*

$$+\infty > f(X) - g(AX - B) + C^T X \geqslant g^c(Y) - f^c(A^T Y - C) + B^T Y > -\infty$$

holds. Moreover, equality

$$f(X_0) - g(AX_0 - B) + C^T X_0 = g^c(Y_0) - f^c(A^T Y_0 - C) + B^T Y_0$$

holds for a pair of feasible solutions X_0 and Y_0 of programs (5.7.1) *and* (5.7.2), *respectively, if and only if*

$$f^c(A^T Y_0 - C) = Y_0^T A X_0 - f(X_0) - C^T X_0$$

and

$$g^c(Y_0) = Y_0^T A X_0 - g(AX_0 - B) - Y_0^T B.$$

Proof. The definitions of f^c and g^c immediately lead to

$$f^c(A^T Y - C) \geqslant Y^T A X - f(X) - C^T X,$$

$$g^c(Y) \leqslant Y^T A X - g(AX - B) - Y^T B$$

for all X and for all Y, which yields the lemma. □

We call program I (5.7.1)

(5.7.4) *(weakly) stably consistent,*

if there are feasible points X_f and X_g of program I such that the function f is (weakly) X_f-stable and g is (weakly) U_g-stable, where $U_g := AX_g - B$. (Weakly) Stable consistency is similarly defined for program (5.7.2).

This notion of stable consistency is slightly more general than Rockafellar's [*1*]: Let $f = f_0 + f_1$ be the sum of a stable convex function f_0 and of another convex function f_1 (and likewise $g = g_0 + g_1$); then, Rockafellar calls program I stably consistent, if there is a point $X_0 \in K(f_0) \cap K(f_1)^I$ such that $AX_0 - B \in K(g_0) \cap K(g_1)^I$. By lemma (5.6.1) and theorem (5.6.3), program I is stably consistent in our sense, if it is in Rockafellar's sense. The following duality theorem (5.7.5) is therefore slightly more general than in Rockafellar's formulation.

(5.7.5) **Duality Theorem of Rockafellar.** *Assume that f is a convex function of R^n and g a concave function on R^m. Then the following statements hold:*

(i) *If program I* (5.7.1) *is weakly stably consistent and if*

$$\inf_X (f(X) - g(AX - B) + C^T X)$$

is finite, then

$$\inf_X (f(X) - g(A X - B) + C^T X) = \sup_Y (g^c(Y) - f^c(A^T Y - C) + B^T Y).$$

Moreover, if program I is stably consistent, then program II (5.7.2) *has an optimal solution* Y_0.

(ii) *Let* f *and* g *be closed functions. If program II* (5.7.2) *is weakly stably consistent and if*

$$\sup_Y (g^c(Y) - f^c(A^T Y - C) + B^T Y)$$

is finite, then

$$\sup_Y (g^c(Y) - f^c(A^T Y - C) + B^T Y) = \inf_X (f(X) - g(A X - B) + C^T X).$$

If, additionally, program II is stably consistent, then program I has an optimal solution X_0.

Proof. We consider only the stably consistent case. The general case is proved similarly. To prove (i), replace in Fenchel's dual programs I and II (5.1.2) $f(X)$ by $f_1(X) := f(X) + C^T X$ and $g(X)$ by $g_1(X) := g(A X - B)$. Then

$$f_1^c(V) = f^c(V - C),$$

and, by the U_g-stability of g

$$g_1^c(V) = \max\{g^c(Y) + Y^T B \mid V = A^T Y\}.$$

Since, by lemma (5.6.1), the function f_1 is X_f-stable and g_1 is X_g-stable, we obtain from duality theorem (5.3.12)

$$\begin{aligned}\inf_X (f(X) - g(A X - B) + C^T X) &= \inf_X (f_1(X) - g_1(X)) = \max_Z (g_1^c(Z) - f_1^c(Z)) \\ &= \max_Z (\max\{g^c(Y) + B^T Y - f^c(Z - C) \mid A^T Y = Z\}) \\ &= \max_Y (g^c(Y) - f^c(A^T Y - C) + B^T Y),\end{aligned}$$

which was to be shown.

(ii) The second part is a consequence of (i), since for closed functions f and g the relations $f = f^{cc}$, $g = g^{cc}$ hold (see theorem (4.7.7)).

The original duality theorem (5.3.12) is clearly contained in the preceding theorem (5.7.5). Therefore, both theorems are equivalent. However, duality theorem (5.3.12) and its corresponding dual programs are more symmetric than their counterparts (5.7.5) and (5.7.1), (5.7.2). On the other hand, the dual programs of this section are more explicit and closer to applications than the original one.

Let, for instance,

$$f(X):=0 \quad \text{for all} \quad X \in R^n,$$

$$g(U):=\begin{cases} 0 & \text{for all} \quad U \geqslant 0,\ U \in R^m, \\ -\infty & \text{otherwise.} \end{cases}$$

Their conjugates are easily calculated:

$$f^c(V)=\sup_X(V^T X - f(X)) = \sup_X V^T X = \begin{cases} 0 & \text{if} \quad V=0, \\ +\infty & \text{if} \quad V \neq 0, \end{cases}$$

$$g^c(Y)=\inf_U(Y^T U - g(U)) = \inf_{U \geqslant 0} Y^T U = \begin{cases} 0 & \text{if} \quad Y \geqslant 0, \\ -\infty & \text{otherwise.} \end{cases}$$

Programs (5.7.1) and (5.7.2) therefore reduce to the dual programs

Minimize $C^T X$ *subject to*

$$A X \geqslant B,$$

Maximize $B^T Y$ *subject to*

$$A^T Y = C, \qquad Y \geqslant 0.$$

Since polyhedral, and in particular linear, functions are stable (theorem (5.5.9)), and so are their conjugates, for they are again polyhedral by (4.8.4), the duality theorem (5.7.5) specializes to the duality theorem of linear programming (1.7.13).

A further direct consequence is a duality theorem by Eisenberg [*1*] for positively homogeneous functions, that is, functions $h\colon R^n \to R \cup \{\pm\infty\}$ which satisfy

$$h(\lambda X) = \lambda h(X) \quad \text{if} \quad \lambda \geqslant 0 \quad \text{and} \quad h(X) \neq \pm\infty.$$

Note that h is closed if it is finite in an R^n-neighborhood of the origin.

Suppose $h(X)$ is a closed positively homogeneous convex function on R^n and $k(Y)$ is a closed positively homogeneous concave function on R^m. Suppose further that A is an $m \times n$-matrix, and define

$$f(X):=h(X), \qquad g(U):=k^c(U).$$

Positively homogeneous functions have been studied in section 4.8, and by (4.8.6) we have:

$$g(U)=\begin{cases} 0 & \text{for} \quad U \in K(k^c), \\ -\infty & \text{otherwise,} \end{cases}$$

$$f^c(Z)=\begin{cases} 0 & \text{for} \quad Z \in K(h^c), \\ +\infty & \text{otherwise.} \end{cases}$$

The closedness of k implies $g^c = k$. Therefore, program I (5.7.1) becomes with $B=0$ and $C=0$:

(5.7.6) *program I: Find* $\inf h(X)$ *subject to*

$$A X \in K(g) = K(k^c),$$

$$\text{i.e., } X \in \{X \mid Y^T A X \geqslant k(Y) \quad \text{for all} \quad Y \in K(k)\}.$$

Similarly, we derive from (5.7.2) the dual program:

(5.7.7) *program II: Find* $\sup k(Y)$ *subject to*

$$A^T Y \in K(f^c) = K(h^c),$$

$$\text{i.e., } Y \in \{Y \mid Y^T A X \leqslant h(X) \quad \text{for all} \quad X \in K(h)\}.$$

With respect to these dual programs (but only in the special case $K(h) = \{X \geqslant 0\}$, $K(k) = \{Y \geqslant 0\}$), Eisenberg [*1*] proved his duality theorem, which is an immediate consequence of (5.7.5):

(5.7.8) **Duality Theorem of Eisenberg.** *Suppose that* $h(X)$ *is a closed positively homogeneous convex function in* R^n, $k(Y)$ *is a closed positively homogeneous concave function in* R^m *and* A *is an* $m \times n$*-matrix. Then the following statements are valid:*

(i) $h(X) \geqslant k(Y)$ *holds for all feasible solutions* X *and* Y *of programs* (5.7.6) *and* (5.7.7), *respectively.*

(ii) *If there is an* $X \in K(h)^I$ *such that* $A X \in K(k^c)^I$, *and if*

$$\inf\{h(X) \mid A X \in K(k^c)\}$$

is finite, then there exists an optimal solution Y_0 *of program* (5.7.7) *such that*

$$\inf\{h(X) \mid A X \in K(k^c)\} = k(Y_0) = \max\{k(Y) \mid A^T Y \in K(h^c)\}.$$

(iii) *If there is a* $Y \in K(k)^I$ *such that* $A^T Y \in K(h^c)^I$, *and if*

$$\sup\{k(Y) \mid A^T Y \in K(h^c)\}$$

is finite, then program (5.7.6) *has an optimal solution* X_0 *satisfying*

$$\min\{h(X) \mid A X \in K(k^c)\} = h(X_0) = \sup\{k(Y) \mid A^T Y \in K(h^c)\}.$$

5.8. Duality Theorems of the Dennis-Dorn Type

Dennis [*2*] and Dorn [*2*] consider minimizing a convex objective function subject to linear constraints. Dorn required the functions involved to be differentiable, whereas Dennis does not. Let us first derive a generalization of Dennis' result from the general duality theorem (5.7.5).

For this purpose, let $f(X)$ be a convex function in R^n, A an $m\times n$-matrix and B a vector of R^m. Further, let $C=0$ in (5.7.1) and (5.7.2), and g as follows:

(5.8.1) $$g(U):=\begin{cases} 0 & \text{for} \quad U\geqslant 0,\ U\in R^m,\\ -\infty & \text{otherwise.}\end{cases}$$

Therefore,

$$g^c(Y)=\inf_U(Y^T U-g(U))=\inf_{U\geqslant 0} Y^T U=\begin{cases} 0 & \text{if} \quad Y\geqslant 0,\\ -\infty & \text{otherwise.}\end{cases}$$

Hence programs (5.7.1) and (5.7.2) reduce to

(5.8.2) *program I: Minimize $f(X)$ subject to*

$$AX\geqslant B, \qquad X\in K(f),$$

program II: Maximize $B^T Y-f^c(A^T Y)$ subject to

$$Y\geqslant 0, \qquad A^T Y\in K(f^c),$$

or, equivalently,

(5.8.3) *program II: Maximize $B^T Y-f^c(Z)$ subject to*

$$Z=A^T Y, \qquad Y\geqslant 0, \qquad Z\in K(f^c).$$

For these programs, lemma (5.7.3) becomes

(5.8.4) **Lemma.** *Any two feasible solutions X_0 and (Y_0, Z_0) of program I (5.8.2) and II (5.8.3), respectively, satisfy the inequality*

$$f(X_0)\geqslant B^T Y_0-f^c(Z_0).$$

They yield equal extrema

$$\begin{aligned} f(X_0)&=\min\{f(X)\,|\,AX\geqslant B,\ X\in K(f)\}\\ &=B^T Y_0-f^c(Z_0)=\max\{B^T Y-f^c(Z)\,|\,Z=A^T Y\in K(f^c),\ Y\geqslant 0\}\end{aligned}$$

if and only if they satisfy the "complementary slackness conditions"

$$Y_0^T A X_0=B^T Y_0=Z_0^T X_0$$

and the "primal-dual coupling relation"

$$f^c(Z_0)=Z_0^T X_0-f(X_0).$$

The corresponding duality theorem is

(5.8.5) **Duality Theorem.** *Let f be a closed convex function on R^n, A an $m \times n$-matrix, and B a vector in R^m. Let further P and C denote the sets*

$$\mathsf{P} := \{X \mid AX \geqslant B\}, \qquad \mathsf{C} := \{Z \mid Z = A^T Y \text{ for some } Y \geqslant 0\},$$

which are a polyhedron and a cone, respectively.

(i) *If $\mathsf{K}(f)^I \cap \mathsf{P} \neq \emptyset$ and if*

$$\inf\{f(X) \mid X \in \mathsf{K}(f) \cap \mathsf{P}\}$$

is finite, then program II (5.8.3) *has an optimal solution (Y_0, Z_0) satisfying*

$$\begin{aligned}\inf\{f(X) \mid X \in \mathsf{K}(f) \cap \mathsf{P}\} &= \max\{B^T Y - f^c(Z) \mid Z = A^T Y \in \mathsf{K}(f^c),\ Y \geqslant 0\}\\ &= B^T Y_0 - f^c(Z_0).\end{aligned}$$

(ii) *If $\mathsf{C} \cap \mathsf{K}(f^c)^I \neq \emptyset$*

and

$$\sup\{B^T Y - f^c(Z) \mid Z = A^T Y \in \mathsf{K}(f^c),\ Y \geqslant 0\}$$

is finite, then program I (5.8.2) *has an optimal solution X_0 satisfying*

$$\begin{aligned}\min\{f(X) \mid X \in \mathsf{K}(f) \cap \mathsf{P}\} &= f(X_0)\\ &= \sup\{B^T Y - f^c(Z) \mid Z = A^T Y \in \mathsf{K}(f^c),\ Y \geqslant 0\}.\end{aligned}$$

Proof. Statement (i) is a consequence of theorem (5.7.5.(i)) and theorem (5.5.9), since the function g (see (5.8.1)) is polyhedral and $\mathsf{K}(f)^I \cap \mathsf{P} \neq \emptyset$ implies the stable consistency of program I.

For the proof of (ii), note that g^c is a polyhedral concave function, which is stable by theorem (5.5.9). Therefore, by lemma (5.6.1), $\mathsf{K}(f^c)^I \cap \mathsf{C} \neq \emptyset$ yields the stable consistency of program II. Application of theorem (5.7.5.(ii)) completes the proof. □

(5.8.6) **Corollary to Theorem (5.8.5).** *If program II* (5.8.3) *has an optimal solution (Y_0, Z_0), then Y_0 can be obtained as an optimal solution of the linear program:*

Maximize $B^T Y$ subject to

$$Z_0 = A^T Y, \qquad Y \geqslant 0.$$

The duality theorem of Dennis [2] obtains, if in (5.8.5) the convex function f is finitely defined for all $X \in \mathsf{R}^n$, but is linear in $n-l$ coordinates. His requirement that f be strictly convex in the remaining variables can be dropped.

Now, we assume the convex function f to be differentiable. Then, by (4.8.5), the conjugate $f^c(Z)$ of the function f in the programs of Dennis, can be replaced by

$$X^T \operatorname{grad} f(X) - f(X) = f^c(\operatorname{grad} f(X))$$

for almost all $Z \in K(f^c)$, provided the regularity condition (4.5.1) is fulfilled. This substitution results in duality theorems, in particular the one by Dorn [2], which are closer to practical applications. The duality features, however, of the dual programs become less apparent. Thus, as has been mentioned in example (5) of section 4.8, the function $X^T \operatorname{grad} f(X) - f(X)$ need not be a convex function of X. The symmetry of the primal and its dual problem was already lost in the formulation by Dennis.

Replacing in programs (5.8.2) and (5.8.3) the vector Z by $\operatorname{grad} f(X)$ and $f^c(Z)$ by $X^T \operatorname{grad} f(X) - f(X)$, we arrive at the following pair of dual programs which have been examined by Dorn [2] under more stringent side conditions:

(5.8.7) *program I: Minimize* $f(X)$ *subject to*

$$AX \geqslant B, \quad X \in K(f).$$

(5.8.8) *program II: Maximize* $B^T Y - X^T \operatorname{grad} f(X) + f(X)$ *subject to*

$$A^T Y = \operatorname{grad} f(X), \quad Y \geqslant 0, \quad X \in K(f)^I.$$

Note that programs (5.8.3) and (5.8.8) are not entirely equivalent as the set $K(f^c)$ is not necessarily equal to the set

$$G := \{\operatorname{grad} f(X) \mid X \in K(f)^I\}.$$

Note furthermore that both programs I and II use the variable X. This departure from the usual terminology, which employs different variable names for the variables of a pair of dual programs respectively, is not incidental. It will be justified by statement (i) of the duality theorem (5.8.13), which expresses an aspect of duality theory not encountered previously, namely that there is not only a relationship between the optimal values of dual programs, but that there are optimal solutions of two dual programs respectively which to some extent coincide. More precisely, if X_0 is an optimal solution of program I, then the same X_0 occurs in some optimal solution of program II.

Just as for programs (5.8.2) and (5.8.3), we are interested in feasible solutions X_0 and (X_1, Y_1) of programs (5.8.7) and (5.8.8), respectively, which satisfy

(5.8.9) $$f(X_0) = B^T Y_1 - X_1^T \operatorname{grad} f(X_1) + f(X_1),$$

the "complementary slackness condition"

(5.8.10)
$$Y_1^T A X_0 = B^T Y_1,$$
$$Y_1^T A X_0 = X_0^T \operatorname{grad} f(X_1),$$

and the "primal-dual coupling relation"

(5.8.11) $$(X_0 - X_1)^T \operatorname{grad} f(X_1) = f(X_0) - f(X_1).$$

Again we have the

(5.8.12) **Lemma.** *Consider the dual programs I (5.8.7) and II (5.8.8) and assume that the function f is convex and differentiable for all* $X \in K(f)^I$. *Let further* X_0 *and* (X_1, Y_1) *be feasible solutions of programs I and II, respectively. Then:*

(i) *The inequality*

$$f(X_0) \geqslant B^T Y_1 - X_1^T \operatorname{grad} f(X_1) + f(X_1)$$

holds for all such X_0 *and* (X_1, Y_1).

(ii) *If* X_0 *and* (X_1, Y_1) *satisfy* (5.8.9), *then they are optimal solutions of the corresponding programs.*

(iii) *The feasible solutions* X_0 *and* (X_1, Y_1) *satisfy* (5.8.9) *if and only if the conditions* (5.8.10) *and* (5.8.11) *are fulfilled.*

(iv) *Moreover, if f fulfills the regularity condition* (4.5.1), *then for any pair of feasible solutions* X_0 *and* (X_1, Y_1) *satisfying* (5.8.9), *the relation*

$$\operatorname{grad} f(X_1) = \operatorname{grad} f(X_0)$$

holds and (X_0, Y_1) *is an optimal solution of program II, also.*

The *proof* is essentially the same as for the corresponding lemma (5.7.3): For any pair of feasible solutions X_0 and (X_1, Y_1) of either problem we have

$$X_0^T \operatorname{grad} f(X_1) = Y_1^T A X_0 \geqslant B^T Y_1.$$

Hence

$$f(X_0) - B^T Y_1 + X_1^T \operatorname{grad} f(X_1) - f(X_1) \geqslant f(X_0) - f(X_1)$$
$$- (X_0 - X_1)^T \operatorname{grad} f(X_1).$$

By (4.4.5), $\operatorname{grad} f(X_1)$ determines a supporting plane. Thus

$$(X_0 - X_1)^T \operatorname{grad} f(X_1) \leqslant f(X_0) - f(X_1) \quad \text{for all} \quad X_0 \in K(f)$$

which proves (i), (ii) and (iii).

In order to prove (iv), we remark that for any pair of feasible solutions X_0 and (X_1, Y_1) satisfying (5.8.9) we have

$$(X_0 - X_1)^T \operatorname{grad} f(X_1) = f(X_0) - f(X_1).$$

Therefore the boundary point $\begin{pmatrix} X_0 \\ f(X_0) \end{pmatrix}$ of $[f]$ lies on the nonvertical supporting plane

$$E := \left\{ \begin{pmatrix} X \\ z \end{pmatrix} \middle| \; X^T \operatorname{grad} f(X_1) - z = X_1^T \operatorname{grad} f(X_1) - f(X_1) \right\}.$$

Now f fulfills the regularity condition (4.5.1). Hence by theorem (4.5.2), we conclude that $X_0 \in K(f)^I$ and $\operatorname{grad} f(X_1) = \operatorname{grad} f(X_0)$. This together with (5.8.9), implies that the pair (X_0, Y_1) is also an optimal solution of program II with

$$f(X_0) = B^T Y_1 - X_0^T \operatorname{grad} f(X_0) + f(X_0). \quad \square$$

As observed before, the sets $G = \{\operatorname{grad} f(X) | X \in K(f)^I\}$ and $K(f^c)$ need not coincide. However, if f is closed convex and satisfies the regularity condition (4.5.1), then by lemma (4.8.5) we have

$$K(f^c) \supseteq G \supseteq K(f^c)^I.$$

In other words, G and $K(f^c)$ are identical except for boundary points. The following duality theorem for programs (5.8.7) and (5.8.8) uses this fact.

(5.8.13) **Duality Theorem.** *Let f be a convex function which is differentiable for all $X \in K(f)^I$. With the notation*

$$P := \{X | AX \geqslant B\},$$

$$C := \{Z | Z = A^T Y \text{ for some } Y \geqslant 0\},$$

$$G := \{\operatorname{grad} f(X) | X \in K(f)^I\},$$

we have:

(i) *If $K(f)^I \cap P \neq \emptyset$ and there is an optimal solution X_0 of program I* (5.8.7), *and if either f satisfies the regularity condition* (4.5.1) *or $X_0 \in K(f)^I$, then the dual program II* (5.8.8) *has an optimal solution $(X_1 = X_0, Y_1)$ which satisfies* (5.8.9).

(ii) *If f is closed and satisfies the regularity condition, $C \cap G^I$ is nonvoid, and the dual program II* (5.8.8) *has an optimal solution (X_1, Y_1), then the primal program I has an optimal solution X_0 satisfying* (5.8.9).

Proof. We modify program II (5.8.8), so as to make it dual to program I (5.8.7) in the sense of Dennis (see (5.8.3)):

(5.8.14) *program II': Maximize $B^T Y - f^c(Z)$ subject to*

$$A^T Y = Z, \qquad Y \geqslant 0, \qquad Z \in K(f^c).$$

(i) By theorem (5.8.5) there exists the optimal solution (Y_1, Z_1) of the modified program (5.8.14), satisfying (see lemma (5.8.4)):

$$f^c(Z_1) = Z_1^T X_0 - f(X_0) \geqslant Z_1^T X - f(X) \quad \text{for all} \quad X \in K(f)$$

and

$$f(X_0) = B^T Y_1 - f^c(Z_1).$$

Then either the regularity condition (4.5.1) or $X_0 \in K(f)^I$, together with the differentiability of f at X_0, implies $Z_1 = \operatorname{grad} f(X_0)$ and $f^c(Z_1) = X_0^T \operatorname{grad} f(X_0) - f(X_0)$. Hence $(X_1 := X_0, Y_1)$ forms also a feasible solution of the original program II (5.8.8) and satisfies (5.8.9).

(ii) By hypothesis, program II (5.8.8) has an optimal solution (X_1, Y_1). We maintain that (Y_1, Z_1) with $Z_1 := \operatorname{grad} f(X_1)$ is an optimal solution of the modified program II′ (5.8.14). In order to prove this statement, we recall

$$K(f^c) \supseteq G \supseteq K(f^c)^I, \tag{5.8.15}$$

which implies

$$\begin{aligned} &\sup\{B^T Y - f^c(A^T Y) \mid Y \geqslant 0,\ A^T Y \in K(f^c)\} \\ &\geqslant \sup\{B^T Y - f^c(A^T Y) \mid Y \geqslant 0,\ A^T Y \in G\} \\ &= B^T Y_1 - f^c(A^T Y_1) = B^T Y_1 - X_1^T \operatorname{grad} f(X_1) + f(X_1). \end{aligned} \tag{5.8.16}$$

In order to prove equality in (5.8.16), let Y_0 be an arbitrary vector satisfying $Y_0 \geqslant 0$ *and* $A^T Y_0 \in K(f^c)$. If we show that

$$B^T Y_0 - f^c(A^T Y_0) \leqslant B^T Y_1 - f^c(A^T Y_1) \tag{5.8.17}$$

then it follows by (5.8.16) that $(Y_1, Z_1 = \operatorname{grad} f(X_1))$ is optimal for program II′ (5.8.14).

Since $C \cap G^I$ is nonvoid by hypothesis, there exists a $\bar{Y} \geqslant 0$ such that $A^T \bar{Y} \in C \cap G^I$. Consider the ray

$$\bar{Y}_\lambda := (1-\lambda) Y_0 + \lambda \bar{Y}.$$

Since, by (5.8.15), $G^I = K(f^c)^I$, $A^T Y_0 \in K(f^c)$ and because of the convexity of $C \cap K(f^c)^I = C \cap G^I$, we get by lemma (3.2.11) that all points $A^T \bar{Y}_\lambda$ with $0 < \lambda \leqslant 1$ belong to $C \cap K(f^c)^I \subseteq C \cap G$. Then by the optimality of (X_1, Y_1),

$$\overline{\lim_{\lambda \to 0}}\, B^T \bar{Y}_\lambda - f^c(A^T \bar{Y}_\lambda) \leqslant B^T Y_1 - f^c(A^T Y_1). \tag{5.8.18}$$

But on the other hand, $\phi(\lambda) := B^T \bar{Y}_\lambda - f^c(A^T \bar{Y}_\lambda)$ is a concave function of λ on $[0,1]$; hence

$$\overline{\lim_{\lambda\to 0}}\, \phi(\lambda) \geqslant \phi(0) = B^T Y_0 - f^c(A^T Y_0).$$

Together with (5.8.18), this proves the inequality (5.8.17). Therefore, program II′ has the optimal solution (Y_1, Z_1), where $Z_1 = \operatorname{grad} f(X_1)$. Thus, by virtue of the duality theorem (5.8.5), part (ii) is proved. □

The following example illustrates that part (i) of the duality theorem (5.8.13) may not hold if neither f is regular in the sense of (4.5.1) nor does any optimal solution of the primal problem lie in $K(f)^I$. Consider

$$f\begin{pmatrix} x_1 \\ x_2 \end{pmatrix} := \begin{cases} x_1 - x_2 & \text{if} \quad x_1 \geqslant x_2, \\ +\infty & \text{otherwise,} \end{cases}$$

where

$$K(f) = \left\{ \begin{pmatrix} x_1 \\ x_2 \end{pmatrix} \middle| \; x_1 \geqslant x_2 \right\}.$$

Clearly, f is convex and differentiable for all $\begin{pmatrix} x_1 \\ x_2 \end{pmatrix} \in K(f)^I$. We further choose

$$A := I = \begin{pmatrix} 1 & 0 \\ 0 & 1 \end{pmatrix}, \qquad B := \begin{pmatrix} 0 \\ 0 \end{pmatrix}.$$

The resulting primal problem reads as follows:

Minimize $x_1 - x_2$ *subject to*

$$x_1 \geqslant 0, \qquad x_2 \geqslant 0, \qquad X = \begin{pmatrix} x_1 \\ x_2 \end{pmatrix} \in K(f).$$

The corresponding dual problem is:

Maximize $0 y_1 + 0 y_2 - (x_1, x_2)\begin{pmatrix} 1 \\ -1 \end{pmatrix} + x_1 - x_2$ *subject to*

$$y_1 = 1, \qquad y_2 = -1, \qquad Y = \begin{pmatrix} y_1 \\ y_2 \end{pmatrix} \geqslant 0, \qquad X = \begin{pmatrix} x_1 \\ x_2 \end{pmatrix} \in K(f)^I.$$

It is seen immediately that the dual problem has no feasible solution at all, whereas $K(f)^I \cap P = \left\{ \begin{pmatrix} x_1 \\ x_2 \end{pmatrix} \middle| \; x_1 > x_2 \geqslant 0 \right\}$ is nonvoid and any $x_1^0 = x_2^0 \geqslant 0$ solves the primal problem optimally.

5.9. Duality Theorems for Quadratic Programs

In this section, we apply the duality theorems of the previous section to quadratic programs. The resulting duality theorems hold in general for ordered fields, and admit algebraic proofs, whereas our derivation holds only for R^n with R being the field of real numbers. Our derivation will, however, show that our formulation of the convex duality theorems of Dennis and Dorn is general enough to include the quadratic duality theorems by the same authors and by Cottle [*1*], at least in the case of real numbers. The original formulations of the convex duality theorems of Dennis [*2*] and Dorn [*2*] are not general enough to include the quadratic duality theorems as special cases.

In order to derive the quadratic duality theorem of Dennis [*1*], let D be a symmetric positive definite (4.8.16) $l \times l$-matrix, C_2 an l-vector, and consider the separable (4.8.13) quadratic function $f: R^k \times R^l \to R$ defined by

$$f(X) := C_1^T X_1 + C_2^T X_2 + \tfrac{1}{2} X_2^T D X_2 = f_1(X_1) + f_2(X_2).$$

This function is everywhere differentiable:

$$\operatorname{grad} f(X) = \begin{pmatrix} C_1 \\ C_2 + D X_2 \end{pmatrix}.$$

It follows therefore from theorem (4.5.3) that the function f fulfills the regularity condition (4.5.1). By (4.8.14) the conjugate of f is again separable:

$$f^c(Z_1, Z_2) = f_1^c(Z_1) + f_2^c(Z_2)$$

where (see example (1) in section 4.8 and formula (4.8.20))

(5.9.1) $$K(f_1^c) = \{C_1\}, \quad f_1^c(C_1) = 0,$$
$$K(f_2^c) = R^l, \quad f_2^c(Z_2) = \tfrac{1}{2}(Z_2 - C_2)^T D^{-1} (Z_2 - C_2).$$

The following pair of programs is then a special case of the dual programs I (5.8.2) and II (5.8.3) of Dennis.

(5.9.2)

program I: *Minimize* $C_1^T X_1 + C_2^T X_2 + \tfrac{1}{2} X_2^T D X_2$ *with* X_1, X_2
subject to $A_1 X_1 + A_2 X_2 \geqslant B$,

program II′: *Maximize* $B^T Y - \tfrac{1}{2}(Z_2 - C_2)^T D^{-1}(Z_2 - C_2)$ *with* Y, Z_1, Z_2,
subject to $A_1^T Y = Z_1$,
$A_2^T Y = Z_2$,
$Y \geqslant 0$,
$Z_1 \in K(f_1^c) = \{C_1\}$,
$Z_2 \in K(f_2^c) = R^l$,

where A_1 is an $m\times k$-matrix, A_2 an $m\times l$-matrix, and $B\in R^m$. The latter program is clearly equivalent to

(5.9.3) *program II:*

$$\begin{aligned} \textit{Maximize}\quad & B^T Y-\tfrac{1}{2}V^T D^{-1}V \text{ with } Y,V \text{ subject to}\\ & A_1^T Y=C_1,\\ & A_2^T Y-V=C_2,\\ & Y\geqslant 0. \end{aligned}$$

Programs I (5.9.2) and II (5.9.3) are Dennis' dual quadratic programs. Since $K(f)^I=K(f)=R^{k+l}$ and $K(f^c)^I=K(f^c)=\{C_1\}\times R^l$ by (5.9.1), theorem (5.8.5) and lemma (5.8.4) yield immediately the

(5.9.4) **Quadratic Duality Theorem of Dennis.** (i) *Any two feasible solutions (X_1,X_2) and (Y,V) of program I* (5.9.2) *and II* (5.9.3), *respectively, satisfy the inequality*

$$C_1^T X_1+C_2^T X_2+\tfrac{1}{2}X_2^T D X_2\geqslant B^T Y-\tfrac{1}{2}V^T D^{-1}V.$$

(ii) *Program I has an optimal solution (X_1,X_2) if and only if program II has an optimal solution (Y,V).*

(iii) *The feasible solutions (X_1,X_2) and (Y,V) of programs I and II, respectively, are optimal solutions of the corresponding programs, if and only if they satisfy*

$$C_1^T X_1+C_2^T X_2+\tfrac{1}{2}X_2^T D X_2= B^T Y-\tfrac{1}{2}V^T D^{-1}V,$$

the complementary slackness conditions

$$Y^T(A_1X_1+A_2X_2)=Y^T B=C_1^T X_1+(V+C_2)^T X_2$$

and the primal-dual coupling relation

$$V=DX_2$$

(which is equivalent to $\tfrac{1}{2}V^T D^{-1}V=V^T X_2-\tfrac{1}{2}X_2^T D X_2$).

Dorn's [1] quadratic dual programs result from the general dual programs (5.8.7) and (5.8.8) by setting

$$f(X):=\tfrac{1}{2}X^T DX+P^T X \quad \text{for} \quad X\in K(f)=R^n,$$

where D is a symmetric positive semidefinite $n\times n$-matrix, and $P\in R^n$. Then again f is everywhere differentiable with $\operatorname{grad} f(X)=DX+P$, and by theorem (4.5.3)

(5.9.5) *f fulfills the regularity condition* (4.5.1).

Furthermore by (4.8.18),

$$K(f^c)=\{DX+P \mid X\in R^n\}=G=\{\operatorname{grad} f(X)\mid X\in K(f)^I=R^n\}.$$

In this case, we obtain the dual programs

(5.9.6) *program I:*

$$\text{Minimize } \tfrac{1}{2}X^T DX + P^T X \quad \text{subject to}$$
$$AX \geqslant B,$$

(5.9.7) *program II:*

$$\text{Maximize } -\tfrac{1}{2}X^T DX + B^T Y \quad \text{with } X, Y \text{ subject to}$$

$$A^T Y - DX = P, \qquad Y \geqslant 0.$$

Here, A denotes an $m \times n$-matrix, and B is a vector of R^m. We are interested in feasible solutions X_0 and (X_1, Y_1) of programs I (5.9.6) and II (5.9.7), respectively, satisfying

$$(5.9.8) \qquad \tfrac{1}{2}X_0^T DX_0 + P^T X_0 = -\tfrac{1}{2}X_1^T DX_1 + B^T Y_1,$$

the complementary slackness condition

$$(5.9.9) \qquad Y_1^T AX_0 = B^T Y_1 = P^T X_0 + X_1^T DX_0$$

and the primal-dual coupling relation

$$(5.9.10) \qquad DX_1 = DX_0.$$

Direct application of lemma (5.8.12) and theorem (5.8.13) yields the

(5.9.11) **Quadratic Duality Theorem of Dorn** [*1*]. *With respect to the dual programs I* (5.9.6) *and II* (5.9.7), *the following statements hold:*

(i) *Any two feasible solutions* X_0 *and* (X_1, Y_1) *of the programs I and II, respectively, satisfy the inequality*

$$\tfrac{1}{2}X_0^T DX_0 + P^T X_0 \geqslant -\tfrac{1}{2}X_1^T DX_1 + B^T Y_1.$$

(ii) *If program I has an optimal solution* X_0, *then there exists a* Y_0, *such that* (X_0, Y_0) *is an optimal solution of program II satisfying* (5.9.8).

(iii) *If the dual program II has an optimal solution* (X_1, Y_1) *then so has the primal program I an optimal solution* X_0 *satisfying* (5.9.8).

(iv) *If the feasible solutions* X_0 *and* (X_1, Y_1) *of the programs I and II, respectively, satisfy* (5.9.8), *then they are optimal. Moreover,* (5.9.8) *is satisfied if and only if* (5.9.9) *and* (5.9.10) *are satisfied.*

The duality theorems for quadratic programs by Dennis and Dorn, described above are asymmetric in that the primal and their respective dual programs are of different structural types. The following pair of dual quadratic programs found by Cottle [*1*] is perfectly symmetric.

Let C and D be symmetric positive semidefinite matrices of orders k and l, respectively. Let further A be an arbitrary $l \times k$-matrix, $B \in R^l$, and $P \in R^k$. Then suppose $(X_1, X_2) \in R^k \times R^l$, $(Y_1, Y_2) \in R^l \times R^k$, and consider the quadratic programs

(5.9.12) *program I:*

$$\textit{Minimize} \quad \tfrac{1}{2} X_1^T C X_1 + \tfrac{1}{2} X_2^T D X_2 + P^T X_1 \quad \textit{subject to}$$
$$A X_1 + D X_2 \geqslant -B, \; X_1 \geqslant 0,$$

(5.9.13) *program II:*

$$\textit{Maximize} \quad -\tfrac{1}{2} Y_1^T D Y_1 - \tfrac{1}{2} Y_2^T C Y_2 - B^T Y_1 \quad \textit{subject to}$$
$$-A^T Y_1 + C Y_2 \geqslant -P, \; Y_1 \geqslant 0.$$

Again we are interested in feasible solutions (X_1, X_2) and (Y_1, Y_2) which satisfy

(5.9.14) $\tfrac{1}{2} X_1^T C X_1 + \tfrac{1}{2} X_2^T D X_2 + P^T X_2 = -\tfrac{1}{2} Y_1^T D Y_1 - \tfrac{1}{2} Y_2^T C Y_2 - B^T Y_1.$

Programs (5.9.12) and (5.9.13) are linked by the

(5.9.15) **Symmetric Duality Theorem** (Cottle [*1*]).

(i) *Any feasible solutions* (X_1, X_2) *and* (Y_1, Y_2) *of programs I* (5.9.12) *and II* (5.9.13), *respectively, satisfy*

$$\tfrac{1}{2} X_1^T C X_1 + \tfrac{1}{2} X_2^T D X_2 + P^T X_1 \geqslant -\tfrac{1}{2} Y_1^T D Y_1 - \tfrac{1}{2} Y_2^T C Y_2 - B^T Y_1.$$

(ii) *If program I has an optimal solution* (X_1, X_2), *then there exists a* Y_1 *such that* (Y_1, Y_2) *with* $Y_2 = X_1$ *is an optimal solution of program II satisfying* (5.9.14).

(iii) *If program II has an optimal solution* (Y_1, Y_2), *then there exists an* X_1 *such that* (X_1, X_2) *with* $X_2 = Y_1$ *is an optimal solution of program I satisfying* (5.9.14).

(iv) *The feasible solutions* (X_1, X_2) *and* (Y_1, Y_2) *of programs I and II, respectively, are optimal solutions if and only if they satisfy* (5.9.14). *In the case of optimality one has moreover*

$$C Y_2 = C X_1, \qquad D Y_1 = D X_2.$$

Proof. Program I (5.9.12) is of the type of Dorn's program (5.8.7). Let

$$f(X_1, X_2) := \tfrac{1}{2} X_1^T C X_1 + \tfrac{1}{2} X_2^T D X_2 + P^T X_1.$$

Then with $Z := (Z_1, Z_2) \in R^k \times R^l$,

$$\begin{aligned} Z^T \operatorname{grad} f(Z) - f(Z) &= Z_1^T (C Z_1 + P) + Z_2^T D Z_2 - \tfrac{1}{2} Z_1^T C Z_1 - \tfrac{1}{2} Z_2^T D Z_2 - P^T Z_1 \\ &= \tfrac{1}{2} Z_1^T C Z_1 + \tfrac{1}{2} Z_2^T D Z_2. \end{aligned}$$

The dual of program I (5.9.12) in the sense of Dorn (5.8.8) thus becomes

$$\begin{aligned}
&\textit{Maximize} \quad -B^T Y_1 - \tfrac{1}{2} Z_1^T C Z_1 - \tfrac{1}{2} Z_2^T D Z_2 \quad \textit{with} \quad Y_1, Y_2, Z_1, Z_2 \quad \textit{subject to} \\
&\qquad A^T Y_1 + Y_2 = C Z_1 + P, \\
&\qquad\qquad D Y_1 = D Z_2, \\
&\qquad (Y_1, Y_2) \geqslant 0, \\
&\qquad (Z_1, Z_2) \in K(f)^I = R^k \times R^l.
\end{aligned}$$

As $D = D^T$, the relation $D Y_1 = D Z_2$ implies

$$Z_2^T D Z_2 = Z_2^T D Y_1 = Y_1^T D Z_2 = Y_1^T D Y_1.$$

Therefore, an equivalent form of the above program is

$$\begin{aligned}
&\textit{Maximize} \quad -\tfrac{1}{2} Y_1^T D Y_1 - \tfrac{1}{2} Z_1^T C Z_1 - B^T Y_1 \quad \textit{subject to} \\
&\qquad\qquad -A^T Y_1 + C Z_1 \geqslant -P, \quad Y_1 \geqslant 0,
\end{aligned}$$

which becomes program II (5.9.13) by renaming Z_1. It is readily seen that the remaining conditions of Dorn's duality theorem (5.8.13) are met.

CHAPTER 6

Saddle Point Theorems

The classical instances of saddle point theorems are the minimax theorems of v. Neumann [*1*] (6.1.1) and Kakutani [*3*]. Since then the saddle point concept has been found to apply in a much larger context. It is also a natural and rich source for duality theorems. In fact, every theorem stating the existence of a "saddle point" can be interpreted as duality theorem for a pair of dual programs.

In this chapter, the generalization by Sion [*1,2*] of Kakutani's theorem will be carried further to include the case of noncompact areas of definitions for "pay-off" functions, and very broad duality theorems will be derived, which permit a weakening of convexity conditions. It is shown that they contain the fundamental saddle point theorem of Kuhn and Tucker [*1*] as a special case.

The saddle point theorem of Kuhn and Tucker is also derived by studying systems of convex inequalities, perhaps the shortest and most self-contained approach (Fan, Glicksberg, and Hoffman [*1*], Bohnenblust, Karlin, and Shapley [*1*]). At the roots of the Kuhn-Tucker theory, however, lies the technique of Lagrange multipliers, and it is attempted to demonstrate this connection to classical calculus.

6.1. The Minimax Theorem of von Neumann

The forerunner of all saddle point theorems is the famous

(6.1.1) **Minimax Theorem of von Neumann.** [*1*]. *For every $m \times n$ matrix A with real coefficients a_{ij} one has*

$$\max_{\substack{X \geqslant 0 \\ \Sigma x_i = 1}} \min_{\substack{Y \geqslant 0 \\ \Sigma y_j = 1}} X A Y = \max_{\substack{X \geqslant 0 \\ \Sigma x_i = 1}} \min_{k} X A_k$$

$$= \min_{\substack{Y \geqslant 0 \\ \Sigma y_j = 1}} \max_{l} {}_l A Y = \min_{\substack{Y \geqslant 0 \\ \Sigma y_j = 1}} \max_{\substack{X \geqslant 0 \\ \Sigma x_i = 1}} X A Y.$$

The theorem is a statement about linear inequalities. In fact it can be regarded as one of the many equivalent formulations of the duality theorem for linear programming (1.7.13). The theorem holds for matrices over any ordered field.

The minimax theorem was motivated by and constitutes the key result of the theory of games. This rich field, initiated by v. Neumann, has close ties to the theory of linear programming. It is, however, too involved to be treated in this book. It may suffice to outline the fundamental concept of

(6.1.2) *matrix games.*

Such a game is defined by some matrix A, and is played by two players, a "row player" and a "column player". The row player selects a row, the column player a column of the matrix A. While making his choice, the row player has no information about the choice of the column player, and vice versa. Suppose the row player selects the rth row, and the column player the sth column of A. The "pay-off" is then given by the value of the element a_{rs}, which is credited to the row player at the expense of the column player. The row player is thus interested in a high value of a_{rs}, whereas the column player prefers low, possibly negative, values.

The question of optimal selections, or "strategies", arises. If the row player is cautious, then he will choose the row with the biggest minimum entry, because he is assured of this pay-off,

$$\underline{c} := \max_i \min_j a_{ij},$$

no matter what the column player decides to do, just as the column player can always avoid loosing more than

$$\bar{c} := \min_j \max_i a_{ij}.$$

One observes

(6.1.3) $$\underline{c} \leqslant \bar{c}.$$

Proof. Let $\bar{c} = a_{rs}$. Since $\min_j a_{ij} \leqslant a_{is}$, one has $\underline{c} = \max_i \min_j a_{ij} \leqslant \max_i a_{is} = a_{rs} = \bar{c}$.

If $\underline{c} < \bar{c}$, then the cautious strategy is unsatisfactory, as it does not induce the other player to play cautiously. Indeed, the caution of the row player is based on the suspicion that the column player might not follow the cautious strategy. If the same game is to be played repeatedly then each player is advised to deviate sometimes from the cautious strategy so as to warn the other player against trying to take advantage of the cautious disposition of his opponent.

Considerations of this nature lead to the notion of a "mixed strategy" in which the row player selects rows ${}_iA$ of A with fixed probabilities $x_i \geqslant 0$, $\sum_{i=1}^{m} x_i = 1$, just as the column player selects columns A_j of A with fixed probabilities $y_j \geqslant 0$, $\sum_{j=1}^{n} y_i = 1$. A

(6.1.4) *mixed strategy*

will then be any row vector $X \geqslant 0$ with $\sum_{i=1}^{m} x_i = 1$ or column vector $Y \geqslant 0$ with $\sum_{j=1}^{n} y_j = 1$. The expected pay-off which results if the row player plays X and the column player plays Y is then given by

$$X\,A\,Y.$$

The question for optimal mixed strategies can now be answered in a meaningful way by virtue of the minimax theorem (6.1.1) of v. Neumann. An optimal strategy $\bar{X}$ is one for which

$$\min_Y \bar{X}\,A\,Y = \max_X \min_Y X\,A\,Y \tag{6.1.5}$$

and, similarly, an optimal strategy $\bar{Y}$ is one for which

$$\max_X X\,A\,\bar{Y} = \min_Y \max_X X\,A\,Y. \tag{6.1.6}$$

The values in (6.1.5) and (6.1.6) coincide according to the minimax theorem, and these optimal strategies, therefore, do not involve the difficulties which arise from a gap between enforceable values. The existence of optimal strategies $\bar{X}$ and $\bar{Y}$ is implicitly stated by the minimax theorem. Moreover

$$\max_X \min_Y X\,A\,Y = \bar{X}\,A\,\bar{Y} = \min_Y \max_X X\,A\,Y$$

holds for any pair of optimal strategies $\bar{X}$, $\bar{Y}$. Indeed

$$\max_X \min_Y X\,A\,Y = \min_Y \bar{X}\,A\,Y \leqslant \bar{X}\,A\,\bar{Y} \leqslant \max_X X\,A\,\bar{Y} = \min_Y \max_X X\,A\,Y,$$

and the minimax theorem yields the equality of the two outermost terms.

Proof of the minimax theorem (6.1.1). First it is easily verified that

$$\inf_Y X\,A\,Y = \min_k X\,A_k, \qquad \sup_X X\,A\,Y = \max_l {}_lA\,Y. \tag{6.1.7}$$

For instance, $\inf_Y XAY \leqslant \min_k XA_k$ is trivial, and the converse is obtained from

$$XAY = \sum_{k=1}^{m} XA_k y_k \geqslant \min_k XA_k,$$

which holds since $y_k \geqslant 0$ and $\sum_k y_k = 1$.

One consequence of (6.1.7) is that

$$\inf_Y XAY = \min_Y XAY, \qquad \sup_X XAY = \max_X XAY.$$

Next we observe that

$$\min_k \hat{X} A_k = \min_Y \hat{X} A Y \leqslant \hat{X} A \hat{Y} \leqslant \max_X XA\hat{Y} = \max_l {}_lA\hat{Y} \tag{6.1.8}$$

for arbitrary $\hat{X} \geqslant 0$ and $\hat{Y} \geqslant 0$ with $\Sigma \hat{x}_i = 1$ and $\Sigma \hat{y}_j = 1$, or in other words, $\sup_X \min_k XA_k \leqslant \inf_Y \max_l {}_lAY$. If equality holds, then the supremum is in fact a maximum, and the infimum a minimum, as stipulated by the theorem.

The existence of a number c and vectors X, Y such that

$$\min_k XA_k = c = \max_l {}_lAY,$$

—the minimax theorem reduces to this statement,—is equivalent to the solvability of the following system of inequalities

$$\begin{aligned}
A^T X^T \qquad & - E_{\text{row}}^T \cdot c \geqslant \ 0,\\
- AY + & E_{\text{col}} \cdot c \geqslant \ 0,\\
X^T \qquad & \qquad\quad \geqslant \ 0,\\
Y & \qquad\quad \geqslant \ 0,\\
-E_{\text{col}}^T X^T \qquad & \qquad\quad = -1,\\
E_{\text{row}} Y & \qquad\quad = \ 1,
\end{aligned} \tag{6.1.9}$$

where E_{row} and E_{col} are row and column vectors, respectively, all entries of which are ones:

$$E_{\text{row}} = \underbrace{(1, \ldots, 1)}_{n \text{ entries}}, \qquad E_{\text{col}} = \left.\begin{pmatrix} 1 \\ \vdots \\ 1 \end{pmatrix}\right\} m \text{ entries}.$$

Assume that the system (6.1.9) were not solvable. Then the theorem (1.1.9) of Kuhn and Fourier yields the existence of vectors U, V, S, T and scalars p, q such that

(6.1.10)
$$\begin{aligned} U^T A^T + S - pE_{\text{col}}^T &= 0, \\ -VA + T + qE_{\text{row}} &= 0, \\ -U^T E_{\text{row}}^T + VE_{\text{col}} &= 0, \\ U, V, S, T &\geqslant 0, \end{aligned}$$

and

$$-p+q>0.$$

We know in addition, that U and V do not vanish. Indeed, $U=0$ would mean that already the following subsystem of (6.1.9),

$$-AY+E_{\text{col}}\cdot c\geqslant 0, \quad X^T,\ Y\geqslant 0, \quad \sum_i x_i=1, \quad \sum_j y_j=1$$

would be unsolvable, which is obviously not the case. Consequently, we may normalize U, V and the other multipliers so that $U^T E_{\text{row}}^T = VE_{\text{col}} = 1$. The condition (6.1.10) are then equivalent to

$$\begin{aligned} &AU \leqslant E_{\text{col}}\cdot p, \\ &VA \geqslant qE_{\text{row}}, \\ &p<q, \\ &U, V\geqslant 0, \quad \Sigma u_j=1, \quad \Sigma v_i=1. \end{aligned}$$

This in turn yields

$$\max_l {}_lAU \leqslant p < q \leqslant \min_k VA_k$$

which contradicts (6.1.8). Hence (6.1.9) must be solvable.

The matrix A is said to have a

(6.1.11) *saddle point* a_{rs}

if $a_{is} \leqslant a_{rs} \leqslant a_{rj}$ for all row indices i and column indices j. In this case,

$$\max_i \min_j a_{ij} = \underline{c} = a_{rs} = \bar{c} = \min_j \max_i a_{ij}.$$

Indeed, $\max_i a_{ij} \geqslant a_{rj} \geqslant a_{rs} = \max_i a_{is}$ for all j, which means $\min_j \max_i a_{ij} = a_{rs}$. The other equation follows analogously.

Therefore $\underline{c}=\bar{c}$ is necessary for a saddle point to exist. That it is also sufficient will be shown in a more general context in the next section (corollary (6.2.11)). Thus saddle points exist precisely if the "cautious" strategies considered at the beginning of this section are optimal strategies. As we have seen, however, the cautious strategies are in general not optimal, and so the existence of saddle points for the matrix A is the exception rather than the rule. In the next section, saddle points will be defined for functions on cartesian products, and the minimax theorem will be seen to establish the existence of saddle points $(\bar{X}, \bar{Y})$ for the "pay-off function" $\Phi(X, Y) := X A Y$.

The minimax theorem of v. Neumann has been generalized in many different directions. One may, for instance, replace the matrix A by a density function defined on a cartesian product $C \times D$, and the mixed strategies X and Y by probability distributions on C and D, respectively (Ville [*1*], Loomis [*1*], Karlin [*1*]). We shall not follow this line of generalization, since it requires measure theory.

6.2. Saddle Points

Chapter 5 has illustrated the central role of Fenchel's duality theorem (5.1.4). In subsequent developments, we shall recognize saddle point theorems as an even more general source of duality in nonlinear programming.

Consider, for instance, Fenchel's dual programs (5.1.2) in the case of a closed function $g = g^{cc}$. Upon replacing $f^c(Y)$ and $g(X)$ by the expressions

$$f^c(Y) = \sup\{Y^T X - f(X) | X \in K(f)\},$$
$$g(X) = \inf\{Y^T X - g^c(Y) | Y \in K(g^c)\},$$

the programs (5.1.2) become

program I: Find $\inf_{X \in K(f)} \sup_{Y \in K(g^c)} (f(X) + g^c(Y) - Y^T X)$,

program II: Find $\sup_{Y \in K(g^c)} \inf_{X \in K(f)} (f(X) + g^c(Y) - Y^T X)$.

The function

$$\Phi(X, Y) := f(X) + g^c(Y) - Y^T X$$

is defined and finite on $K(f) \times K(g^c)$. It is a convex function in X and a convace function in Y.

By lemma (5.1.3), all pairs (X_0, Y_0) of optimal solutions to Fenchel's problems satisfy

$$\inf_{X \in K(f)} \sup_{Y \in K(g^c)} \Phi(X, Y) = \Phi(X_0, Y_0) = \sup_{Y \in K(g^c)} \inf_{X \in K(f)} \Phi(X, Y)$$

and are characterized by the relations

$$\Phi(X_0, Y) \leqslant \Phi(X_0, Y_0) \leqslant \Phi(X, Y_0)$$

for all $X \in K(f)$ and $Y \in K(g^c)$. In other words, each solution pair (X_0, Y_0) is a

(6.2.1) *saddle point*

of Φ and Fenchel's duality theorems (5.1.4) and (5.3.12) can be considered as existence theorems for saddle points of Φ.

More generally, let $\Phi: C \times D \to R$ be a real function defined for all $(X, Y) \in C \times D$. Then Φ is called a

(6.2.2) *(quasi-, pseudo-) convex-concave function*

if $C \subseteq E$, $D \subseteq F$ are convex subsets of real topological vector spaces E and F, respectively, and if for each $Y \in D$, $\Phi(., Y)$ is a (quasi-, pseudo-) convex function on C and if for each $X \in C$, $\Phi(X, .)$ is a (quasi-, pseudo-) concave function on D (see (4.9.1), (4.9.2), (4.9.6)).

Another generalization of convex-concave functions is due to Fan [3]. This definition is remarkable inasmuch as it does not require the regions of definition C and D to carry a linear structure: A function $\Phi: C \times D \to R$ defined on the cartesian product of two arbitrary sets C and D is called a

(6.2.3) *convex-concavelike function,*

if for any $X_1, X_2 \in C$, $\lambda \geqslant 0$, $\mu \geqslant 0$, $\lambda + \mu = 1$ there is an $X_0 \in C$ such that

$$\Phi(X_0, Y) \leqslant \lambda \Phi(X_1, Y) + \mu \Phi(X_2, Y) \quad \text{for all} \quad Y \in D,$$

if for any $Y_1, Y_2 \in D$, $\lambda \geqslant 0$, $\mu \geqslant 0$, $\lambda + \mu = 1$ there is a $Y_0 \in D$ with

$$\Phi(X, Y_0) \geqslant \lambda \Phi(X, Y_1) + \mu \Phi(X, Y_2) \quad \text{for all} \quad X \in C.$$

Clearly, not every convex-concavelike function Φ is quasi-convex-concave. On the other hand, Nikaidô [2] gave an example of a function Φ which is quasi-convex-concave but not convex-concavelike:

$$\Phi(x, y) := \frac{-y^2}{(y-x)^2 + 1} \quad \text{for} \quad -1 \leqslant x, \quad y \leqslant 1.$$

There is no $x \in [-1, 1]$ such that

$$\Phi(x, y) \leqslant \tfrac{1}{2}(\Phi(1, y) + \Phi(-1, y))$$

for all $y \in [-1, 1]$.

If C and D are topological spaces, then a function Φ on $C \times D$ is called

(6.2.4) *semicontinuous*

if $\Phi(., Y)$ is lower semicontinuous (l.s.c.) for all $Y \in D$ and $\Phi(X, .)$ is upper semicontinuous (u.s.c.) for all $X \in C$.

Saddle points (6.2.1) have been defined for arbitrary functions $\Phi: C \times D \to R$. With any such function Φ, we may associate moreover the following two programs, which are dual to each other

(6.2.5) *program I: Find* $\inf_{X \in C} \sup_{Y \in D} \Phi(X, Y)$,

program II: Find $\sup_{Y \in D} \inf_{X \in C} \Phi(X, Y)$.

A point $(X_1, Y_1) \in C \times D$ is regarded as an

(6.2.6) *optimal solution*

of program I if

$$\sup_{Y \in D} \Phi(X, Y) \geqslant \Phi(X_1, Y_1) = \sup_{Y \in D} \Phi(X_1, Y) = \max_{Y \in D} \Phi(X_1, Y) \quad \text{for all} \quad X \in C.$$

In this case

$$\Phi(X_1, Y_1) = \min_{X \in C} \sup_{Y \in D} \Phi(X, Y).$$

Likewise, an optimal solution (X_2, Y_2) of program II is defined by the relations

$$\inf_{X \in C} \Phi(X, Y) \leqslant \Phi(X_2, Y_2) = \inf_{X \in C} \Phi(X, Y_2) = \min_{X \in C} \Phi(X, Y_2) \quad \text{for all} \quad Y \in D,$$

with the consequence

$$\max_{Y \in D} \inf_{X \in C} \Phi(X, Y) = \Phi(X_2, Y_2).$$

In view of the analogy between the function $\Phi: C \times D \to R$ and the pay-off matrix A for games discussed in the previous section, we call Φ a

(6.2.7) *pay-off function*.

Obviously

$$\sup_{Y \in D} \inf_{X \in C} \Phi(X, Y) \leqslant \inf_{X \in C} \sup_{Y \in D} \Phi(X, Y) \tag{6.2.8}$$

since the inequality

$$\Phi(X, Z) \leqslant \sup_{Y \in D} \Phi(X, Y) \quad \text{for} \quad Z \in D$$

implies

$$\inf_{X \in C} \Phi(X, Z) \leqslant \inf_{X \in C} \sup_{Y \in D} \Phi(X, Y) \quad \text{for} \quad Z \in D,$$

and therefore (6.2.8).

Now, if (X_0, Y_0) is a saddle point of Φ, then by definition (6.2.1)

$$\Phi(X_0, Y) \leqslant \Phi(X_0, Y_0) \leqslant \Phi(X, Y_0),$$

and therefore

$$\max_{Y \in D} \Phi(X_0, Y) = \Phi(X_0, Y_0) = \min_{X \in C} \Phi(X, Y_0).$$

Thus every saddle point of Φ yields a solution of both programs I and II (6.2.5), simultaneously. Conversely, if (X_0, Y_0) is an optimal solution of both programs I and II, then it is obviously a saddle point.

Note that an optimal solution of program I need not be a saddle point of Φ:

Consider the game defined by the matrix

$$A := \begin{pmatrix} 1 & 1 \\ 1 & 0 \end{pmatrix}.$$

The elements of the matrix constitute a pay-off function $\Phi: C \times D \to R$ with $C = D = \{1,2\}$ standing for the column and the row indices of A, respectively. The associated program I,

$$\inf_{X \in C} \sup_{Y \in D} \Phi(X, Y),$$

has the optimal solution $(X=1, Y=2)$. Indeed, $a_{21} = \max_i a_{i1}$ and $\max_i a_{i1} \leqslant \max_i a_{ij}$ for all j. However, $(1,2)$ is not a saddle point as $\Phi(1,2) = a_{21} > a_{22} = \Phi(2,2)$.

(6.2.9) **Theorem.** *Let $\Phi(X, Y)$ be an arbitrary pay-off function. Then the following statements hold:*

(i) *If (X_0, Y_0) is a saddle point of $\Phi(X, Y)$, then (X_0, Y_0) is an optimal solution of both programs I and II* (6.2.5).

(ii) *If (X_1, Y_1) is an optimal solution of program I and (X_2, Y_2) is optimal for program II with $\Phi(X_1, Y_1) = \Phi(X_2, Y_2)$, then (X_1, Y_2) is a saddle point of Φ.*

Proof. The first part has been shown above. To prove (ii), let $(X_1\, Y_1)$ be optimal for program I, $(X_2\, Y_2)$ optimal for program II and $\Phi(X_1, Y_1) = \Phi(X_2, Y_2)$. Then

$$\sup_{Y \in D} \Phi(X, Y) \geqslant \Phi(X_1, Y_1) = \sup_{Y \in D} \Phi(X_1, Y) = \Phi(X_2, Y_2)$$
$$= \inf_{X \in C} \Phi(X, Y_2) \geqslant \inf_{X \in C} \Phi(X, Y).$$

This implies at once

$$\Phi(X_1, Y_1) = \sup_{Y \in D} \Phi(X_1, Y) \geqslant \Phi(X_1, Y_2) \geqslant \inf_{X \in C} \Phi(X, Y_2) = \Phi(X_2, Y_2)$$

and therefore

$$\Phi(X_1, Y_1) = \Phi(X_1, Y_2) = \Phi(X_2, Y_2).$$

Together with the inequalities

$$\Phi(X_1, Y_1) \geqslant \Phi(X_1, Y), \quad \Phi(X_2, Y_2) \leqslant \Phi(X, Y_2) \quad \text{for all} \quad X \in C, \quad Y \in D$$

this gives the saddle point relation for (X_1, Y_2):

$$\Phi(X_1, Y) \leqslant \Phi(X_1, Y_2) \leqslant \Phi(X, Y_2) \quad \text{for all} \quad X \in C, \quad Y \in D.$$

As a corollary, we note

(6.2.10) **Corollary to Theorem (6.2.9).** *Let (X_0, Y_0) be a saddle point of $\Phi(X, Y)$. If (X_1, Y_1) is an optimal solution of program I, then (X_1, Y_0) is also a saddle point of $\Phi(X, Y)$. If (X_2, Y_2) is an optimal solution of program II, then (X_0, Y_2) is also a saddle point of $\Phi(X, Y)$.*

(6.2.11) **Corollary to Theorem (6.2.9).** *If $\underline{c} = \bar{c}$ holds for a matrix game, then it has a saddle point* (6.1.11).

6.3. Minimax Theorems for Compact Sets

Existence theorems for saddle points are commonly called "minimax theorems". By theorem (6.2.9), every saddle point (X_0, Y_0) of a pay-off function Φ provides an optimal solution to programs I and II (6.2.5). Thus minimax theorems can be viewed as duality theorems for the corresponding dual programs (6.2.5). In this section, the minimax theorem of Sion [*2*] for quasi-convex-concave functions and a theorem of Fan [*3*] for convex-concavelike functions shall be established. Both theorems generalize the fundamental result of Kakutani [*3*] for continuous quasi-convex-concave functions.

The pay-off functions $\Phi: C \times D \to R$ to be considered are defined on compact topological spaces C and D. In this case a necessary and sufficient criterion for the existence of saddle points is available:

(6.3.1) **Lemma** (Fan [*3*]). *Let the real function $\Phi: C \times D \to R$ be defined on the cartesian product $C \times D$ of two compact (Hausdorff-)spaces C and D and let Φ be semicontinuous* (6.2.4), *i.e. $\Phi(., Y)$ is l.s.c. for all $Y \in D$ and $\Phi(X, .)$ is u.s.c. for all $X \in C$. Then:*

(i) $\max_{Y \in D} \min_{X \in C} \Phi(X, Y) = \min_{X \in C} \max_{Y \in D} \Phi(X, Y)$

holds if and only if

(ii) *for any finite sets $\{X_1, \ldots, X_m\} \subseteq C$, $\{Y_1, \ldots, Y_n\} \subseteq D$ there exist $X_0 \in C$ and $Y_0 \in D$ such that*

(6.3.2) $\Phi(X_0, Y_k) \leqslant \Phi(X_i, Y_0)$ for $k = 1, 2, \ldots, n, \quad i = 1, 2, \ldots, m.$

Proof. Note first that for all $X \in C$, $\max_{Y \in D} \Phi(X, Y)$ is attained since D is compact and $\Phi(X,.)$ is u.s.c.. Moreover, $F(X) := \max_{Y \in D} \Phi(X, Y)$ is l.s.c. on C as it is a maximum of l.s.c. functions, and therefore $\min_{X \in C} F(X) = \min_{X \in C} \max_{Y \in D} \Phi(X, Y)$ is attained, since C is compact. Similarly, $\max_{Y \in D} \min_{X \in C} \Phi(X, Y)$ is meaningful. Also we note that

$$\max_{Y \in D} \min_{X \in C} \Phi(X, Y) \leqslant \min_{X \in C} \max_{Y \in D} \Phi(X, Y), \tag{6.3.3}$$

as a consequence of (6.2.8).

Condition (i) implies (ii), since

$$\begin{aligned}\max_{Y \in D} \min_{X_i} \Phi(X_i, Y) \geqslant \max_{Y \in D} \min_{X \in C} \Phi(X, Y) &= \min_{X \in C} \max_{Y \in D} \Phi(X, Y) \\ &\geqslant \min_{X \in C} \max_{Y_k} \Phi(X, Y_k).\end{aligned}$$

If, on the other hand, (i) does not hold, then by (6.3.3) there is a number c such that

$$\max_{Y \in D} \min_{X \in C} \Phi(X, Y) < c < \min_{X \in C} \max_{Y \in D} \Phi(X, Y).$$

Therefore, the sets $A_X := \{Y \mid \Phi(X, Y) < c\}$ cover D,

$$D \subseteq \bigcup_{X \in C} A_X,$$

and are open, since $\Phi(X,.)$ is u.s.c. Now, D is compact, hence there is a finite set $\{X_1, \ldots, X_m\} \subseteq C$ such that

$$D \subseteq \bigcup_{i=1}^{m} A_{X_i},$$

which implies

$$\max_{Y \in D} \min_{1 \leqslant i \leqslant m} \Phi(X_i, Y) < c. \tag{6.3.4}$$

Analogously, there is a finite set $\{Y_1, \ldots, Y_n\} \subseteq D$ such that $C \subseteq \bigcup_{k=1}^{n} B_{Y_k}$ with $B_Y := \{X \mid \Phi(X, Y) > c\}$, and

$$\min_{X \in C} \max_{1 \leqslant k \leqslant n} \Phi(X, Y_k) > c. \tag{6.3.5}$$

Since (6.3.4) and (6.3.5) preclude the existence of X_0, Y_0 satisfying (6.3.2) for the above pair of finite sets, (ii) cannot hold.

For the proof of the announced theorem of Sion [*2*] we first need a lemma on quasi-convex-concave functions. This lemma will follow from a theorem of Berge (3.7.5) on combinatorial properties of convex sets.

(6.3.6) **Lemma** (Sion [2]). *Let C be a closed*[1] *convex set, $\tilde{D}=\{Y_1,\ldots,Y_n\}$ a finite set, and $\Phi\colon C\times\tilde{D}\to R$ a real function such that the function $\Phi(.,Y_k)$ is quasi-convex and l.s.c. on C for each $Y_k\in\tilde{D}$. Suppose the set $\tilde{D}$ is minimal with respect to the following property: For every $X\in C$ there is a $Y\in\tilde{D}$ such that $\Phi(X,Y)>c$. Then there exists an $X_0\in C$ such that $\Phi(X_0,Y_k)>c$ for all $Y_k\in\tilde{D}$.*

Proof. The sets

$$K_k:=\{X\in C\mid\Phi(X,Y_k)\leqslant c\}$$

are closed convex sets, since $\Phi(.,Y_k)$ is l.s.c. and C is closed[1]. Moreover, $C\cap\bigcap_{k=1}^{n}K_k=\emptyset$ holds, and the minimality of $\tilde{D}$ implies

$$C\cap\bigcap_{\substack{k=1\\k\neq j}}^{n}K_k\neq\emptyset\quad\text{for}\quad j=1,2,\ldots,n.$$

Hence, theorem (3.7.5), when applied to the sets $M:=C$, K_k, $k=1,\ldots,n$, gives

$$C\not\subseteq\bigcup_{k=1}^{n}K_k.$$

There is, therefore, a point $X_0\in C$ such that $X_0\notin\bigcup_{k=1}^{m}K_k$, i.e., $\Phi(X_0,Y)>c$ holds for all $Y_k\in\tilde{D}$.

Now it is easy to prove Sion's generalization [2] of Kakutani's minimax theorem [3]:

(6.3.7) **Minimax Theorem of Sion-Kakutani.** *Let C and D be compact convex sets and $\Phi\colon C\times D\to R$ a semicontinuous quasi-convex-concave pay-off function. Then*

$$\max_{Y\in D}\ \min_{X\in C}\Phi(X,Y)=\min_{X\in C}\ \max_{Y\in D}\Phi(X,Y).$$

The original theorem of Kakutani was restricted to continuous quasi-convex-concave functions. It is an elegant application of Kakutani's fixed point theorem (3.9.9). This approach has been further developed by Glicksberg [1].

Proof of minimax theorem (6.3.7). We have only to establish condition (ii) of lemma (6.3.1). Assume that there are finite sets $\tilde{C}_1\subseteq C$, $\tilde{D}_1\subseteq D$ and a number c such that

$$\max_{Y\in D}\ \min_{X\in\tilde{C}_1}\Phi(X,Y)<c<\min_{X\in C}\ \max_{Y\in\tilde{D}_1}\Phi(X,Y).$$

[1] The theorem and its proof remains valid if C is only convex, but all level sets $\{X\in C\mid\Phi(X,Y_k)\leqslant\alpha\}$, $k=1,2,\ldots,n$, $\alpha\in R$, are closed.

Then
$$\max_{Y\in\mathscr{H}\tilde{D}_1}\ \min_{X\in\tilde{C}_1}\Phi(X,Y)<c<\min_{X\in\mathscr{H}\tilde{C}_1}\ \max_{Y\in\tilde{D}_1}\Phi(X,Y).$$
Let $\tilde{D}_2$ denote the smallest subset of $\tilde{D}_1$ such that
$$c<\min_{X\in\mathscr{H}\tilde{C}_1}\ \max_{Y\in\tilde{D}_2}\Phi(X,Y)$$
still holds. Clearly, we have also
$$\max_{Y\in\mathscr{H}\tilde{D}_2}\ \min_{X\in\tilde{C}_1}\Phi(X,Y)<c.$$
Then let $\tilde{C}_2$ denote the smallest subset of $\tilde{C}_1$ such that
$$\max_{Y\in\mathscr{H}\tilde{D}_2}\ \min_{X\in\tilde{C}_2}\Phi(X,Y)<c$$
still holds. Again, we have also
$$c<\min_{X\in\mathscr{H}\tilde{C}_2}\ \max_{Y\in\tilde{D}_2}\Phi(X,Y).$$
Define in the same way sets $\tilde{D}_3\subseteq\tilde{D}_2$, $\tilde{C}_3\subseteq\tilde{C}_2$, etc. Since $\tilde{D}_1,\tilde{C}_1$ are finite sets, this process terminates after finitely many steps with sets $\tilde{C}\subseteq\tilde{C}_1$, $\tilde{D}\subseteq\tilde{D}_1$, which are minimal with respect to the following property
$$\min_{X\in\mathscr{H}\tilde{C}}\ \max_{Y\in\tilde{D}}\Phi(X,Y)>c>\max_{Y\in\mathscr{H}\tilde{D}}\ \min_{X\in\tilde{C}}\Phi(X,Y).$$
Lemma (6.3.6), and its quasi-concave analog, then yield the existence of $X_0\in\mathscr{H}\tilde{C}$ and $Y_0\in\mathscr{H}\tilde{D}$ with
$$\Phi(X,Y_0)<c<\Phi(X_0,Y)$$
for all $Y\in\tilde{D}$, $X\in\tilde{C}$. Since $\Phi(X_0,.)$ is quasi-concave and $\tilde{D}$ is a finite set, this implies
$$c<\Phi(X_0,Y)\quad\text{for all}\quad Y\in\mathscr{H}\tilde{D}.$$
Symmetrically,
$$\Phi(X,Y_0)<c\quad\text{for all}\quad X\in\mathscr{H}\tilde{C},$$
and the contradiction
$$\Phi(X_0,Y_0)<c<\Phi(X_0,Y_0)$$
is obvious. □

The minimax theorem (6.1.1) of v. Neumann is the special case of the Sion-Kakutani theorem (6.3.7) where
$$C:=\left\{X\in R^m \mid X\geqslant 0\quad\text{and}\ \sum_{i=1}^{m}x_i=1\right\},$$
$$D:=\left\{Y\in R^n \mid Y\geqslant 0\quad\text{and}\ \sum_{k=1}^{n}y_k=1\right\},$$
$$\Phi(X,Y):=X^T A Y.$$

The minimax theorem of v. Neumann also generalizes to semicontinuous (6.2.4) convex-concavelike (6.2.3) pay-off functions:

(6.3.8) **Minimax Theorem of Fan** [3]. *Let C and D be compact Hausdorff-spaces and $\Phi: C\times D\to R$ a semicontinuous convex-concavelike function. Then*

$$\min_{X\in C}\max_{Y\in D}\Phi(X,Y)=\max_{Y\in D}\min_{X\in C}\Phi(X,Y).$$

Proof. We only have to verify condition (ii) of lemma (6.3.1): If $\{X_1,\dots,X_n\}\subseteq C$, $\{Y_1,\dots,Y_m\}\subseteq D$ are arbitrary finite subsets of C and D, respectively, and A is the $n\times m$-matrix

$$A:=(\Phi(X_i,Y_k)),$$

then the minimax theorem of v. Neumann (6.1.1) immediately shows that there are numbers $\xi_1\geqslant 0,\dots,\xi_m\geqslant 0$, $\sum_{i=1}^{m}\xi_i=1$, $\eta_1\geqslant 0,\dots,\eta_n\geqslant 0$, $\sum_{k=1}^{n}\eta_k=1$ such that

$$\text{(6.3.9)}\qquad \max_k\sum_{i=1}^{m}\xi_i\Phi(X_i,Y_k)=\min_i\sum_{k=1}^{n}\Phi(X_i,Y_k)\eta_k.$$

Since Φ is convex-concavelike (6.2.3), there is an $X_0\in C$ and a $Y_0\in D$ such that

$$\Phi(X_0,Y_k)\leqslant\sum_{i=1}^{m}\xi_i\Phi(X_i,Y_k)\quad\text{for all}\quad k=1,2,\dots,n,$$

$$\Phi(X_i,Y_0)\geqslant\sum_{k=1}^{n}\Phi(X_i,Y_k)\eta_k\quad\text{for all}\quad i=1,2,\dots,m.$$

Together with (6.3.9) this implies

$$\Phi(X_0,Y_k)\leqslant\Phi(X_i,Y_0)\quad\text{for all}\quad i=1,\dots,m,\qquad k=1,\dots,n.$$

that is, condition (ii) of lemma (6.3.1) is satisfied.

6.4. Minimax Theorems for Noncompact Sets

In many applications, the sets C, D on which the pay-off function Φ is defined are not compact. Additional conditions must then be imposed on Φ to guarantee the existence of saddle points. Such conditions essentially reduce the noncompact case to the compact case. They can also be regarded as generalized "constraint qualifications" (see sections 6.6 and 6.7) conditions of this kind have been introduced by Stoer [1], [2], and in slightly weaker form by Mangasarian and Ponstein [1]. We follow the terminology of these authors, while retaining full generality, and say that $\Phi: C \times D \to R$, where $C \subseteq R^m$ and $D \subseteq R^n$, has the

(6.4.1) *low-value property at* $(\bar{X}, \bar{Y}) \in C \times D$

if there is an R^m-neighborhood $U(\bar{X})$ of $\bar{X}$ and a compact set $B \subseteq D$ such that for each $X \in U(\bar{X}) \cap C$ there exists a $Y \in B$ with $\Phi(X, Y) \geqslant \Phi(\bar{X}, \bar{Y})$.

Similarly, we say that Φ has the

(6.4.2) *high-value property at* $(\bar{X}, \bar{Y}) \in C \times D$

if there is a neighborhood $U(\bar{Y})$ of $\bar{Y}$ and a compact set $B \subseteq C$ such that for each $Y \in U(\bar{Y}) \cap D$ there exists an $X \in B$ with $\Phi(X, Y) \leqslant \Phi(\bar{X}, \bar{Y})$.

(6.4.3) **Theorem.** *Suppose Φ is a semicontinuous pseudoconvex-concave function on $C \times D$, $C \subseteq R^m$, $D \subseteq R^n$. Then*

(i) *Φ has a saddle point $(\bar{X}, Y_0) \in C \times D$ if and only if there is a $\bar{Y} \in D$ such that Φ has the low-value property* (6.4.1) *at $(\bar{X}, \bar{Y})$ and $\Phi(\bar{X}, \bar{Y}) = \sup_{Y \in D} \Phi(\bar{X}, Y)$.*

(ii) *Φ has a saddle point $(X_0, \bar{Y}) \in C \times D$ if and only if there is a $\bar{X} \in C$ such that Φ has the high-value property* (6.4.2) *at $(\bar{X}, \bar{Y})$ and $\Phi(\bar{X}, \bar{Y}) = \inf_{X \in C} \Phi(X, \bar{Y})$.*

Proof. Because of symmetry it suffices to prove (i). If $(\bar{X}, Y_0)$ is a saddle point of Φ then the choice

$$\bar{Y} := Y_0, \qquad U(\bar{X}) := R^m, \qquad B := \{Y_0\}$$

establishes the low-value property of Φ at $(\bar{X}, \bar{Y})$ and the relation $\Phi(\bar{X}, \bar{Y}) = \sup_{Y \in D} \Phi(\bar{X}, Y)$.

To prove the "if"-part of (i), suppose that Φ has the low-value property at $(\bar{X}, \bar{Y})$ for some $\bar{Y} \in D$ with

$$\Phi(\bar{X}, \bar{Y}) = \sup_{Y \in D} \Phi(\bar{X}, Y). \tag{6.4.4}$$

That is, there is a neighborhood $U(\bar{X})$ of $\bar{X}$ and a compact set $B \subseteq D$ such that for all $X \in U(\bar{X}) \cap C$ there is a $Y \subseteq B$ with $\Phi(X, Y) \geqslant \Phi(\bar{X}, \bar{Y})$. We may assume that $U(\bar{X})$ is closed and convex. Otherwise we replace it by a closed convex subneighborhood. We may also assume that B is convex, because otherwise we replace it by its convex hull, which is again compact.

Now consider an arbitrary compact convex subset K of C which contains $\bar{X}$:

$$\bar{X} \in K \subseteq C.$$

Then

$$U_K := U(\bar{X}) \cap K$$

is a nonvoid compact convex set, and the minimax theorem of Sion-Kakutani applies to the restriction of Φ to $U_K \times B$, whence there exists a saddle point (X_K, Y_K) in $U_K \times B$:

$$\text{(6.4.5)} \qquad \begin{array}{llll} \text{(i)} & \Phi(X_K, Y_K) \geqslant \Phi(X_K, Y) & \text{for all} & Y \in B, \\ \text{(ii)} & \Phi(X_K, Y_K) \leqslant \Phi(X, Y_K) & \text{for all} & X \in U_K. \end{array}$$

The low-value property (6.4.1) implies

$$\Phi(\bar{X}, \bar{Y}) \leqslant \Phi(X_K, Y_1)$$

for some $Y_1 \in B$, and by (6.4.5.(i)),

$$\Phi(X_K, Y_1) \leqslant \Phi(X_K, Y_K),$$

whence

$$\Phi(\bar{X}, \bar{Y}) \leqslant \Phi(X_K, Y_K).$$

Since $\bar{X} \in U_K$, we obtain from (6.4.5.(ii)) and (6.4.4) that

$$\Phi(X_K, Y_K) \leqslant \Phi(\bar{X}, Y_K) \leqslant \Phi(\bar{X}, \bar{Y}),$$

whence

$$\Phi(X_K, Y_K) = \Phi(\bar{X}, Y_K) = \Phi(\bar{X}, \bar{Y}).$$

From this we get by (6.4.4) that

$$\Phi(\bar{X}, Y) \leqslant \Phi(\bar{X}, \bar{Y}) = \Phi(\bar{X}, Y_K) \qquad \text{for all} \qquad Y \in D.$$

Moreover, (6.4.5.(ii)) leads to

$$\Phi(\bar{X}, Y_K) = \Phi(X_K, Y_K) \leqslant \Phi(X, Y_K) \qquad \text{for all} \qquad X \in U_K.$$

We conclude that

(6.4.6) *If* (X_K, Y_K) *is a saddle point of* Φ *in* $U_K \times B$, *then* $(\bar{X}, Y_K)$ *is a saddle point of* Φ *in the larger set* $U_K \times D$.

Our next observation concerns the set

$$\mathcal{Y}_K$$

of all points $Y \in B$ such that $(\bar{X}, Y)$ is a saddle point of Φ in $U_K \times D$. It was seen above, that $\mathcal{Y}_K$ is not empty. In addition, we claim that

(6.4.7) $\mathcal{Y}_K$ *is closed.*

Indeed, suppose that

$$\{Y_i\}_{i=1,2,\ldots} \to Y_0 \in B, \qquad Y_i \in \mathcal{Y}_K \qquad \text{for all} \qquad i.$$

By definition of Y_K, all points $(\bar{X}, Y_i)$ are saddle points on $U_K \times D$, giving rise to the same value of Φ, $\Phi(\bar{X}, Y_i) = \Phi(\bar{X}, \bar{Y})$. Since $\Phi(\bar{X}, .)$ is u.s.c., we have therefore

$$\Phi(\bar{X}, \bar{Y}) = \lim_{i \to \infty} \Phi(\bar{X}, Y_i) \leqslant \Phi(\bar{X}, Y_0),$$

and by (6.4.4),

$$\Phi(\bar{X}, \bar{Y}) = \lim_{i \to \infty} \Phi(\bar{X}, Y_i) = \Phi(\bar{X}, Y_0).$$

Thus (6.4.4) yields the left saddle inequality for $(\bar{X}, Y_0)$:

$$\Phi(\bar{X}, Y) \leqslant \Phi(\bar{X}, Y_0) \qquad \text{for all} \qquad Y \in D.$$

Using again that all $(\bar{X}, Y_i)$ are saddle points, and that $\Phi(X, .)$ is u.s.c., we obtain

$$\Phi(\bar{X}, Y_0) = \lim_{i \to \infty} \Phi(\bar{X}, Y_i) \leqslant \overline{\lim_{i \to \infty}}\, \Phi(X, Y_i) \leqslant \Phi(X, Y_0)$$

for all $X \in U_K$, establishing the other saddle inequality. Thus $(\bar{X}, Y_0)$ is a saddle point of Φ in $U_K \times D$, and therefore $Y_0 \in Y_K$. This proves (6.4.7).

Now consider any finite set of compact convex sets

$$K_1, \ldots, K_s$$

with $\bar{X} \in K_j \subseteq C$ for $j = 1, \ldots, s$. Clearly

$$Y_{K_1} \cap \cdots \cap Y_{K_s} \supseteq Y_{\mathscr{H}(K_1 \cup \cdots \cup K_s)} \neq \emptyset.$$

Since all Y_K are closed subsets of a compact set, namely B, and since as was just seen every finite family of sets Y_K has nonempty intersection, it follows that

$$Y_C := \bigcap_{\bar{X} \in K \subseteq C} Y_K \neq \emptyset.$$

Any

$$Y_0 \in Y_C$$

is such that $(\bar{X}, Y_0)$ is a saddle point of Φ on $(U(\bar{X}) \cap C) \times D$. Indeed, for each $X \in U(\bar{X}) \cap C$ there exists a compact convex set K with $\{X, \bar{X}\} \subseteq K \subseteq C$. Since $Y_0 \in Y_K$ and $X \in U_K$,

$$\Phi(\bar{X}, Y_0) \leqslant \Phi(X, Y_0).$$

Thus

$$\Phi(\bar{X}, Y_0) \leqslant \Phi(X, Y_0) \qquad \text{for all} \qquad X \in U(\bar{X}) \cap C.$$

But this inequality holds in fact for all $X \in C$, since local minima are global for pseudoconvex functions.

The left saddle inequality

$$\Phi(\bar{X}, Y_0) \geqslant \Phi(\bar{X}, Y) \quad \text{for all} \quad Y \in D$$

is again an immediate consequence of (6.4.4). □

If the low-value property (6.4.1) is strengthened by requiring the existence of a compact $B \subseteq D$ such that for each $X \in C$, not just for each X in a neighborhood of $\bar{X}$, there exists $Y \in B$ with $\Phi(X, Y) \geqslant \Phi(\bar{X}, \bar{Y})$, then the above proof carries through for semicontinuous quasi-convex saddle functions (see Karamardian [*1*] and Mehndiratta [*1*] for a similar result).

If program I (6.2.5) has an optimal solution $(\bar{X}, \bar{Y})$, then clearly

$$\Phi(\bar{X}, \bar{Y}) = \sup_{Y \in D} \Phi(\bar{X}, Y).$$

Therefore theorem (6.4.3) is essentially equivalent to the following theorem, which emphasizes the viewpoint of duality:

(6.4.8) **Duality Theorem.** *Suppose Φ is a semicontinuous pseudoconvex-concave function on $C \times D$, $C \subseteq R^m$, $D \subseteq R^n$. Then the following statements hold:*

(i) *If program I* (6.2.5) *has an optimal solution $(\bar{X}, \bar{Y})$, then there is an optimal solution (X_0, Y_0) of programm II with $\Phi(\bar{X}, \bar{Y}) = \Phi(X_0, Y_0)$ if and only if Φ has the low-value property at $(\bar{X}, \bar{Y})$.*

(ii) *If programm II* (6.2.5) *has an optimal solution (X, Y), then program I has an optimal solution (X_0, Y_0) with $\Phi(\bar{X}, \bar{Y}) = \Phi(X_0, Y_0)$ if and only if Φ has the high-value property at $(\bar{X}, \bar{Y})$.*

Note that if $(\bar{X}, \bar{Y})$ and (X_0, Y_0) are optimal for programs I and II, respectively, and satisfy $\Phi(\bar{X}, \bar{Y}) = \Phi(X_0, Y_0)$, then by theorem (6.2.9) $(\bar{X}, Y_0)$ is a saddle point of Φ and is optimal for both programs I and II.

The following theorem gives convenient sufficient conditions for the low-value property (6.4.1).

(6.4.9) **Theorem.** *Suppose that Φ is a closed continuous quasiconvex-concave function on $C \times D$, where C and D are convex sets in R^m and R^n, respectively. Let $(\bar{X}, \bar{Y})$ be an optimal solution of program I* (6.2.5). *Then Φ has the low-value property at $(\bar{X}, \bar{Y})$ if the set*

$$B_X := \{Y \mid \Phi(X, Y) = \max_{Y \in D} \Phi(X, Y)\}$$

is bounded for $X = \bar{X}$.

Proof. By hypothesis, the level set

$$B_{\bar{X}} = \{Y \in D \mid \Phi(\bar{X}, Y) \geqslant \Phi(\bar{X}, \bar{Y})\}$$

of the closed quasiconcave function $\Phi(\bar{X},.)$ is bounded and nonempty, $\bar{Y} \in B_{\bar{X}}$. Lemma (4.9.7) therefore gives the existence of an $\varepsilon > 0$ such that

$$B := \{Y \in D \mid \Phi(\bar{X}, Y) \geqslant \max_{Y \in D} \Phi(\bar{X}, Y) - \varepsilon\}$$
$$= \{Y \in D \mid \Phi(\bar{X}, Y) \geqslant \Phi(\bar{X}, \bar{Y}) - \varepsilon\}$$

is a compact convex set. The low value property will be established, if there exists an R^m-neighborhood $U(\bar{X})$ of $\bar{X}$ such that

(6.4.10) $\quad B_X \neq \emptyset \quad$ and $\quad B_X \subseteq B \quad$ for all $\quad X \in U(\bar{X}) \cap C.$

Indeed this implies in view of the definition of B_X

$$\max_{Y \in B} \Phi(X, Y) \geqslant \max_{Y \in B_X} \Phi(X, Y) = \max_{Y \in D} \Phi(X, Y) \geqslant \Phi(\bar{X}, \bar{Y})$$

for $X \in U(\bar{X}) \cap C$, and therefore the low value property of Φ at $\bar{X}$ with respect to the sets $U(\bar{X}), B$.

To prove (6.4.10) we show first that there is a R^m-neighborhood $U_2(\bar{X})$ of $\bar{X}$ such that $U_2(\bar{X}) \cap C$ is a compact convex set. Indeed, let L be the level set

$$L := \{X \in C \mid \Phi(X, \bar{Y}) \leqslant \Phi(\bar{X}, \bar{Y}) + \varepsilon\}.$$

As $\bar{X} \in L$ and $\Phi(., \bar{Y})$ is continuous, L contains a neighborhood V of $\bar{X}$,

$$V := \{X \in C \mid \Phi(X, \bar{Y}) < \Phi(\bar{X}, \bar{Y}) + \varepsilon\},$$

which is relative open in C; that is, there exists an R^m-neighborhood $U_1(\bar{X})$ of $\bar{X}$ such that

$$V = U_1(\bar{X}) \cap C = U_1(\bar{X}) \cap L.$$

Now choose any compact convex neighborhood $U_2(\bar{X})$ of $\bar{X}$ with $U_2(\bar{X}) \subseteq U_1(\bar{X})$.

Then $U_2(\bar{X}) \cap C \supseteq U_2(\bar{X}) \cap L$; but on the other hand, $U_2(\bar{X}) \cap C \subseteq U_1(\bar{X}) \cap C = U_1(\bar{X}) \cap L$ and therefore also, $U_2(\bar{X}) \cap C = U_2(\bar{X}) \cap L$.

This shows that $U_2(\bar{X}) \cap C = U_2(\bar{X}) \cap L$ is a compact convex set, as L is a closed set in R^m by the quasiconvexity of $\Phi(., \bar{Y})$.

Now, Φ is continuous, hence uniformly continuous on the compact set $(U_2(\bar{X}) \cap C) \times B$. Therefore, there exists an R^m-neighborhood $U(\bar{X}) \subseteq U_2$ of $\bar{X}$ such that for all $X \in U(\bar{X}) \cap C$ and $Y \in B$,

(6.4.11) $$|\Phi(\bar{X}, Y) - \Phi(X, Y)| \leqslant \frac{\varepsilon}{4}.$$

Then choose a $Y_1 \in B^I$ such that

(6.4.12) $$\Phi(\bar{X}, X_1) > \Phi(\bar{X}, \bar{Y}) - \frac{\varepsilon}{4}.$$

This is possible because of the continuity of $\Phi(\bar{X},.)$ and $\bar{Y} \in B$. Now, suppose (6.4.10) is not true. Then there exists an $X_0 \in U(\bar{X}) \cap C$ and a $Y_0 \in D \sim B$ such that

(6.4.13) $$\Phi(X_0, Y_0) \geqslant \Phi(X_0, Y_1).$$

Let $Y' = \lambda Y_1 + (1-\lambda) Y_0$, $0 < \lambda < 1$, be the boundary point of B, $Y' \in B$, lying on the segment (Y_0, Y_1). Then by $Y_0 \notin B$, $Y_1 \in B^I$, the definition of B, and the continuity of $\Phi(\bar{X}, .)$, we have

(6.4.14) $$\Phi(\bar{X}, Y') = \Phi(\bar{X}, \bar{Y}) - \varepsilon.$$

Now it follows from (6.4.11) and (6.4.14)

(6.4.15) $$\Phi(X_0, Y') \leqslant \Phi(\bar{X}, Y') + \frac{\varepsilon}{4} = \Phi(\bar{X}, \bar{Y}) - \frac{3}{4}\varepsilon.$$

Again by (6.4.11) and by (6.4.12)

$$\Phi(X_0, Y_1) \geqslant \Phi(\bar{X}, Y_1) - \frac{\varepsilon}{4} \geqslant \Phi(\bar{X}, \bar{Y}) - \frac{\varepsilon}{2},$$

and therefore by (6.4.13) and (6.4.15)

$$\Phi(X_0, Y_0) \geqslant \Phi(X_0, Y_1) > \Phi(X_0, Y').$$

But this contradicts the quasiconcavity of $\Phi(X_0, .)$ (Y' lies between Y_0 and Y_1). ▯

The following theorem has thus been proved:

(6.4.16) **Theorem.** *If, in addition to the hypotheses of theorem* (6.4.9) *the function* $\Phi(\bar{X}, Y)$ *is a strictly concave function of* Y *in some neighborhood of* $\bar{Y}$, *then* $(\bar{X}, \bar{Y})$ *is a saddle point of* $\Phi(X, Y)$.

6.5. Lagrange Multipliers

Consider the problems of minimizing a function $f: R^n \to R$ subject to "side conditions" expressed as equations:

(6.5.1) *Minimize* $f(X)$ *subject to*

$$F(X) = 0,$$

where F stands for a row vector of m functions $f_i: R^n \to R$

$$F := \begin{vmatrix} f_1 \\ \vdots \\ f_m \end{vmatrix}.$$

The unconstrained minimum of a differentiable function $f: R^n \to R$ is necessarily assumed at a "stationary point" X_0, i.e. at a point with

$$\operatorname{grad} f(X_0) = 0.$$

While far from being sufficient, this condition usually narrows down the search to a finite number of points, which then can be examined individually, provided their determination is not too difficult. Problem (6.5.1) can sometimes be reduced to an unconstrained minimization problem, by using the constraints $f_i(X)=0$ for eliminating some of the variables. In many cases, in which this elimination either cannot be carried out in closed form or would destroy inherent symmetries, it is desirable to have necessary conditions for minima in terms of the original functions f, f_i and their derivatives. The following "Lagrange multiplier" technique aims at providing such conditions.

Let X_0 be any feasible point, i.e. any point satisfying $F(X_0)=0$. For an infinitesimal analysis, each constraint $f_i(X)=0$ is replaced by the linear constraint

$$\bar{f}_i(X) := \operatorname{grad} f_i(X_0)^T (X - X_0), \qquad i = 1, \ldots, m,$$

which represents the tangent plane of the surface $\{X \mid f_i(X)=0\}$ at the point X_0, provided $\operatorname{grad} f_i(X_0) \neq 0$. Under fairly general conditions, which will be stated below, the restriction of f to $\{X \mid F(X)=0\}$ is locally minimal at X_0 only if its restriction to the linear manifold

$$M := \bigcap_{i=1}^{m} \{X \mid \bar{f}_i(X)=0\}$$

is locally minimal at X_0. This in turn requires the gradient of F at X_0 to be orthogonal to M. As the orthogonal complement to M is spanned by the vectors $\operatorname{grad} f_i(X_0)$, the desired necessary condition takes the form

$$\operatorname{grad} f(X_0) \in \mathscr{L}\{\operatorname{grad} f_1(X_0), \ldots, \operatorname{grad} f_m(X_0)\}.$$

There exists therefore a vector $Y \in R^m$ so that

$$\operatorname{grad} f(X_0)^T + Y^T \operatorname{Grad} F(X_0)^T = 0, \tag{6.5.2}$$

where $\operatorname{Grad} F$ stands for the transpose of the

$$\text{Jacobian matrix} \quad \begin{vmatrix} \dfrac{\partial f_1}{\partial x_1}, \ldots, \dfrac{\partial f_1}{\partial x_n} \\ \vdots \qquad\quad \vdots \\ \dfrac{\partial f_m}{\partial x_1}, \ldots, \dfrac{\partial f_m}{\partial x_n} \end{vmatrix}. \tag{6.5.3}$$

In other words,

$$\operatorname{Grad} F = (\operatorname{grad} f_1, \ldots, \operatorname{grad} f_m).$$

The components of Y in (6.5.2) are called

(6.5.4) *Lagrange multipliers.*

Condition (6.5.2) can be written as

$$\operatorname{grad}_X \Phi(X, Y)\big|_{X=X_0} = 0,$$

where

$$\Phi(X, Y) := f(X) + Y^T F(X)$$

is the

(6.5.5) *Lagrangian function*

of the problem (6.5.1).

As was mentioned above, there are exceptions to the rule (6.5.2). Consider for instance the problem

$$\textit{Minimize} \quad f\begin{pmatrix} x_1 \\ x_2 \end{pmatrix} \quad \textit{subject to}$$

$$x_1^2 - x_2 = 0,$$
$$x_1^2 + x_2 = 0.$$

The point $\begin{pmatrix} 0 \\ 0 \end{pmatrix}$ is the only feasible solution, and therefore the optimal solution no matter what value is taken by $\operatorname{grad} f$. However, since

$$\operatorname{grad} f_1\begin{pmatrix} 0 \\ 0 \end{pmatrix} = \begin{pmatrix} 0 \\ -1 \end{pmatrix}, \quad \operatorname{grad} f_2\begin{pmatrix} 0 \\ 0 \end{pmatrix} = \begin{pmatrix} 0 \\ +1 \end{pmatrix},$$

(6.5.2) requires $\frac{\partial f}{\partial x_1}\begin{pmatrix} 0 \\ 0 \end{pmatrix} = 0$, contradicting the generality of $\operatorname{grad} f\begin{pmatrix} 0 \\ 0 \end{pmatrix}$.

This and similar pathological situations are excluded by the following "constraint qualification":

(6.5.6) *The rank of the Jacobian matrix* (6.5.3) *at* X_0 *should be maximal, i.e.*

$$\operatorname{rank}(\operatorname{Grad} F(X_0)) = m.$$

This constraint qualification is implicit in the following formulation of the main theorem on Lagrange multipliers:

(6.5.7) **Theorem.** *Suppose the functions f and f_i, $i=1,\dots,m$ in problem* (6.5.1) *have continuous partial derivatives in R^n. Necessary for f to assume a local minimum at the feasible solution X_0 is that the gradient vectors*

$$\operatorname{grad} f(X_0), \qquad \operatorname{grad} f_1(X_0), \dots, \operatorname{grad} f_m(X_0)$$

be linearly dependent.

Indeed, if (6.5.6) is satisfied, then the existence of Lagrange multipliers Y satisfying (6.5.2) is necessary for a local minimum at X_0.

An elegant proof of theorem (6.5.7), capable of generalization to more general linear vector spaces, can be based on the "implicit function theorem" of analysis (see Hadley [*3*], Luenberger [*2*]). We shall not describe this proof, since we are mainly interested in problems or programs of the form

(6.5.8) $$\text{Minimize } f(X) \text{ subject to} \quad F(X) \leqslant 0, \quad X \in H,$$

where the objective function f and the constraint functions f_i are defined on an open region

$$H \subseteq R^n.$$

Necessary and sufficient conditions for a local minimum are then provided by the following fundamental (see also Mangasarian and Fromowitz [*1*]):

(6.5.9) **Theorem of John** [*1*][1]. *Suppose the function f and f_i, $i=1,\dots,m$, in problem* (6.5.8) *have continuous partial derivatives in H. Necessary for f to assume a local minimum at a feasible solution X_0 is the existence of nonnegative numbers $y_0 \geqslant 0$, $Y \geqslant 0$, which do not all vanish, satisfying*

(i) $$y_0 \operatorname{grad} f(X_0)^T + Y^T \operatorname{Grad} F(X_0)^T = 0,$$

and the "complementary slackness" condition

(ii) $$Y^T F(X_0) = 0.$$

---

[1] John [*1*] proved a somewhat more general result, allowing for infinitely many constraints which form a compact metric space. Similar results had been obtained before within the calculus of variations (e.g. Valentine [*1*]).

Sufficient for a local minimum at X_0 *is that, in addition to* (i) *and* (ii),

$$\text{(iii)} \qquad \operatorname{rank} \begin{vmatrix} y_0 \dfrac{\partial f}{\partial x_1}, \dots, y_0 \dfrac{\partial f}{\partial x_n} \\ y_1 \dfrac{\partial f_1}{\partial x_1}, \dots, y_1 \dfrac{\partial f_1}{\partial x_n} \\ \vdots \\ y_m \dfrac{\partial f_m}{\partial x_1}, \dots, y_m \dfrac{\partial f_m}{\partial x_n} \end{vmatrix} = n \quad at \quad X_0.$$

Condition (ii) is called "complementary slackness" condition, for if $f_i(X_0)<0$, then it implies $y_i=0$, and if $y_i>0$, then it implies $f_i(X_0)=0$; in other words, at most one of the two inequalities $f_i(X_0)\leqslant 0$ and $y_i\geqslant 0$ may show "slack". The third condition is essentially a condition on that submatrix of the Jacobian for which the coefficients y_i are positive.

Proof of theorem (6.5.9). Suppose $X_0 \in H$ is feasible and a local minimum of (6.5.8). Consider the sequence of indices

$$I=\{i_1, i_2, \dots, i_s\}$$

characterized by

$$f_{i_k}(X_0)=0, \qquad k=1,\dots,s.$$

We proceed to show that the system of inequalities in Z,

$$\text{(6.5.10)} \qquad \begin{aligned} \operatorname{grad} f(X_0)^T Z &< 0, \\ \operatorname{Grad} F_I(X_0)^T Z &< 0, \end{aligned}$$

has no solution.

Assume that Z were a solution. Since H is an open set, there exist positive numbers $\theta>0$ such that $X_0+\rho Z \in H$ for $\rho \in [0,\theta]$. Since the functions f, f_i are continuously differentiable, there exist numbers ρ, ρ_i between 0 and θ such that

$$\begin{aligned} f(X_0+\theta Z) &= f(X_0)+\theta \operatorname{grad} f(X_0+\rho Z)^T Z, \\ f_i(X_0+\theta Z) &= f_i(X_0)+\theta \operatorname{grad} f_i(X_0+\rho_i Z)^T Z. \end{aligned}$$

For all sufficiently small $\theta>0$, we have in addition $\operatorname{grad} f(X_0+\rho Z)^T Z<0$, and $\operatorname{grad} f_i(X_0+\rho_i Z)^T Z<0$ for $i \in I$. Furthermore, if $i \notin I$, then $f_i(X_0)<0$, and again by continuity, $f_i(X_0+\theta Z)<0$. Thus

$$f(X_0+\theta Z)<f(X_0) \quad \text{and} \quad F(X_0+\theta Z)<0$$

for sufficiently small $\theta>0$. But X_0 was supposed to be a local minimum. Hence there is no Z satisfying (6.5.10).

The transposition theorem of Gordan (1.6.3) can now be applied to (6.5.10). It yields the existence of nonnegative multipliers $y_0, y_{i_1}, \dots, y_{i_s}$, not all of them zero, such that

$$y_0 \operatorname{grad} f(X_0)^T + y_{i_1} \operatorname{grad} f_{i_1}(X_0)^T + \cdots + y_{i_s} \operatorname{grad} f_{i_s}(X_0)^T = 0.$$

If the values y_{i_k} are extended to a vector $Y \in R^m$ so that $y_i = 0$ if $i \notin I$, then y_0 and Y are as required by (i) and (ii). This proves the necessity of these two conditions.

To prove the sufficiency of conditions (i), (ii) and (iii), suppose $f(X_0)$ is not a local minimum. Then there exist a sequence of positive numbers $\{\theta_r\}_{r=1,2,\dots} \to 0$ and a sequence of directions $\{Z_r\}_{r=1,2,\dots}, \|Z_r\| = 1$, such that for all r,

$$f(X_0 + \theta_r Z_r) < f(X_0),$$
$$F(X_0 + \theta_r Z_r) \leqslant 0.$$

The mean value theorem yields the existence of $\rho_r \in [0, \theta_r]$ such that

$$\theta_r \operatorname{grad} f(X_0 + \rho_r Z_r)^T Z_r = f(X_0 + \theta_r Z_r) - f(X_0) < 0$$

and therefore

$$\operatorname{grad} f(X_0 + \rho_r Z_r)^T Z_r < 0.$$

Similarly, there exist numbers $\rho_{ir} \in [0, \theta_r]$ such that

$$\operatorname{grad} f_i(X_0 + \rho_{ir} Z_r)^T Z_r \leqslant 0 \quad \text{for} \quad y_i > 0 \quad \text{and} \quad i > 0.$$

For a suitable subsequence, the Z_r converge towards a vector $Z \neq 0$, for which then (putting $\operatorname{grad} f_0(X_0) := \operatorname{grad} f(X_0)$)

$$\operatorname{grad} f_i(X_0)^T Z \leqslant 0 \quad \text{for} \quad y_i > 0.$$

But in view of (i), Z must in fact solve

$$\operatorname{grad} f_i(X_0)^T Z = 0 \quad \text{for} \quad y_i > 0,$$

which contradicts (iii). □

Again, a "constraint qualification", stipulating that

(6.5.11) *the gradient vectors* $\{\operatorname{grad} f_i(X_0) \mid f_i(X_0) = 0\}$ *admit no linear dependence with nonnegative coefficients;*

guarantees, that $y_0 > 0$ in (6.5.9.(i)) and that therefore nonnegative Lagrange multipliers exist (see Cottle [*2*]).

The sufficient condition of John's theorem is a rather weak result, since it applies only to points for which at least n constraints are satisfied as equations. In this next section, we shall describe conditions which are both necessary and sufficient for minima under the additional assumption of convexity.

The technique of Lagrange multipliers can be extended to more general linear spaces, for instance, function spaces. Side conditions may then take the form of differential equations or inequalities, the objective functions may be expressed by an integral. Optimization of this kind is the domain of Variational Calculus and its modern offspring, the theory of Optimal Control. The central result of the latter, Pontryagin's celebrated Maximum Principle (Pontryagin, Boltyanskii, Gamkrelidze, and Mischenko [*1*] involves generalized Lagrange Multipliers (see for instance Bliss [*1*], Luisternik and Sobolev [*1*], Lee and Markus [*1*], Luenberger [*2*], Neustadt [*1*]).

6.6. Kuhn-Tucker Theory for Differentiable Functions

We are condernded again with programs of the type

(6.6.1) $$\textit{Minimize} \quad f(X) \quad \textit{subject to} \quad F(X) \leqslant 0$$

where, as in section 6.5, F stands for an m-vector of constraint functions

$$F = \begin{vmatrix} f_1 \\ \vdots \\ f_m \end{vmatrix}.$$

In addition, we shall assume that the objective function $f\colon R^n \to R \cup \{+\infty\}$ and the constraint functions $f_i\colon R^n \to R \cup \{+\infty\}$ are convex. The program (6.6.1) is then an instance of a

convex program.

Its feasible region

$$C_f := \{X \in K(f) \mid F(X) \leqslant 0\}$$

is a convex set.

Irregularities in the behavior of solutions to (6.6.1) are frequently caused by constraints which are satisfied as equations by all feasible solutions. We call such constraints $f_i(X) \leqslant 0$

(6.6.2) *singular.*

If there are no feasible solutions X, i.e. if $C_f = \emptyset$, then all constraints are considered to be singular (compare (1.6.5), (2.3.5), and (3.4.10) for analogous definitions). Note that

(6.6.3) *if* $U \in C_f^I$, *then*

$$f_i(U) < 0$$

for all nonsingular constraints. In particular, the existence of some point U *such that* $F(U) < 0$ *is necessary and sufficient for the absence of singular constraints.*

Proof. Suppose $U \in C_f^I$ and $f_i(X) < 0$ for some $X \in C_f$. Then by lemma (3.2.9),

$$X_\varepsilon := U + \varepsilon(U - X) \in C_f^I$$

for sufficiently small $\varepsilon > 0$. By convexity, and since $f_i(X_\varepsilon) \leqslant 0$,

$$f_i(U) \leqslant \frac{1}{1+\varepsilon} f_i(X_\varepsilon) + \frac{\varepsilon}{1+\varepsilon} f_i(X) < 0.$$

The second part of (6.6.3) is straightforward if one keeps in mind that all constraints are singular whenever C_f is empty. □

The

(6.6.4) *absence of singular constraints in* $F(X) \leqslant 0$ (Slater [*1*])

can now serve as constraint qualification for a Lagrange multiplier theorem. This constraint qualification is frequently given the form

$$F(U) < 0 \quad \textit{for some} \quad U \in R^n,$$

which was seen above to be equivalent to (6.6.4).

(6.6.5) **Theorem of Kuhn and Tucker** [*1*]. *Suppose the functions* f *and* f_i, $i = 1, \ldots, m$, *are differentiable in some* R^n*-neighborhood of the feasible region* C_f, *and that the constraint qualification* (6.6.4) *is met. Then* $X_0 \in C_f$ *is minimum if and only if there exist nonnegative Lagrange multipliers* $Y \geqslant 0$ *such that the following "Kuhn-Tucker conditions" hold:*

$$\operatorname{grad} f(X_0)^T + Y^T \operatorname{Grad} F(X_0)^T = 0 \quad \textit{and} \quad Y^T F(X_0) = 0.$$

Proof. Recall that a convex differentiable function is also continuously differentiable by theorem (4.4.7). The necessity of the above Kuhn-Tucker conditions will therefore follow from the theorem of John (6.5.9), after the constraint qualification (6.5.11) will have been verified.

Since all constraints are nonsingular, there exists a U such that $F(U) < 0$. Putting $Z := U - X_0$, we have by relations (4.4.3) and (4.3.2) for convex functions

$$\operatorname{grad} f_i(X_0)^T Z = f_i'(X_0; Z) \leqslant f_i(X_0 + Z) - f_i(X_0) = f_i(U) - f_i(X_0).$$

Thus

$$\operatorname{grad} f_i(X_0)^T Z < 0$$

for all i with $f_i(X_0)=0$. Any nonnegative linear dependence of these gradient vectors $\operatorname{grad} f_i(X_0)$ would immediately lead to the contradiction $0<0$. Hence (6.5.11) must hold.

To prove sufficiency, suppose $F(X_0+Z)\leqslant 0$ for some Z. Then again by (4.4.3) and (4.3.2),

$$\operatorname{grad} f_i(X_0)^T Z = f_i'(X_0; Z) \leqslant f_i(X_0+Z) - f_i(X_0).$$

In particular,

$$\operatorname{grad} f_i(X_0)^T Z \leqslant 0$$

for all i with $f_i(X_0)=0$. Similarly

$$f(X_0+Z)-f(X_0) \geqslant f'(X_0; Z) = \operatorname{grad} f(X_0)^T Z$$

and in view of the Kuhn-Tucker conditions

$$f(X_0+Z)-f(X_0) \geqslant \operatorname{grad} f(X_0)^T Z = -y_i \operatorname{grad} f_i(X_0)^T Z \geqslant 0. \quad \square$$

6.7. Saddle Points of the Lagrangian

Slater's constraint qualification, that there be no singular constraints, does not require differentiability for its formulation. The question arises, whether the Kuhn-Tucker theorem can be extended to nondifferentiable functions. This is indeed the case, and one way to accomplish this is to formulate it as a saddle point theorem. With program (6.6.1), Kuhn and Tucker [*1*] therefore associated the convex-concave Lagrangian

$$\Phi(X, Y) := f(X) + Y^T F(X),$$

defined on the set $C \times R^n_+$, where C stands for the intersection of the domains of finiteness of all functions involved:

$$C := K(f) \cap \bigcap_{i=1}^{m} K(f_i).$$

Consider then the two programs I and II (6.2.5) associated with $\Phi(X, Y)$. Since

$$\sup_{Y\geqslant 0} \Phi(X, Y) = \sup_{Y\geqslant 0} (f(X) + Y^T F(X)) = \begin{cases} +\infty & \text{if } X\in C \text{ and } F(X)\nleqslant 0, \\ f(X) & \text{if } X\in C \text{ and } F(X)\leqslant 0, \end{cases}$$

we have

$$\inf_{X\in C} \sup_{Y\geqslant 0} \Phi(X, Y) = \inf_{X\in C_f} f(X).$$

Thus program I is equivalent to our convex program (6.6.1), in which we are interested: each optimal solution $\bar{X}$ of (6.6.1) gives rise to an optimal solution $(\bar{X},0)$ of program I, and vice versa, if $(\bar{X}, Y_0)$ is optimal for program I, then $\bar{X}$ is optimal for (6.6.1).

On the other hand, if $(\bar{X},\bar{Y})$ is a saddle point of the Lagrangian function $\Phi(X,Y)=f(X)+Y^T F(X)$, then $\bar{X}$ is an optimal solution of the convex program (6.6.1), since in this case $(\bar{X},\bar{Y})$ solves program I. Then $(\bar{X},0)$ also solves program I, which implies the complementary slackness condition

$$\bar{Y}^T F(\bar{X})=0.$$

Theorem (6.6.5), which stated necessary and sufficient conditions for optimality of $\bar{X}$ in terms of Lagrange multipliers, can therefore be formulated as a saddle point theorem. It then becomes a consequence of the general duality theorem (6.4.8), which requires only that the function f and f_i in (6.6.1) are closed. The low-value property of the Lagrangian $\Phi(X,Y)=f(X)+Y^T F(X)$ at the point $(\bar{X},0)$ acts as a constraint qualification. It is theoretically the most powerful one, for it is necessary and sufficient. Unfortunately, it is in general not easily verified since it requires knowledge of the optimal solution $\bar{X}$ of (6.6.1). Slater's constraint qualification is an instance of a constraint qualification that is easily verifiable in particular applications. It is, however, too restrictive in many instances since it excludes the simultaneous occurrence of linear constraints $l(X)\leqslant 0$, $-l(X)\leqslant 0$, which together act as a linear equation and are therefore singular. The following constraint qualification is a slight generalization of one given by Uzawa [*1*]. It requires that

(6.7.1) *all nonlinear constraints* $f_i(X)\leqslant 0$ *in the system* $F(X)\leqslant 0$ *are nonsingular and that*

(6.7.2) $$C_f \cap C^I \neq \emptyset,$$

where $C_f := \{X\in C \mid F(X)\leqslant 0\}$ *is the set of feasible solutions of the convex program* (6.6.1).

We observe that

(6.7.3) $$C_f \cap C^I \neq \emptyset \iff C_f^I \subseteq C^I.$$

Proof. Let $U\in C_f^I$. If U is the only point in $C_f\cap C^I$, then $U\in C^I$ by (6.7.2). Suppose therefore that there exists $V\in C_f\cap C^I$ with $V\neq U$. By lemma (3.2.9), there exists $W\in C_f^I$ such that U belongs to the open segment (W,V). But $(W,V)\subseteq C^I$ by the accessibility lemma (3.2.11). □

The following is but a reformulation of (6.7.1):

(6.7.4) *There exists* $U \in C_f \cap C^I$ *such that*

$$f_i(U) < 0$$

for all nonlinear constraints in $F(X) \leqslant 0$.

Indeed, if $C_f \cap C^I \neq \emptyset$, then by (6.7.3) and (6.6.3) any $U \in C_f^I$ will do.

Slater's constraint qualification (6.6.4) is a special case of Uzawa's constraint qualification (6.7.1). We have to show that $F(U) < 0$ implies $C_f \cap C^I \neq \emptyset$. To this end, let $V \in C^I$. Then for sufficiently small $\lambda > 0$, $\lambda < 1$,

$$F(\lambda V + (1-\lambda) U) \leqslant \lambda F(V) + (1-\lambda) F(U) < 0.$$

For these λ, $\lambda V + (1-\lambda) U \in C_f \cap C^I$ by the accessibility lemma (3.2.11).

Among other things, condition (6.7.2) will allow us to drop the l.s.c. requirement for the functions f and f_i. Consider the convex-concave function

$$\underline{\Phi}(X, Y) := \underline{f}(X) + Y^T \underline{F}(X), \tag{6.7.5}$$

where $\underline{f}$ and $\underline{F}$ are the l.s.c. closures on C (not necessarily on R^n) of f and F, respectively. Then $\underline{\Phi}$ is l.s.c. in X and u.s.c. in Y, and therefore semicontinuous (6.2.4). If $\bar{X}$ is optimal for (6.6.1), then

$$C_f \cap C^I \neq \emptyset \Rightarrow \underline{f}(\bar{X}) = f(\bar{X}) = \min\{\underline{f}(X) \mid \underline{F}(X) \leqslant 0\}. \tag{6.7.6}$$

In other words, if $(\bar{X}, 0)$ solves program I for Φ, then it solves program I also for $\underline{\Phi}$.

Proof of (6.7.6). Clearly, $\overline{C_f} \supseteq \underline{C}_f \supseteq C_f$, where

$$\underline{C}_f := \{X \in C \mid \underline{F}(X) \leqslant 0\}.$$

Hence, $\underline{C}_f^I = C_f^I$ by (3.2.13). According to (6.7.3), $C_f^I \subseteq C^I$. Therefore $\underline{f}(X) = f(X)$ for $X \in C_f^I$ (theorem (4.1.5)). Consider a line segment $\mathscr{H}\{X, U\}$, where $X \in \underline{C}_f$ and $U \in C_f^I$. Then $\underline{f}$ is continuous on $\mathscr{H}\{X, U\}$, and there exists a sequence $\{X_k\}_{k=1,2,\ldots}$ of points in C_f^I such that $X_k \to X$ and $f(X_k) = \underline{f}(X_k) \to \underline{f}(X)$ as $k \to \infty$. Consequently, $\underline{f}(X) \geqslant f(\bar{X})$ for all $X \in \underline{C}_f$.

(6.7.7) *If* $C_f \cap C^I \neq \emptyset$, *and if* $\bar{X}$ *is optimal for* (6.6.1), *then* $(\bar{X}, Y_0)$ *is a saddle point of* Φ *if and only if it is also a saddle point of* $\underline{\Phi}$.

Proof. Since $(\bar{X}, 0)$ solves program I,

$$\Phi(\bar{X}, Y) \leqslant f(\bar{X}) \quad \text{for all} \quad Y \geqslant 0.$$

Since $\underline{f}(\bar{X}) + Y_0^T \underline{F}(\bar{X})$ is an upper bound of $\underline{f}(\bar{X}) + Y^T \underline{F}(\bar{X})$ for all $Y \geqslant 0$, we must have complementary slackness

$$Y_0^T \underline{F}(\bar{X}) = 0,$$

in view of $\underline{F}(\bar{X}) \leqslant F(\bar{X}) \leqslant 0$. In fact, we must even have

$$Y_0^T F(\bar{X}) = 0,$$

since $y_{0i} > 0$ implies $\underline{f}_i(\bar{X}) = 0 \geqslant f_i(\bar{X}) \geqslant \underline{f}_i(\bar{X})$ whence $f_i(\bar{X}) = 0$. Consequently, for all $X \in C$ and $Y \geqslant 0$,

$$\Phi(\bar{X}, Y) \leqslant f(\bar{X}) = \Phi(\bar{X}, Y_0) = \underline{\Phi}(\bar{X}, Y_0) \leqslant \underline{\Phi}(X, Y_0) \leqslant \Phi(X, Y_0).$$

This proves the "if" direction. The "only if" direction is quite easy to show, as

$$\Phi(\bar{X}, Y_0) = \min_{X \in C} \Phi(X, Y_0)$$

implies

$$\Phi(\bar{X}, Y_0) = \underline{\Phi}(\bar{X}, Y_0) = \min_{X \in C} \underline{\Phi}(X, Y_0).$$

Indeed, this gives immediately

$$\underline{\Phi}(\bar{X}, Y) \leqslant \Phi(\bar{X}, Y) \leqslant \Phi(\bar{X}, Y_0) = \underline{\Phi}(\bar{X}, Y_0) \leqslant \underline{\Phi}(X, Y_0)$$

for all $X \in C$ and $Y \geqslant 0$. □

As an application of the duality theorem (6.4.8) we now obtain the

(6.7.8) **Saddle Point Theorem of Kuhn and Tucker** [*1*]. *Let f be the objective function, and F the vector of constraint functions of the convex program* (6.6.1). *Put* $C := K(f) \cap \bigcap_{i=1}^{m} K(f_i)$. *Then the following statements hold for the Lagrangian* $\Phi(X, Y) := f(X) + Y^T F(X)$:

(i) *If* $(\bar{X}, \bar{Y})$ *is a saddle point of* Φ *in* $C \times R_+^m$, *then* $\bar{X}$ *is an optimal solution of program* (6.6.1).

(ii) *If* $\bar{X}$ *is an optimal solution of* (6.6.1), *then there exists* $Y_0 \geqslant 0$ *such that* $(\bar{X}, Y_0)$ *is a saddle point of* Φ *on* $C \times R_+^m$, *provided the constraint qualification* (6.7.1) *is met.*

Proof. In view of (6.7.6) and (6.7.7) it suffices to establish the low-value property of $\underline{\Phi}$ at $(\bar{X}, 0)$ so that the duality theorem (6.4.8) applies to $\underline{\Phi}$. That is, we have to find an R^n-neighborhood $U(\bar{X})$ and a compact set $B \subseteq R_+^m$ such that for each $X \in U(\bar{X}) \cap C$ there exists $Y \in B$ with

$$\underline{\Phi}(X, Y) \geqslant \Phi(\bar{X}, 0) = \underline{f}(\bar{X}) = f(\bar{X}).$$

To this end we introduce the vector

of all indices i for which the constraints $f_i(X) \leqslant 0$ are singular within the system $F(X) \leqslant 0$, and the vector

$$N$$

of all indices i for which the constraints $f_i(X) \leqslant 0$ are nonsingular. The constraint qualification (6.7.1) ensures that all constraints in the subsystem ${}_SF(X) \leqslant 0$ are linear.

Suppose $X \in \mathscr{M}C$ and ${}_SF(X) \leqslant 0$, and let $U \in C_f^I \subseteq C^I$. Then ${}_SF(U)=0$ and ${}_NF(U)<0$ by (6.6.3).

Since ${}_NF$ is continuous at $U \in C^I$, there exists a point W on the line segment (X,U) such that ${}_NF(W)<0$. In addition, ${}_SF(W) \leqslant 0$, whence $W \in C_f$. Consequently $X \in \mathscr{M} C_f$, and we have shown that

$$\mathscr{M} C_f = \{X \in R^n \mid {}_SF(X) \leqslant 0\} \cap \mathscr{M} C.$$

Without restriction of generality, we may assume that

$$\mathscr{M} C_f = \{X \in R^n \mid {}_SF(X) \leqslant 0\}. \tag{6.7.9}$$

Indeed, we may add linear inequalities $G(X) \leqslant 0$ such that $\mathscr{M}C = \{X \mid G(X) \leqslant 0\}$. This will extend the domain of Φ, but the reader will see at once that any saddle point of the extended Lagrangian gives rise to a saddle point of the original one by reducing the dummy multipliers which are associated with $G(X) \leqslant 0$ to zero.

We proceed to show the low-value property holds if $\underline{\Phi}$ is restricted to

$$(C \cap \mathscr{M} C_f) \times R^n_+,$$

i.e. if only points in the affine hull $\mathscr{M} C_f$ of the set of feasible solutions are considered. Note in this context that $\mathscr{M} \underline{C}_f = \mathscr{M} \underline{C}_f^I = \mathscr{M} C_f^I = \mathscr{M} C_f$.

For some $U \in C_f^I$, let

$$m := \max_{i \in N} \frac{\underline{f}(\bar{X}) - \underline{f}(U)}{\underline{f}_i(U)}.$$

We claim that the low-value property holds for

$$U(\bar{X}) := R^n \supseteq C \cap \mathscr{M} C_f, \qquad B := \{Y \mid 0 \leqslant y_i \leqslant m \quad \text{for} \quad i=1,2,\ldots,m\}.$$

Suppose that $X \in C \cap \mathscr{M} C_f$. If $X \in \underline{C}_f$, then $\Phi(X,0) = f(X) \geqslant \underline{f}(\bar{X})$ by (6.7.6). If $X \notin \underline{C}_f$, then consider the last point $\tilde{X} = \lambda U + \mu X$, $\lambda, \mu > 0$, $\lambda + \mu = 1$, with $\tilde{X} \in \underline{C}_f$ on the line segment $\mathscr{H}\{U, X\}$. Such a point exists since the functions $\underline{f}$ and $\underline{f}_i$ are l.s.c., and by (4.1.10) continuous when restricted to $\mathscr{H}(U, \tilde{X})$. $\tilde{X}$ is determined by an index k such that $\underline{f}_k(\tilde{X}) = 0$. Now define $Y \in B$ by putting

$$y_i := \begin{cases} m & \text{if} \quad i=k, \\ 0 & \text{otherwise.} \end{cases}$$

By convexity,

$$\underline{f}_k(X) \geqslant \frac{\underline{f}_k(\tilde{X}) - \lambda \underline{f}_k(U)}{\mu} = -\frac{\lambda}{\mu}\underline{f}_k(U),$$

$$\underline{f}(X) \geqslant \frac{\underline{f}(\tilde{X}) - \lambda \underline{f}(U)}{\mu} \geqslant \frac{\underline{f}(\bar{X}) - \lambda \underline{f}(U)}{\mu}.$$

Consequently

$$\begin{aligned}\underline{\Phi}(X,Y) &= \underline{f}(X) + m\underline{f}_k(X)\\ &\geqslant \frac{\underline{f}(\bar{X}) - \lambda \underline{f}(U)}{\mu} - \left(\frac{\underline{f}(\bar{X}) - \underline{f}(U)}{\underline{f}_k(U)}\right)\left(\frac{\lambda}{\mu}\underline{f}_k(U)\right) = \underline{f}(\bar{X}).\end{aligned}$$

We can thus apply the duality theorem (6.4.8) to the restriction of $\underline{\Phi}$ to $(C \cap \mathscr{M} C_f) \times R_+^m$. We obtain the existence of $Y_1 \geqslant 0$ with

$$\phi(X) := \underline{\Phi}(X, Y_1) \geqslant \underline{f}(\bar{X}) = \underline{f}(\bar{X}) \quad \text{for all} \quad X \in C \cap \mathscr{M} C_f$$

and

$$\phi(\bar{X}) = f(\bar{X}).$$

Now consider the epigraph $[\phi]$, where ϕ is defined as above for $X \in C$ and ∞ otherwise. This epigraph is a convex set which is supported by the manifold

$$M := \left\{\begin{pmatrix} X \\ z \end{pmatrix} \in R^{n+1} \;\middle|\; X \in \mathscr{M} C_f,\ z = f(\bar{X})\right\}.$$

By the general supporting plane theorem (3.4.11), there exists a supporting plane

$$E = \left\{\begin{pmatrix} X \\ z \end{pmatrix} \in R^{n+1} \;\middle|\; z - f(\bar{X}) = A^T(X - \bar{X})\right\} \supseteq M$$

so that

$$\phi(X) - f(\bar{X}) \geqslant A^T(X - \bar{X}) \quad \text{for all} \quad X. \tag{6.7.10}$$

Since $E \supseteq M$, the equations $A^T(X - \bar{X}) = 0$, and a fortiori

$$-A^T(X - \bar{X}) \leqslant 0,$$

are consequences of the homogeneous linear inequalities

$${}_SF(X) - {}_SF(\bar{X}) \leqslant 0.$$

Hence, by the Farkas lemma (1.4.8), there exists a vector $Z \geqslant 0$ with

$$Z^T({}_SF(X)) = -A^T(X - \bar{X}).$$

Let Y_2 be the m-vector with ${}_N Y_2 := 0$ and ${}_S Y_2 := Z$. Then

$$Y_0 := Y_1 + Y_2 \geqslant 0$$

gives rise to a saddle point $(\bar{X}, Y_0)$ of $\underline{\Phi}$. Indeed, since $\bar{X} \in C \cap \mathcal{M} C_f$,

$$\underline{\Phi}(\bar{X}, Y_0) = \phi(\bar{X}) = f(\bar{X}),$$

and by (6.7.10)

$$\Phi(\bar{X}, Y) \leqslant \Phi(\bar{X}, 0) = f(\bar{X}) = \underline{\Phi}(\bar{X}, Y_0) \leqslant \phi(X) - A^T(X - \bar{X}) = \underline{\Phi}(X, Y_0).$$

This completes the proof of the theorem. □

Finally, we remark that in constraint qualification (6.7.1), the condition (6.7.2) is not superfluous, even if there are only linear constraints: Take in R^1

$$f(x) := -\sqrt{x} \quad \text{with} \quad K(f) = \{x \geqslant 0\} =: C,$$
$$F(x) := x.$$

Program (6.6.1) then has the optimal solution $x_0 := 0$. But

$$\Phi(x, y) := -\sqrt{x} + y \cdot x$$

has no saddle point (x, y) in $C \times D$, where $D := \{y \geqslant 0\}$.

Rockafellar [1] determined the conjugates δ_i^c of the convex characteristic functions

$$\delta_i(X) := \begin{cases} 0 & \text{if} \quad f_i(X) \leqslant 0, \\ +\infty & \text{otherwise,} \end{cases}$$

and applied Fenchel's duality theorem in order to derive the Kuhn-Tucker theorem. Another proof is based directly on the properties of systems of convex inequalities. It will be described in section 6.10.

Multiplier theorems for quasiconvex functions have been given by Luenberger [1].

6.8. Duality Theorems and Lagrange Multipliers

Interpreting the Kuhn-Tucker theorem as a saddle point theorem leads to associating with it the two dual program I and II (6.2.5), which we restate without essential change as follows:

(6.8.1) *minimize* $f(X)$ *subject to* $F(X) \leqslant 0$

and

(6.8.2) *maximize* $\inf_{X \in C} (f(X) + Y^T F(X))$ *subject to* $Y \geqslant 0$.

Here C denotes again the intersection of the domains of finiteness of all functions concerned

$$C := K(f) \cap \bigcap_{i=1}^{m} K(f_i).$$

The "primal program" (6.8.1) is in fact our original convex program (6.6.1).

Theorem (6.7.8) states already one part of a duality theorem: if $\bar{X}$ is an optimal solution of program (6.8.1), then there exists $Y_0 \geq 0$ such that $(\bar{X}, Y_0)$ is a saddle point of Φ and therefore an optimal solution of program (6.8.2):

$$f(\bar{X}) + Y_0^T F(\bar{X}) = \min_{X \in C} (f(X) + Y_0^T F(X)) = \max_{Y \geq 0} \inf_{X \in C} (f(X) + Y^T F(X)).$$

This statement was subject to a constraint qualification as, for instance, (6.7.1) or the one by Slater (6.6.4).

The other part of the duality theorem, which leads from an optimal solution $(\bar{X}, \bar{Y})$ of the dual program (6.8.2) to a saddle point $(X_1, \bar{Y})$ of the Lagrangian function $\Phi(X, Y) := f(X) + Y^T F(X)$, and therefore to an optimal solution X_1 of program (6.8.1) is merely an application of the general theorems (6.4.8), (6.4.9), and (6.4.16):

(6.8.3) **Theorem.** (Stoer [*1*], [*2*]). *Let*

$$\Phi(X, Y) := f(X) + Y^T F(X)$$

be the Lagrangian of the convex program (6.8.1), *where the objective function f and the n constraint functions f_i in F are closed and continuous. Suppose that $(\bar{X}, \bar{Y})$ is an optimal solution of the dual program* (6.8.2). *Then the following statements hold:*

(i) *There exists $X_0 \in C$ such that $(X_0, \bar{Y})$ is a saddle point of Φ in $C \times R_+^m$ if and only if Φ has the high-value property at $(\bar{X}, \bar{Y})$.*

(ii) *There exists $X_0 \in C$ such that $(X_0, \bar{Y})$ is a saddle point of Φ on $C \times R_+^m$ if the set*

$$A_Y := \{X \mid \Phi(X, \bar{Y}) = \Phi(\bar{X}, \bar{Y}) = \inf_{Z \in C} \Phi(Z, \bar{Y})\}$$

is bounded.

(iii) *If $\Phi(\cdot, \bar{Y})$ is strictly convex in some neighborhood of $\bar{X}$, then $(\bar{X}, \bar{Y})$ is a saddle point of Φ on $C \times R_+^m$.*

(iv) *In cases (i) and (ii), the points X_0 and $\bar{X}$ need not be equal. They are, however, connected by the relation*

$$f(X_0) = \Phi(X_0, \bar{Y}) = f(X_0) + \bar{Y}^T F(X_0) = \Phi(\bar{X}, \bar{Y}) = f(\bar{X}) + \bar{Y}^T F(\bar{X}).$$

The formulation of the second program (6.8.2) can be further developed if the functions f and f_i are differentiable in C^I. Under fairly general conditions, the minimum of $\Phi(\cdot, Y)$ is actually assumed in C^I. It then must be a stationary point with

$$\operatorname{grad}_X \Phi(X, Y) = \operatorname{grad} f(X)^T + Y^T \operatorname{Grad} F(X)^T = 0.$$

The dual program then takes the form:

$$\text{(6.8.4)} \qquad \textit{maximize} \quad f(X) + Y^T F(X) \quad \textit{subject to}$$
$$\operatorname{grad} f(X)^T + Y^T \operatorname{Grad} F(X)^T = 0, \qquad Y \geqslant 0, \qquad X \in C^I.$$

We repeat that every optimal solution of (6.8.4) is also an optimal solution of the dual program (6.8.2). An optimal solution $(\bar{X}, \bar{Y})$ of the dual program (6.8.2), however, solves (6.8.4) if and only if

$$\bar{X} \in C^I.$$

This will be automatically the case if C is an open subset of R^n, in particular, if $C = R^n$. It will also be enforced, if $C = K(f)$ and f satisfies the regularity condition (4.5.1).

In the special case $C = R^n$, Wolfe [*2*], Huard [*1*], Hanson [*1*] and Mangasarian [*1*] have investigated the dual programs (6.8.1) and (6.8.4). The Kuhn-Tucker Theorem considered as a duality theorem leading from optimal solutions of the primal program (6.8.1) to those of the dual program (6.8.4) is exactly the duality theorem of Wolfe. A converse theorem, leading from the dual to the primal program has been established by Huard, Hanson and Mangasarian, however, under considerably more restrictive conditions.

Note finally that the dual programs (6.8.1) and (6.8.4) become the (generalized) dual problems of Dorn [*2*] (5.8.7), (5.8.8) if there are only linear constraints.

6.9. Constrained Minimax Programs

Let $\Phi(X, Y)$ be a convex-concave function defined on the set $C \times D$, where $C \in R^n$ and $D \in R^m$ are convex sets. Suppose further that

$$F = \begin{bmatrix} f_1 \\ \vdots \\ f_s \end{bmatrix}, \qquad G = \begin{bmatrix} g_1 \\ \vdots \\ g_t \end{bmatrix},$$

are vectors of convex functions f_i and concave functions g_i such that $K(f_i) \supseteq C$, $i = 1, \ldots, s$, and $K(g_i) \supseteq D$, $i = 1, \ldots, t$. Consider the following pair of constrained dual programs:

(6.9.1) *program I:* *Find* $\inf_{\substack{X \in C \\ F(X) \leqslant 0}} \sup_{\substack{Y \in D \\ G(Y) \geqslant 0}} \Phi(X, Y)$,

program II: *Find* $\sup_{\substack{Y \in D \\ G(Y) \geqslant 0}} \inf_{\substack{X \in C \\ F(X) \leqslant 0}} \Phi(X, Y)$.

Clearly, if we define the convex sets of feasible solutions

$$C_f := \{X \in C \mid F(X) \leqslant 0\},$$
$$D_f := \{Y \in D \mid G(Y) \geqslant 0\},$$

and the convex-concave function Φ_f as the restriction of Φ to $C_f \times D_f$, then these constrained programs are only a special case of the programs I and II (6.2.5) with Φ_f, C_f, D_f replacing Φ, C, D, respectively. Now define the function

$$\Psi(X, U; Y, V) := \Phi(X, Y) + V^T F(X) + U^T G(Y)$$

for all $X \in C$, $U \geqslant 0$, $Y \in D$, $V \geqslant 0$. We call Ψ the

(6.9.2) *Lagrangian*

function of the constrained programs I and II (6.9.1). It is a convex-concave function on $C_1 \times C_2 \subseteq R^{n+t} \times R^{m+s}$, where

$$C_1 := C \times R^t_+, \qquad D_1 := D \times R^s_+ .$$

The function Ψ gives rise to a pair of unconstrained dual minimax programs I and II of the type (6.2.5):

(6.9.3) *program I:* *Find* $\inf_{(X,U) \in C_1} \sup_{(Y,V) \in D_1} \Psi(X, U; Y, V)$,

program II: Find $\sup_{(Y,V) \in D_1} \inf_{(X,U) \in C_1} \Psi(X, U; Y, V)$.

We call these programs the

Lagrangian programs associated with (6.9.1).

Finally, we introduce a third convex-concave function, the restriction Ψ_f of Ψ to the convex set $C_2 \times D_2$, where

$$C_2 := C_f \times R^t_+, \qquad D_2 := D_f \times R^s_+,$$

and call Ψ_f the

(6.9.4) *restricted Lagrangian*

of the programs (6.8.1). To Ψ_f belong the

restricted Lagrangian programs

(6.9.5) *program I:* *Find* $\inf_{(X,U)\in C_2} \sup_{(Y,V)\in D_2} \Psi_f(X,U;Y,V)$,

program II: *Find* $\sup_{(Y,V)\in D_2} \inf_{(X,U)\in C_2} \Psi_f(X,U;Y,V)$.

We proceed to examine the relationship between dual problems (6.9.1), (6.9.3), and (6.9.5), in particular the relationship between saddle points of the respective Lagrangian functions Φ_f, Ψ, and Ψ_f.

(6.9.6) **Theorem.** (i) *Φ_f has a saddle point $(X_0,Y_0)\in C_f\times D_f$ if and only if Ψ_f has a saddle point $(X_0,U_0;Y_0,V_0)\in C_2\times D_2$. In the latter case*

$$V_0^T F(X_0)=U_0^T G(Y_0)=0.$$

(ii) *A saddle point $(X_0,U_0;Y_0,V_0)\in C_1\times D_1$ of Ψ is also a saddle point of Ψ_f.*

Proof. Suppose that $(X_0,Y_0)\in C_f\times D_f$ is a saddle point of Φ_f:

$$\Phi(X_0,Y)\leqslant\Phi(X_0,Y_0)\leqslant\Phi(X,Y_0) \quad \text{for all} \quad X\in C_f \quad \text{and} \quad Y\in D_f.$$

Then

$$\begin{aligned}\Psi(X_0,0;Y_0,0)&=\Phi(X_0,Y_0)\leqslant\Phi(X,Y_0)\leqslant\Phi(X,Y_0)+U^T G(Y_0)\\ &=\Psi(X,U;Y_0,0) \quad \text{for} \quad (X,U)\in C_2\end{aligned}$$

and

$$\begin{aligned}\Psi(X_0,0;Y_0,0)&=\Phi(X_0,Y_0)\geqslant\Phi(X_0,Y)\geqslant\Phi(X_0,Y)+V^T F(X_0)\\ &=\Psi(X_0,0;Y,V) \quad \text{for} \quad (Y,V)\in D_2.\end{aligned}$$

This proves one direction of (i).

To prove the other direction of (i) (and simultaneously part (ii) of the theorem), suppose that $(X_0,U_0;Y_0,V_0)$ is a saddle point of Ψ_f (Ψ):

$$\begin{aligned}\Phi(X_0,Y)+V^T F(X_0)+U_0^T G(Y)&\leqslant\Phi(X_0,Y_0)+V_0^T F(X_0)+U_0^T G(Y_0)\\ &\leqslant\Phi(X,Y_0)+V_0^T F(X)+U^T G(Y_0)\end{aligned}$$

for all $(X,U)\in C_2\,(C_1)$ and $(Y,V)\in D_2\,(D_1)$. It follows at once that $G(Y_0)\geqslant 0$, because otherwise the right hand term would not be bounded below as U ranges in R_+^t. Similarly $F(X_0)\leqslant 0$. Therefore $(X_0,Y_0)\in C_f\times D_f$ (and $(X_0,U_0;Y_0,V_0)$ is seen to be a saddle point of Ψ_f, which proves (ii)). Furthermore, $U_0^T G(Y_0)$ and $V_0^T F(X_0)$ must both vanish for $(X_0, U_0; Y_0, V_0)$ to be a saddle point. Therefore

$$\Phi(X_0, Y) \leqslant \Phi(X_0, Y) + U_0^T G(Y) \leqslant \Phi(X_0, Y_0) \leqslant \Phi(X, Y_0) + V_0^T F(X)$$
$$\leqslant \Phi(X, Y_0)$$

for all $X \in C_f$ and $Y \in D_f$. □

The relationship between saddle points of Φ_f and Ψ is not as simple. As may be expected in view of the behavior of the standard Lagrangian function $f(X) + Y^T F(X)$ (sections 6.6, 6.7), some constraint qualification is needed. As such we shall use the analogue to Uzawa's constraint qualification (6.7.1):

(6.9.7) *All nonlinear constraints $f_i(X) \leqslant 0$ and $g_i(X) \geqslant 0$ are nonsingular within the systems $F(X) \leqslant 0$ and $G(X) \geqslant 0$, respectively. In addition, $C_f \cap C^I \neq \emptyset$ and $D_f \cap D^I \neq \emptyset$.*

Again, the latter conditions actually imply $C_f^I \subseteq C^I$ and $D_f^I \subseteq D^I$. Furthermore, $f_i(X) < 0$ and $g_j(Y) > 0$ holds for all nonlinear constraints, if $X \in C_f^I$ and $Y \in D_f^I$.

(6.9.8) **Theorem.** (i) *If the Lagrangian Ψ (6.9.2) has a saddle point $(X_0, U_0; Y_0, V_0) \in C_1 \times D_1$, i.e. if*

(6.9.9)

(a) $\Phi(X_0, Y_0) + V_0^T F(X_0) + U_0^T G(Y_0) \leqslant \Phi(X, Y_0) + V_0^T F(X) + U^T G(Y_0)$,

(b) $\Phi(X_0, Y_0) + V_0^T F(X_0) + U_0^T G(Y_0) \geqslant \Phi(X_0, Y) + V^T F(X_0) + U_0^T G(Y)$,

for all $X \in C$, $U \geqslant 0$, $Y \in D$, $V \geqslant 0$, then $(X_0, Y_0) \in C_f \times D_f$ is a saddle point of Φ_f:

(6.9.10) $\quad \Phi(X_0, Y) \leqslant \Phi(X_0, Y_0) \leqslant \Phi(X, Y_0)$ *for all* $X \in C_f$, $Y \in D_f$.

(ii) *If Φ_f has a saddle point $(X_0, Y_0) \in C_f \times D_f$, that is, if (6.9.10) holds, and if the constraint qualification (6.9.7) is met, then there are $U_0 \geqslant 0$, $V_0 \geqslant 0$ such that $(X_0, U_0; Y_0, V_0) \in C_1 \times D_1$ is a saddle point of Ψ, that is (6.9.9) holds.*

(iii) *If $(X_0, U_0; Y_0, V_0)$ is a saddle point of Ψ, then complementary slackness holds:*

$$U_0^T G(Y_0) = V_0^T F(X_0) = 0.$$

Proof. (i) Follows from (6.9.6).

(ii) Suppose that (6.9.10) holds. Then $X_0 \in C_f$ is the optimal solution of the program

$$\text{Minimize} \quad \Phi(X, Y_0) \quad \text{subject to} \quad X \in C, \quad F(X) \leqslant 0.$$

As there is an $X_1 \in C^I \cap C_f$ such that $f_i(X_1) < 0$ for all nonlinear f_i, it follows from the saddle point theorem (6.7.8) of Kuhn and Tucker that there is a $V_0 \geqslant 0$ such that $V_0^T F(X_0) = 0$ and

$$\Phi(X_0, Y_0) \leqslant \Phi(X, Y_0) + V_0^T F(X)$$

for all $X \in C$. Dually, it follows in the same way that there is a $U_0 \geqslant 0$ with

$$U_0^T G(Y_0) = 0$$

and

$$\Phi(X_0, Y_0) \geqslant \Phi(X_0, Y) + U_0^T G(Y)$$

for all $Y \in D$. Hence, we have

$$\begin{aligned}\Phi(X_0, Y_0) &\doteq \Phi(X_0, Y_0) + V_0^T F(X_0) + U_0^T G(Y_0)\\ &\leqslant \Phi(X, Y_0) + V_0^T F(X) \qquad \text{for all} \quad X \in C\\ &\leqslant \Phi(X, Y_0) + V_0^T F(X) + U^T G(Y_0) \quad \text{for all} \quad X \in C \quad \text{and} \quad U \geqslant 0,\end{aligned}$$

as $Y_0 \in D_f$ and $G(Y_0) \geqslant 0$, that is, (6.9.9.(a)) holds. A symmetrical argument yields (6.9.9.(b)).

(iii) is a trivial consequence of (i). □

The above results on saddle points of general Lagrangian functions relate to a duality theorem which has been obtained by Dantzig, Eisenberg, and Cottle [*1*] for differentiable convex functions. Suppose that

$$\operatorname{grad}_X \Phi(X, Y), \qquad \operatorname{Grad} F(X)$$

exist for all $X \in C^I$ and all $Y \in D$, and that

$$\operatorname{grad}_Y \Phi(X, Y), \qquad \operatorname{Grad} G(Y)$$

exist for $Y \in D^I$ and all $X \in C$. Here Grad F stands again for the transpose of the Jacobian matrix (6.5.3), and Grad G is understood correspondingly.

Then the pairs of dual programs (6.9.3) and (6.9.5) can be restated in a way that is analogous to the transformation of program (6.8.2) into program (6.8.4). Consider program (6.9.3) and suppose that $(X_0, U_0; Y_0, V_0)$ is optimal for program I (6.9.3) with $X_0 \in C^I$ and $Y_0 \in D^I$. The latter conditions will be automatically satisfied if $\Phi(.,Y)$ and $\Phi(X,.)$ both fulfill the regularity condition (4.5.1). Now by definition (6.9.2) of Ψ,

$$\inf_{(X,U)\in C_1} \sup_{(Y,V)\in D_1} (\Phi(X,Y)+V^T F(X)+U^T G(Y))$$
$$= \inf_{(X,U)\in C_2} \sup_{Y\in D} (\Phi(X,Y)+U^T G(Y))$$
$$= \inf\{\Phi(X,Y)+U^T G(Y) \mid \operatorname{grad}_Y\Phi(X,Y)^T + U^T \operatorname{Grad} G(Y)^T = 0,\quad (X,U)\in C_2,\ Y\in D^I\}.$$

Thus, if $(X_0, U_0; Y_0, V_0)$ is optimal for both programs I and II (6.9.3) and satisfies $X_0 \in C^I$, $Y_0 \in D^I$, then by (6.9.6) $(X_0, Y_0) \in C_f \times D_f$, and $(X_0, U_0; Y_0, V_0)$ is also a common optimal solution of

(6.9.11) *program I:* *Minimize* $\Phi(X,Y)+U^T G(Y)$ *subject to*

$\operatorname{grad}_Y \Phi(X,Y)^T + U^T \operatorname{Grad} G(Y)^T = 0,$

$X\in C, \quad Y\in D^I,$

$F(X)\leqslant 0, \quad G(Y)\geqslant 0, \quad U\geqslant 0.$

program II: *Maximize* $\Phi(X,Y)+V^T F(X)$ *subject to*

$\operatorname{grad}_X \Phi(X,Y) + V^T \operatorname{Grad} F(X)^T = 0,$

$Y\in D, \quad X\in C^I,$

$G(Y)\geqslant 0, \quad F(X)\leqslant 0, \quad V\geqslant 0.$

For the special case, $C := R^n$, $D := R^m$, $F(X) := -X$, $G(Y) = Y$, these programs reduce to the dual programs studied by Dantzig, Eisenberg, and Cottle [*1*]:

(6.9.12) *program I:* *Minimize* $\Phi(X,Y) - Y^T \operatorname{grad}_Y\Phi(X,Y)$ *subject to*

$X\geqslant 0, \quad Y\geqslant 0, \quad \operatorname{grad}_Y\Phi(X,Y)\leqslant 0.$

Program II: *Maximize* $\Phi(X,Y) - X^T \operatorname{grad}_X\Phi(X,Y)$ *subject to*

$X\geqslant 0, \quad Y\geqslant 0, \quad \operatorname{grad}_X\Phi(X,Y)\geqslant 0.$

Duality theorems for these programs, as well as their generalized versions (6.9.11) are obtained in obvious fashion by combining the general theorems (6.4.8), (6.4.9), (6.4.16) with the theorems (6.9.6) and (6.9.8). They will not be listed.

Rockafellar [*3*], [*8*] has linked the theory of convex-concave functions $\Phi(X,Y)$ to the theory of conjugate functions: Permitting infinite values, a function $\Phi: R^{m+n} \to R\cup\{\pm\infty\}$ is called a convex-concave saddle-function, if $\Phi(.,Y)$ is convex in the wider sense (4.2.5) for all $Y\in R^n$ and $\Phi(X,.)$ is concave in the wider sense for all $X\in R^m$. Concave-convex saddle-functions Ψ are defined similarly. By skew-conjugacy, a convex-concave saddle-function Φ gives rise to the convex (in the wider sense) function

$$F(X,V) := \sup_Y (\Phi(X,V) + V^T Y)$$

and the concave (in the wider sense) function

$$G(U,Y):=\inf_X(\Phi(X,Y)-U^T Y),$$

called its "convex parent", and "concave parent", respectively. Φ is called closed if F and G are (skew-) conjugate to each other:

$$F(X,V)=\sup_Y \sup_U(G(U,Y)+U^T X+V^T Y),$$

$$G(U,Y)=\inf_X \inf_V (F(X,V)-Y^T V-X^T U)$$

(then F and G are strongly closed (4.7.5), (4.7.6)). Let $[\Phi]$ denote the equivalence class of all (closed) convex-concave saddle-functions with the same skew-conjugate parents F and G. Equivalence classes $[\Psi]$ of closed concave-convex functions are defined analogously. Then, by Rockafellar [8], one has a four way one-to-one correspondence between strongly closed convex functions F, strongly closed concave functions G, equivalence classes $[\Phi]$ of closed convex-concave saddle-functions Φ and equivalence classes $[\Psi]$ of closed concave-convex saddle-functions, given by the above mappings and

$$F\to[\Phi]:\quad \Phi(X,Y):=\inf_V(F(X,V)-V^T Y),$$

$$[\Phi]\to[\Psi]:\quad \bar{\Psi}(U,V):=\inf_X \sup_Y(\Phi(X,Y)-U^T X+V^T Y),$$

$$\underline{\Psi}(U,V):=\sup_Y \inf_X(\Phi(X,Y)-U^T X+V^T Y),$$

$$\bar{\Psi},\underline{\Psi}\in[\Psi]$$

$$[\Psi]\to G:\quad G(U,Y)=\inf_V(\Psi(U,V)-Y^T V),$$

and similar mappings.

This four way correspondence gives rise to the following programming problems:

(I) Find $\bar{X}\geqslant 0$, $\bar{V}\geqslant 0$ such that $F(\bar{X},\bar{V})$ is finite and $F(X,V)\geqslant F(\bar{X},\bar{V})$ for $X\geqslant 0$, $V\geqslant 0$.

(II) Find $\bar{U}\geqslant 0$, $\bar{Y}\geqslant 0$ such that $G(\bar{U},\bar{Y})$ is finite and $G(U,Y)\leqslant G(\bar{U},\bar{Y})$.

(III) Find $\bar{X}\geqslant 0$, $\bar{Y}\geqslant 0$ such that $\Phi(\bar{X},\bar{Y})$ is finite and $\Phi(X,\bar{Y})\geqslant\Phi(\bar{X},\bar{Y})\geqslant\Phi(\bar{X},Y)$ for all $X\geqslant 0, Y\geqslant 0$.

(IV) Find $\bar{U}\geqslant 0$, $\bar{V}\geqslant 0$ such that $\Psi(\bar{U},\bar{V})$ is finite and $\Psi(U,\bar{V})\leqslant\Psi(\bar{U},\bar{V})\leqslant\Psi(\bar{U},V)$ for all $U\geqslant 0$, $V\geqslant 0$.

Rockafellar proved in [8] the following duality theorem: If $K(F)^I\cap\{X\geqslant 0, V\geqslant 0\}\neq\emptyset$ and $K(G)^I\cap\{U\geqslant 0, Y\geqslant 0\}\neq\emptyset$ then programs (I)–(IV) have a common optimal solution $\bar{X}\geqslant 0$, $\bar{Y}\geqslant 0$, $\bar{U}\geqslant 0$, $\bar{V}\geqslant 0$ with $F(\bar{X},\bar{V})=G(\bar{U},\bar{Y})=\Phi(\bar{X},\bar{Y})=\Psi(\bar{U},\bar{V})$.

Another recent generalization of the dual problems (6.9.12) and the corresponding duality theorem is due to Balas [3]. He considers convex-concave functions $\Phi(s,t;.,.)\colon R^m\times R^n\to R$ depending on two parameters $s\in\Omega_1$, $t\in\Omega_2$, ranging in their respective parameter domains Ω_1,Ω_2. That is, for each $(s,t)\in\Omega_1\times\Omega_2$, $Y\in R^n$, $\Phi(s,t;.,Y)$ is a convex function on R^m and for each $(s,t)\in\Omega_1\times\Omega_2$, $X\in R^m$, $\Phi(s,t;X,.)$ is a concave function on R^n. Further he requires that these convex and concave functions are everywhere twice differentiable and that Φ is either separable with respect to s or t, that is Φ can be written either in the form

$$\Phi(s,t;X,Y)=\Phi_1(s)+\Phi_2(t;X,Y)$$

or in the form

$$\Phi(s,t;X,Y)=\Phi_1(t)+\Phi_2(s;X,Y)$$

with certain functions Φ_1, Φ_2.

The dual problems of Balas then are (compare (6.9.12)).

(6.9.13) *Program I:*

Find $\sup_t \inf_s \inf_{X,Y}\{\Phi(s,t;X,Y)-Y^T \operatorname{grad}_Y \Phi(s,t;X,Y) \mid (s,t;X,Y)\in Z_1\}$

where

$$Z_1:=\{(s,t;X,Y) \mid s\in\Omega_1,\ t\in\Omega_2,\ X\geqslant 0,\ Y\geqslant 0,\ \operatorname{grad}_Y\Phi(s,t;X,Y)\leqslant 0\};$$

(6.9.14) *Program II:*

Find $\inf_s \sup_t \sup_{X,Y}\{\Phi(s,t;X,Y)-X^T \operatorname{grad}_X \Phi(s,t;X,Y) \mid (s,t;X,Y)\in Z_2\}$

where

$$Z_2:=\{s,t;X,Y) \mid s\in\Omega_1,\ t\in\Omega_2,\ X\geqslant 0,\ Y\geqslant 0,\ \operatorname{grad}_X\Phi(s,t;X,Y)\geqslant 0\}.$$

If, for instance, Φ is separable with respect to s,

$$\Phi(s,t;X,Y)=\Phi_1(s)+\Phi_2(t;X,Y),$$

and $(\bar{s},\bar{t},\bar{X},\bar{Y})$ is optimal for (6.9.13), then it is easily seen that $(\bar{X},\bar{Y})$ is an optimal solution of the following program of type (6.9.12) for the convex-concave function $\Phi_2(\bar{t};.,.)$:

(6.9.15) *Program I': Minimize* $\Phi_2(\bar{t};X,Y)-Y^T \operatorname{grad}_Y \Phi_2(\bar{t};X,Y)$ *subject to*

$$X\geqslant 0, \qquad Y\geqslant 0, \qquad \operatorname{grad}_Y\Phi_2(\bar{t};X,Y)\leqslant 0.$$

In this way, any duality theorem for the programs (6.9.12) gives rise to a duality theorem for the programs (6.9.13), (6.9.14). Again, they will not be listed. It should be noted that the dual programs of Balas are very general, since the hypotheses concerning the parameters $s\in\Omega_1$, $t\in\Omega_2$ are rather weak (separability of Φ). By suitable specialization, e.g. $\Omega_1=\{s\in R^n \mid s\geqslant 0,\ s_i \text{ integer}\}$ etc., one can find duality theorems in integer programming, as explained in Balas [*1*], [*2*], [*3*].

6.10. Systems of Convex Inequalities

This section deals with systems of convex inequalities, i.e. systems of the form

$$f_i(X)\leqslant 0, \qquad i\in I,$$

originating from a,—possibly infinite—, set of convex functions $f_i(X)$. Several results of chapter 1 on systems of linear inequalities carry over to the general convex case. This is particularly true for finite systems of strict convex inequalities. In the presence, however, of inequalities which are not strict, difficulties arise. Additional hypotheses are necessary, which turn out to be nothing else but the constraint qualifications encountered in sections 6.6 and 6.7. The theory leads to a further proof of the saddle point theorem of Kuhn-Tucker (6.7.8).

One has in analogy to Stiemke's transposition theorem (1.6.4)

(6.10.1) **Theorem.** *Let $f_i(X)$, $i=1,2,\ldots,m$, be convex functions in R^n with $C:=\bigcap_i K(f_i)\neq\emptyset$. The system*

$$F(X):=\begin{bmatrix} f_1(X) \\ f_2(X) \\ \cdots\cdots \\ f_m(X) \end{bmatrix}<0,$$

is unsolvable if and only if there is a nonnegative vector $Z\geqslant 0$, $Z\in R^m$, $Z\neq 0$ such that

$$Z^T F(X)\geqslant 0 \quad \textit{for all} \quad X\in C.$$

Proof. The "if"-part being trivial let us turn to the "only if"-part of (6.10.1). Define in R^m the set

$$A:=\{V\in R^m \mid V>F(X) \text{ for some } X\in C\}.$$

Clearly, A is a nonvoid convex set, since the functions f_i are convex. Moreover, A does not contain the origin 0 of R^m, as there is no $X\in C$ such that $F(X)<0$. By the separation theorem (3.3.9), there is a plane

$$E=\{V \mid Z^T V=b\}, \quad Z\neq 0$$

separating 0 from A in R^m:

$$Z^T V\geqslant 0=Z^T 0 \quad \text{for all} \quad V\in A.$$

The construction of A implies at once $Z\geqslant 0$. Now assume $Z^T F(X_0)<0$ for some $X_0\in C$, and let $V>F(X_0)$. Then $V\in A$ and therefore $Z^T V\geqslant 0$. Also

$$V_\lambda:=F(X_0)+\lambda(V-F(X_0))\in A$$

for all $\lambda>0$, with contradicts the fact that $Z^T V_\lambda<0$ for sufficiently small λ. Thus $Z^T F(X)\geqslant 0$ for all $X\in C$, which proves the theorem in view of $Z\neq 0$. □

Theorem (6.10.1) can be sharpened by invoking Helly's first theorem (3.7.1). By hypothesis, the intersection $\bigcap_{1\leqslant i\leqslant m} K_i$ of the convex sets $K_i:=\{X \mid f_i(X)<0 \text{ and } X\in C\}$ is void. Hence there is a subsystem

$$f_{i_j}(X)<0, \quad j=1,2,\ldots,k, \quad k\leqslant n+1$$

of at most $n+1$ inequalities having no solution $X\in C$. Therefore it is no restriction of generality to assume that

(6.10.2) *the vector Z in theorem* (6.10.1) *has at most* $n+1$ *nonvanishing components.*

(6.10.3) **Corollary to Theorem** (6.10.1) (*Bohnenblust, Karlin and Shapley [1]*). *Let* $f_i(X)$, $i \in I$, *be a (possibly infinite) family of closed convex functions in* R^n *such that the intersection* $\bigcap_{i \in I} K(f_i) =: C$ *is a compact convex set. If the system of inequalities*

$$f_i(X) \leqslant 0, \qquad i \in I$$

has no solution X *in* C, *there are* $m \leqslant n+1$ *indices* $i_j \in I$, $j=1,2,\dots,m$, *and nonnegative numbers* z_j, $j=1,2,\dots,m$, *such that*

$$\inf_{X \in C} \{z_1 f_{i_1}(X) + \cdots + z_m f_{i_m}(X)\} > 0.$$

Proof. Define for each $i \in I$ and each $\varepsilon > 0$ the set

$$K_{i,\varepsilon} := \{X \in R^n \mid f_i(X) \leqslant \varepsilon \quad \text{and} \quad X \in C\},$$

which is closed, convex, and compact since it is contained in the compact set C. By hypothesis, the intersection

$$\bigcap_{\substack{i \in I \\ \varepsilon > 0}} K_{i,\varepsilon} = \emptyset$$

is empty. By Helly's second theorem (3.7.4) there are finitely many sets

$$K_{i_j,\varepsilon_j}, \qquad j=1,2,\dots,m, \qquad m \leqslant n+1,$$

among the $K_{i,\varepsilon}$ such that $\bigcap_{j=1}^{m} K_{i_j,\varepsilon_j} = \emptyset$. Therefore, the finite system

$$f_{i_j}(X) - \varepsilon_j < 0, \qquad j=1,2,\dots,m, \qquad m \leqslant n+1,$$

has no solution $X \in C$. By theorem (6.10.1), there are numbers $z_j \geqslant 0$, not all of them zero, with

$$\sum_{j=1}^{m} z_j f_{i_j}(X) \geqslant \sum_{j=1}^{m} z_j \varepsilon_j > 0 \qquad \text{for all} \qquad X \in C. \quad \square$$

Consider now the following system of convex inequalities and linear equations

(6.10.5)
$$f(X) < 0, \qquad F(X) \equiv \begin{bmatrix} f_1(X) \\ f_2(X) \\ \vdots \\ f_l(X) \end{bmatrix} \leqslant 0, \qquad G(X) \equiv AX - B = 0,$$

where A is an $m \times n$-matrix, and $B \in R^m$. Because of the nonstrict inequalities in this system, a constraint qualification is needed. We choose the following one, which is related to Uzawa's constraint qualification (6.7.1):

(6.10.6) *There exists* $U \in C^I$ *with* $F(U) < 0$ *and* $G(U) = 0$, *where*

$$C := K(f) \cap \bigcap_{i=1}^{l} K(f_i).$$

We then have the

(6.10.7) **Theorem.** *Suppose the system of convex inequalities* (6.10.5) *meets the constraint qualification* (6.10.6). *Then it has no solution if and only if there exist vectors* $Y_1 \geqslant 0$, Y_2 *with*

$$f(X) + Y_1^T F(X) + Y_2^T G(X) \geqslant 0 \quad \textit{for all} \quad X \in C.$$

Proof. In analogy to the proof of theorem (6.10.1), we consider the nonempty convex set

$$A := \left\{ \begin{bmatrix} v_0 \\ V_1 \\ V_2 \end{bmatrix} \in R^{1+l+m} \;\middle|\; v_0 > f(X),\ V_1 \geqslant F(X),\ V_2 = G(X) \text{ for some } X \in C \right\}.$$

Again nonsolvability of the system (6.10.5) entails that the origin of R^{1+l+m} is not contained in A. By the separation theorem (3.3.9), there exists a plane

$$E = \left\{ \begin{bmatrix} v_0 \\ V_1 \\ V_2 \end{bmatrix} \;\middle|\; z_0 v_0 + Z_1^T V_1 + Z_2^T V_2 = b \right\}, \quad \begin{bmatrix} z_0 \\ Z_1 \\ Z_2 \end{bmatrix} \neq \begin{bmatrix} 0 \\ 0 \\ 0 \end{bmatrix},$$

which separates the origin from A,

$$(6.10.8) \qquad z_0 v_0 + Z_1^T V_1 + Z_2^T V_2 \geqslant b \geqslant 0 \quad \text{for all} \quad \begin{bmatrix} v_0 \\ V_1 \\ V_2 \end{bmatrix} \in A,$$

without containing both A and the origin. If $b = 0$, this implies at once $E \not\supseteq A$. Hence

$$(6.10.9) \qquad \text{there exists} \quad \begin{bmatrix} \bar{v}_0 \\ \bar{V}_1 \\ \bar{V}_2 \end{bmatrix} \in A \quad \text{with} \quad z_0 \bar{v}_0 + Z_1^T \bar{V}_1 + Z_2^T \bar{V}_2 > 0.$$

The same holds clearly if $b > 0$. As in the proof of theorem (6.10.1), it is readily seen that

(6.10.10)

$$z_0 \geqslant 0, \quad Z_1 \geqslant 0, \quad \textit{and} \quad z_0 f(X) + Z_1^T F(X) + Z_2^T G(X) \geqslant 0 \quad \textit{for} \quad X \in C.$$

It remains to be shown that $z_0>0$, because then $Y_1:=\frac{1}{z_0}Z_1$ and $Y_2:=\frac{1}{z_0}Z_2$ are suitable.

Assume $z_0=0$. Let then U be according to (6.10.6). By (6.10.10), $Z_1^T F(U)\geqslant 0$. Since both $F(U)<0$ and $Z_1\geqslant 0$,

$$Z_1=0.$$

By (6.10.9), $Z_2^T \bar{V}_2>0$. By definition of A, $\bar{V}_2=G(\bar{X})$ for some $\bar{X}\in C$. By lemma (3.2.9),

$$U-\varepsilon(\bar{X}-U)\in C$$

for some $\varepsilon>0$. By the linearity of G,

$$G(U-\varepsilon(\bar{X}-U))=(1+\varepsilon)G(U)-\varepsilon G(\bar{X})=-\varepsilon G(\bar{X}),$$

and by (6.10.10),

$$Z_2^T G(U-\varepsilon(\bar{X}-U))=-\varepsilon Z_2^T G(\bar{X})\geqslant 0,$$

contradicting $Z_2^T G(\bar{X})=Z_2^T \bar{V}_2>0$. Thus $z_0>0$, and the theorem is proved. ▯

We shall need a slight modification of theorem (6.10.7) in order to arrive at our formulation of the saddle-point theorem (6.7.8) of Kuhn-Tucker. Consider the system of convex inequalities

(6.10.11) $$f(X)<0, \qquad F(X)\equiv\begin{bmatrix} f_1(X)\\ f_2(X)\\ \vdots\\ f_m(X)\end{bmatrix}\leqslant 0$$

and choose the constraint qualification (6.7.1) of Uzawa, letting the function f play the role of the objective function in (6.7.1). More precisely, if we split the system $F(X)\leqslant 0$ into its nonsingular part

$${}_N F(X)\leqslant 0$$

and its singular part

$${}_S F(X)\leqslant 0,$$

the latter consisting of those inequalities $f_i(X)\leqslant 0$ which are singular in $F(X)\leqslant 0$, then there exists $U\in C^I$ so that ${}_N F(U)<0$ and ${}_S F(U)=0$, where again

$$C:=K(f)\cap\bigcap_{i=1}^{m}K(f_i).$$

(6.10.12) **Theorem.** *Suppose the system of convex inequalities* (6.10.11) *meets the constraint qualification* (6.7.1). *Then it has no solution if and only if there exists a vector* $Y \geqslant 0$ *with*

$$f(X) + Y^T F(X) \geqslant 0 \quad \textit{for all} \quad X \in C.$$

Proof. The system

$$f(X) < 0, \quad {}_N F(X) \leqslant 0, \quad {}_S F(X) = 0$$

is of the form (6.10.5), meeting the constraint qualification (6.10.6). By theorem (6.10.7), there exists Y such that ${}_N Y \geqslant 0$ and

$$f(X) + Y^T F(X) \geqslant 0 \quad \text{for all} \quad X \in C. \tag{6.10.13}$$

Let k be the number of singular constraints. If $k=0$, then $Y = {}_N Y \geqslant 0$ and nothing has to be shown. We proceed then by induction over k. Suppose that the theorem has been proved for $k-1$ singular inequalities, and that k singular inequalities

$$f_{i_j}(X) \leqslant 0, \quad j = 1, \ldots, k$$

are present. By singularity, the k systems, $s = 1, \ldots, k$,

$$\begin{aligned} &{}_N F(X) \leqslant 0, \\ &f_{i_j}(X) \leqslant 0, \quad j \neq s, \\ &f_{i_s}(X) < 0, \end{aligned}$$

are nonsolvable. Then the induction hypothesis yields vectors $Y^{(s)} \geqslant 0$, $s = 1, \ldots, k$, such that

$$Y^{(s)T} F(X) \geqslant 0, \quad y_{i_s}^{(s)} = 1.$$

Adding suitable multiples of these vectors $Y^{(s)}$ to the vector Y in (6.10.13) will make the latter nonnegative. Note that one can always achieve $y_i = 0$ for at least one of the singular constraints $f_i(X) \leqslant 0$. □

Theorem (6.10.12) is but a rewording of part (ii) of the saddle point theorem (6.7.8) of Kuhn-Tucker. Indeed if

$$f(\bar{X}) = \min\{f(X) \mid F(X) \leqslant 0\}$$

for some vector F of convex functions which satisfy the constraint qualification (6.7.1), then

$$f(X) - f(\bar{X}) < 0, \quad F(X) \leqslant 0,$$

is not solvable, and by theorem (6.10.12) there exists $Y \geqslant 0$ such that

$$f(X) - f(\bar{X}) + Y^T F(X) \geqslant 0 \quad \text{for all} \quad X \in C.$$

Bibliography

Abadie, J.: [*1*] The dual to Fourier's method for solving linear inequalities. International Symposium on Mathematical Programming. London 1964.

— [*2*] (ed.) Nonlinear programming. Amsterdam: North-Holland Publ. Co. 1967.

— [*3*] On the Kuhn-Tucker theorem. In Abadie [*2*] (ed.) 19—36 (1967).

Alexandroff, A. D.: [*1*] Almost everywhere existence of the second differential of a convex function and some properties of convex surfaces connected with it. Leningrad State University Annals, Uchenje Zapiski, Math. Ser. **6,** 3—35 (1939).

— [*2*] Konvexe Polyeder. (German translation. Russian version publ. 1950). Berlin: VEB Deutsch. Verl. d. Wiss. 1958.

Alexandroff, P. S.: [*1*] Combinatorial Topology; English translation. Vol. 1. Rochester, N. Y.: Graylock 1956.

—, und H. Hopf: [*1*] Topologie. Berlin: Springer 1935. New York: Chelsea 1965.

Anderson, R. D., and V. L. Klee, Jr.: [*1*] Convex functions and upper semicontinuous collections. Duke Math. J. **19,** 349—357 (1952).

Antosiewicz, H. A.: [*1*] (ed.) Proc. second symposium in linear programming. 2 vols. (mimeographed). Washington: U.S. Air Force 1955.

— [*2*] A theorem on alternatives for pairs of matrices. Pacific J. Math. **5,** 641—642 (1955).

Arrow, K. J., and A. C. Enthoven: [*1*] Quasi-concave programming. Econometrica **29,** 779—800 (1961).

—, and L. Hurwicz: [*1*] Reduction of constrained maxima to saddle-point problems. In Neyman [*1*] (ed.) vol. V, 1—20 (1956).

—, —, and H. Uzawa: [*1*] (eds.) Studies in linear and non-linear programming. Stanford, Calif.: Stanford Univ. Press 1958.

— [*2*] Constraint qualifications in maximization problems. Naval Res. Logist. Quart. **8,** 175—191 (1961).

Asplund, E.: [*1*] Fréchet differentiability of convex functions. Acta Math. **121,** 31—47 (1968).

—, and R.T. Rockafellar: [*1*] Gradients of convex functions. Trans. Amer. Math. Soc. **139,** 443—467 (1969).

Aumann, G.: [*1*] Reelle Funktionen. Berlin-Göttingen-Heidelberg: Springer 1954.

Balakrishnan, A.V.: [*1*] Control theory and the calculus of variations. New York: Academic Press 1969.

Balas, E.: [*1*] Duality in discrete programming. Technical report No. 67—5. Dept. of Operations Res., Stanford University 1967.

Balas, E.: [*2*] Duality in discrete programming II. The quadratic case. Management Sciences Research Report No. 116. Carnegie-Mellon University, Pittsburgh 1967.

— [*3*] Duality in discrete programming III. Nonlinear objective functions and constraints. Management Sciences Research Report No. 121. Carnegie-Mellon University, Pittsburgh 1968.

— [*4*] Duality in discrete programming IV. Applications. Management Sciences Research Report No. 145. Carnegie-Mellon University, Pittsburgh 1968.

Balinski, M. L.: [*1*] On the graph structure of convex polyhedra in *n*-space. Pacific J. Math. **11,** 431—434 (1961).

Baer, R.: Linear algebra and projective geometry. New York: Academic Press 1952.

Beckenbach, E. F.: [*1*] Convex Functions. Bull. Amer. Math. Soc. **54,** 439—460 (1948).

—, and R. Bellman: [*1*] Inequalities. Berlin-Göttingen-Heidelberg: Springer 1961.

Bellman, R., and K. Fan: [*1*] On systems of linear inequalities in hermitian matrix variables. In Klee [*22*] (ed.) 1—11 (1963).

—, and M. Hall, Jr.: [*1*] (eds.) Combinatorial Analysis. Proc. Symp. Appl. Math. **10.** Providence, R. I.: Amer. Math. Soc. 1960.

—, and W. Karush: [*1*] Mathematical programming and the maximum transform. SIAM J. Appl. Math. **10,** 550—567 (1962).

Ben-Israel, A.: [*1*] Notes on linear inequalities 1: The intersection of the non-negative orthant with complementary orthogonal subspaces. J. Math. Anal. Appl. **9,** 303—314 (1964).

Berge, C.: [*1*] Théorie générale des jeux à n personnes. Memoir. Sci. Math. **138.** Paris: Gauthiers-Villars 1957.

— [*2*] Espaces topologiques et fonctions multivoques. Paris: Dunod 1959.

— [*3*] Sur une propriété combinatoire des ensembles convexes. C. R. Acad. Sci. Paris **248**, 2698 (1959).

—, et A. Ghouila-Houri: [*1*] Programmes, jeux et réseaux de transport. Paris: Dunod 1962.

Bernau, S. J.: On bare points. J. Austral. Math. Soc. **9,** 25—28 (1969).

Bernholtz, B.: [*1*] A new derivation of the Kuhn-Tucker conditions. Operations Res. **12,** 295—299 (1964).

Besicovitch, A. S.: [*1*] On singular points of convex surfaces. In Klee [*22*] (ed.) 21—23 (1963).

Birkhoff, G.: [*1*] Lattice theory. 3rd ed. Amer. Math. Soc. Colloq. Publ. **25.** Providence, R. I.: Amer. Math. Soc. 1967.

—, and S. MacLane: [*1*] A survey of modern algebra. 3rd ed. New York: Macmillan 1965.

Bishop, E., and R. R. Phelps: [*1*] The support functionals of a convex set. In Klee [*22*] (ed.) 27—35 (1963).

Björck, G.: [*1*] The set of extreme points of a compact convex set. Ark. Mat. **3,** 463—468 (1958).

Blaschke, W.: [*1*] Kreis und Kugel. Leipzig: Veit 1916. New York: Chelsea 1949.

Bliss, G. A.: [*1*] Lectures on the calculus of variations. Chicago: Univ. of Chicago Press 1946.

Blumenthal, L. M.: [1] Metric methods in linear inequalities. Duke Math. J. **15,** 955—966 (1948).

— [2] Two existence theorems for systems of linear inequalities. Pacific J. Math. **2,** 523—530 (1952).

Bohnenblust, H. F., S. Karlin, and L. S. Shapley: [1] Games with continuous convex pay-off. In Kuhn and Tucker [2] (eds.) 181—192 (1950).

Bonnesen, T., und W. Fenchel: [1] Theorie der konvexen Körper. Berlin: Springer 1934. New York: Chelsea 1948.

Bonnice, W., and V. L. Klee: [1] The generation of convex hulls. Math. Ann. **152,** 1—29 (1963).

Botts, T.: [1] Convex Sets. Amer. Math. Monthly **49,** 527—535 (1942).

Bourbaki, N.: [1] Éléments de mathématique. Livre II: Algèbre. Chapitre II: Algèbre linéaire. Act. Sci. et Ind. Nr. **1032.** Paris: Hermann 1947.

— [2] Éléments de mathématique. Livre III: Topologie générale. Chapitres I, II. Act. Sci. et Ind. Nr. **1142.** 4th ed. Paris: Hermann 1965.

— [3] Éléments de mathématique. Livre V: Espaces vectoriels topologiques. Act. Sci. et Ind. Nr. **1189.** Paris: Hermann 1953.

Bourgin, D. G.: [1] Un indice dei punti uniti. Nota 1, 2, 3. Atti Acad. Naz. dei Lincei **19,** 435—440 (1955), **20,** 43—48 (1956), **21,** 395—400 (1956).

Bram, J.: [1] The Lagrange multiplier theorem for max-min with several constrains. SIAM J. Appl. Math. **14,** 665—667 (1966).

Braunschweiger, C. C.: [1] An extension of the nonhomogeneous Farkas theorem. Amer. Math. Monthly **69,** 969—975 (1962).

—, and H. E. Clark: [1] An extension of the Farkas theorem. Amer. Math. Monthly **69,** 272—277 (1962).

Brøndsted, A.: [1] Conjugate convex functions in topological vector spaces. Mat.-Fys. Medd. Danske Vid. Selsk. **34,** No. 2, 1—26 (1964).

—, and R. T. Rockafellar: [1] On the subdifferentiability of convex functions. Proc. Amer. Math. Soc. **16,** 605—611 (1965).

Brouwer, L. E. J.: [1] Über Abbildungen von Mannigfaltigkeiten. Math. Ann. **71,** 97—115 (1912).

— [2] Über den natürlichen Dimensionsbegriff. J. Reine Angew. Math. **142,** 146—152 (1913).

Browder, F. E.: [1] On a generalization of the Schauder fixed point theorem. Duke Math. J. **26,** 291—303 (1959).

— [2] Another generalization of the Schauder fixed point theorem. Duke Math. J. **32,** 399—406 (1965).

— [3] A further generalization of the Schauder fixed point theorem. Duke Math. J. **32,** 575—578 (1965).

— [4] Nonexpansive nonlinear operators in a Banach space. Proc. Nat. Acad. Sci. U.S.A. **54,** 1041—1044 (1965).

Burger, E.: Über homogene lineare Ungleichungssysteme. Z. angew. Math. Mech. **36,** 135—139 (1956).

Busemann, H.: [1] Convex Surfaces. New York: Interscience 1958.

— [2] Convexity on Grassmann manifolds. Enseign. Math. ser. 2, **7,** 139—152 (1961—1962).

—, and W. Feller: [1] Krümmungseigenschaften konvexer Flächen. Acta Math. **66,** 1—47 (1936).

—, G. Ewald, and G. C. Shephard: [1] Convex bodies and convexity on Grassmann cones. Parts I to IV. Math. Ann. **151,** 1—41 (1963).

Canon, M., C. Cullum, and E. Polak: [*1*] Constrained minimization problems in finite-dimensional spaces. J. SIAM Control **4,** 528—547 (1966).

Carathéodory, C.: [*1*] Über den Variabilitätsbereich der Koeffizienten von Potenzreihen, die gegebene Werte nicht annehmen. Math. Ann. **64,** 95—115 (1907).

— [*2*] Über den Variabilitätsbereich der Fourierschen Konstanten von positiven harmonischen Funktionen. Rend. Circ. Mat. Palermo **32,** 193—217 (1911).

Carver, W. B.: [*1*] Systems of linear inequalities. Ann. of Math. **23,** 212—220 (1921—1922).

Černikov, S. N.: [*1*] Systems of linear inequalities. (Russian) Uspehi Mat. Nauk., N. S. **8,** No. 2, 7—73 (1953).

Charnes, A., and W. W. Cooper: [*1*] Management models and industrial applications of linear programming. Vol. I, II. New York: John Wiley 1961.

— [*2*] Programming with linear fractional functionals. Naval Res. Log. Quart. **9,** 181—186 (1962).

—, —, and K. Kortanek: [*1*] A duality theory for convex programs with convex constraints. Bull. Amer. Math. Soc. **68,** 605—608 (1962).

— [*2*] Duality, Haar programs, and finite sequence spaces. Proc. Nat. Acad. Sci. U.S.A. **48,** 783—786 (1962).

— [*3*] Duality in semi-infinite programs and some works of Haar and Carathéodory. Management Sci. **9,** 209—228 (1963).

— [*4*] Semi-infinite programs which have no duality gap. Management Sci. **12,** 113—121 (1965).

Choquet, G.: [*1*] Theory of capacities. Ann. Inst. Fourier (Grenoble) **5,** 131—295 (1953—1954).

— [*2*] Unicité des représentations intégrales au moyen des points extrémaux dans les cônes convexes. C. R. Acad. Sci. Paris **243,** 555—557 (1956).

— [*3*] Existence des représentations intégrales au moyen des points extrémaux dans les cônes convexes. C. R. Acad. Sci. Paris **243,** 699—702 (1956).

— [*4*] Le théorème de représentation intégrale dans les ensembles convexes compacts. Ann. Inst. Fourier (Grenoble) **10,** 333—344 (1960).

—, H. Corson, and V. Klee: [*1*] Exposed points of convex sets. Pacific. J. Math. **17,** 33—44 (1966).

Corson, H. H.: [*1*] A compact convex set in E^3 whose exposed points are of the first category. Proc. Amer. Math. Soc. **16,** 1015—1021 (1965).

—, and V. Klee: [*1*] Topological classification of convex sets. In Klee [*22*] (ed.) 37—51 (1963).

Cottle, R. W.: [*1*] Symmetric dual programs. Quart. Appl. Math. **21,** 237—243 (1963).

— [*2*] A theorem of Fritz John in mathematical programming. Memorandum RM-3P58-PR. Santa Monica, Calif.: RAND Corp. 1963.

— [*3*] Note on a fundamental theorem in quadratic programming. SIAM J. Appl. Math. **12,** 663—665 (1964).

— [*4*] Nonlinear programs with positively bounded Jacobians. SIAM J. Appl. Math. **14,** 147—158 (1966).

Danskin, J. M.: [*1*] The theory of max-min, with applications. SIAM J. Appl. Math. **14,** 641—664 (1966).

— [*2*] The theory of max-min and its applications to weapons allocation problems. Lecture Notes in Operations Research and Mathematical Economics **5.** Berlin-Heidelberg-New York: Springer 1967.

Dantzig, G. B.: [1] Linear programming and extensions. Princeton: Princeton Univ. Press 1963.
—, and A. F. Veinott, Jr.: [1] (eds.) Mathematics of the decision sciences. Part I. Lect. Appl. Math. **11.** Providence, R. I.: Amer. Math. Soc. 1968.
—, E. Eisenberg, and R. W. Cottle: [1] Symmetric dual non-linear programs. Pacific J. Math. **15,** 809—812 (1965).

Danzer, L., B. Grünbaum, and V. Klee: [1] Helly's theorem and its relatives. In Klee [22] (ed.) 101—180 (1963).

Davis, C.: [1] The intersection of a linear subspace with the positive orthant. Michigan Math. J. **1,** 163—168 (1952).
— [2] Remarks on a previous paper. Michigan Math. J. **2,** 23—25 (1953—1954).
— [3] Theory of positive linear dependence. Amer. J. Math. **76,** 733—746 (1954).
— [4] The set of non-linearity of a convex piecewise-linear function. Scripta. Math. **24,** 219—228 (1959).

Day, M. M.: [1] Normed linear spaces. Berlin-Göttingen-Heidelberg: Springer 1958. 2nd printing corr. Berlin: Springer 1962.

Dennis, J. B.: [1] A dual problem for a class of quadratic programs. M.I.T. Res. Note **1,** Nov. 1957.
— [2] Mathematical Programming and Electrical Networks. Cambridge, Mass.: Technology Press 1959.

De Santis, R.: A generalization of Helly's theorem. Proc. Amer. Math. Soc. **8,** 336—340 (1957).

Dieter, U.: [1] Dualität bei konvexen Optimierungs-(Programmierungs-)Aufgaben. Z. Unternehmensforschung **9,** 91—111 (1965).
— [2] Optimierungsaufgaben in topologischen Vektorräumen I: Dualitätstheorie. Z. Wahrscheinlichkeitstheorie verw. Geb. **5,** 89—117 (1966).

Dieudonné, J.: [1] Foundations of modern analysis. New York: Academic Press 1960.

Dines, L. L.: [1] Systems of linear inequalities. Ann. of Math. **20,** 191—199 (1918—1919).
— [2] Convex extension and linear inequalities. Bull. Amer. Math. Soc. **42,** 353—365 (1936).
—, and N. H. McCoy: [1] On linear inequalities. Trans. Roy. Soc. Canada, Sect. III **27,** 37—70 (1933).

Dorn, W. S.: [1] Duality in quadratic programming. Quart. Appl. Math. **18,** 155—162 (1960).
— [2] A duality theorem for convex programs. IBM J. of Res. and Devel. **4,** 407—413 (1960).
— [3] A symmetric dual theorem for quadratic programs. J. Operations Res. Soc. Japan **2,** 93—97 (1960).
— [4] Self-dual quadratic programs. J. Soc. Indust. Appl. Math. **9,** 51—54 (1961).

Dubins, L. E.: [1] On extreme points of convex sets. J. Math. Anal. Appl. **5,** 237—244 (1962).

Duffin, R. J.: [1] Infinite programs. In Kuhn-Tucker [4] (eds.) 157—171 (1956).
— [2] Dual programs and minimum cost. J. Soc. Indust. Appl. Math. **10,** 119—123 (1962).
—, and E. L. Peterson: [1] Duality theory for geometric programming. SIAM J. Appl. Math. **14,** 1307—1349 (1966).

Dunford, N., and J. T. Schwartz: [1] Linear operators, Part I: General theory. New York: Interscience 1966.

Eggleston, H. G.: [*1*] Convexity. Cambridge: Cambridge University Press 1958.

—, B. Grünbaum, and V. Klee: [*1*] Some semicontinuity theorems for convex polytopes and cell-complexes. Comment. Math. Helv. **39**, 165—188 (1964).

Eidelheit, M.: [*1*] Zur Theorie der konvexen Mengen in linearen normierten Räumen. Studia Math. **6**, 104—111 (1936).

Eilenberg, S., and D. Montgomery: [*1*] Fixed point theorems for multivalued transformations. Amer. J. Math. **68**, 214—222 (1946).

Eisenberg, E.: [*1*] Duality in homogeneous programming. Proc. Amer. Math. Soc. **12**, 783—787 (1961).

— [*2*] On cone functions. In Graves-Wolfe [*1*] (eds.) 26—33 (1963).

Ellis, J. W.: [*1*] A general set-separation theorem. Duke Math. J. **19**, 417—421 (1952).

Ewald, G., and G. C. Shephard: [*1*] Normed vector spaces consisting of classes of convex sets. Math. Z. **91**, 1—19 (1966).

Falk, J. E.: [*1*] Lagrange multipliers and nonlinear programming. J. Math. Anal. Appl. **19**, 141—159 (1967).

Fan, K.: [*1*] A generalization of Tucker's combinatorial lemma with topological applications. Ann. of Math. **56**, 431—437 (1952).

— [*2*] Fixed-point and minimax theorems in locally convex topological linear spaces. Proc. Nat. Acad. Sci. U.S.A. **38**, 121—126 (1952).

— [*3*] Minimax theorems. Proc. Nat. Acad. Sci. U.S.A. **39**, 42—47 (1953).

— [*4*] On systems of linear inequalities. In Kuhn and Tucker [*4*] (eds.) 99—156 (1956).

— [*5*] Existence theorems and extreme solutions for inequalities concerning convex functions of linear transformations. Math. Z. **68**, 205—216 (1957—1958).

— [*6*] On the equilibrium value of a system of convex and concave functions. Math. Z. **70**, 271—280 (1958—1959).

— [*7*] Convex sets and their applications. Argonne Natl. Lab., Appl. Math. Div. Summer Lectures 1959.

— [*8*] A generalization of Tychonoff's fixed point theorem. Math. Ann. **142**, 305—310 (1961).

— [*9*] On the Krein-Milman theorem. In Klee [*22*] (ed.) 211—219 (1963).

—, I. Glicksberg, and A. J. Hoffman: [*1*] Systems of inequalities involving convex functions. Proc. Amer. Math. Soc. **8**, 617—622 (1957).

Farkas, J.: [*1*] Theorie der einfachen Ungleichungen. J. Reine Angew. Math. **124**, 1—27 (1902).

Fenchel, W.: [*1*] On conjugate convex functions. Canad. J. Math. **1**, 73—77 (1949).

— [*2*] Convex cones, sets and functions. Lecture notes by D. W. Blackett, Princeton University 1953.

Fenske, C.: [*1*] Lokales Fixpunktverhalten bei stetigen Abbildungen in kompakten konvexen Mengen. Forschungsberichte des Landes Nordrhein-Westfalen **1931**, 17—52, Köln: Westdeutscher Verlag 1968.

Fourier, J.-B. J.: [*1*] Solution d'une question particulière du calcul des inégalités. 1826. Oeuvres II. 317—328.

Gaddum, J. W.: [*1*] A theorem on convex cones with applications to linear inequalities. Proc. Amer. Math. Soc. **3**, 957—960 (1952).

Gale, D.: [*1*] Convex polyhedral cones and linear inequalities. In Koopmans [*1*] (ed.) 287—297 (1951).

Gale, D.: [2] Linear combinations of vectors with non-negative coefficients. Amer. Math. Monthly **59**, 46—47 (1952).
— [3] On convex polyhedra. Bull. Amer. Math. Soc. **61**, 556 (1955).
— [4] Neighboring vertices on a convex polyhedron. In Kuhn-Tucker [4] (eds.) 255—263 (1956).
— [5] Theory of linear economic models. New York: McGraw-Hill 1960.
— [6] Neighborly and cyclic polytopes. In Klee [22] (ed.) 225—232 (1963).
— [7] On the number of faces of a convex polytope. Canad. J. Math. **16**, 12—17 (1964).
— [8] A geometric duality theorem with economic application. Review of Econ. Studies **34**, 19—24 (1967).
—, and V. Klee: [1] Continuous convex sets. Math. Scand. **7**, 379—391 (1959).
—, H. W. Kuhn, and A. W. Tucker: [1] Linear programming and the theory of games. In Koopmans [1] (ed.) 317—329 (1951).

Gerstenhaber, M.: [1] Theory of convex polyhedral cones. In Koopmans [1] (ed.) 298—316 (1951).

Ghouila-Houri, A.: [1] Sur l'étude combinatoire des familles de convexes. Compt. Rend. Acad. Sci. Paris **252**, 494—496 (1961).

Glicksberg, I.: [1] A further generalization of the Kakutani fixed point theorem, with applications to Nash equilibrium points. Proc. Amer. Math. Soc. **3**, 170—174 (1952).

Goldman, A. J.: [1] Resolution and separation theorems for polyhedral convex sets. In Kuhn-Tucker [4] (eds.) 41—51 (1956).
—, and A. W. Tucker: [1] Polyhedral convex cones. In Kuhn-Tucker [4] (eds.) 19—40 (1956).
— [2] Theory of linear programming. In Kuhn-Tucker [4] (eds.) 53—97 (1956).

Gol'stein, E. G.: [1] Dual problems of convex and fractionally convex programming in function spaces. Dokl. Akad. Nauk SSR **172**, 1007—1010 (1967). Engl. translation: Soviet Math. Dokl. **8**, 212—216 (1967).

Good, R. A.: [1] Systems of linear relations. SIAM Review **1**, 1—31 (1959).

Gordan, P.: [1] Über die Auflösung linearer Gleichungen mit reellen Coefficienten. Math. Ann. **6**, 23—28 (1873).

Gr(a)eub, W. (H.): [1] Lineare Algebra. Berlin-Göttingen-Heidelberg: Springer 1960.
— [2] Linear algebra. 3rd edition. Berlin-Heidelberg-New York: Springer 1967.
— [3] Multilinear algebra. Berlin-Heidelberg-New York: Springer 1967.

Graves, R. L., and P. Wolfe: [1] (ed.) Recent advances in mathematical programming. New York: McGraw-Hill 1963.

Grünbaum, B.: [1] On a theorem of L. A. Santaló. Pacific J. Math. **5**, 351—359 (1955).
— [2] The dimension of intersections of convex sets. Pacific J. Math. **12**, 197—202 (1962).
— [3] A generalization of theorems of Kirszbraun and Minty. Proc. Amer. Math. Soc. **13**, 812—814 (1962).
[4] Common secants for families of polyhedra. Arch. Math. **15**, 76—80 (1964).
— [5] On the facial structure of convex polytopes. Bull. Amer. Math. Soc. **71**, 599—600 (1965).
— [6] Convex Polytopes. London: Interscience 1967.
—, and T. S. Motzkin: [1] On polyhedral graphs. In Klee [22] (ed.) 285—290 (1963).

Gustin, W.: [*1*] On the interior of the convex hull of a euclidean set. Bull. Amer. Math. Soc. **53,** 299—301 (1947).

Haar, A.: [*1*] Über lineare Ungleichungen. Acta. Math. Szeged **2,** 1—14 (1924).

Hadley, G.: [*1*] Linear algebra. Reading, Mass.: Addison-Wesley 1961.

— [*2*] Linear programming. Reading, Mass.: Addison-Wesley 1962.

— [*3*] Nonlinear and dynamic programming. Reading, Mass.: Addison-Wesley 1964.

Hadwiger, H.: [*1*] Altes und Neues über konvexe Körper. Basel: Birkhäuser 1955.

Halkin, H.: [*1*] Necessary and sufficient condition for a convex set to be closed. Amer. Math. Monthly **73,** 628—630 (1966).

— [*2*] A property of nonseparated convex sets. Proc. Amer. Math. Soc. **17,** 1389—1395 (1966).

— [*3*] Convexity and Control theory. Functional Analysis and Optimization. New York: Academic Press 1966.

Hammer, P. C.: [*1*] Convex bodies associated with a convex body. Proc. Amer. Math. Soc. **2,** 781—793 (1951).

— [*2*] Maximal convex sets. Duke Math. J. **22,** 103—106 (1955).

— [*3*] Semispaces and the topology of convexity. In Klee [*22*] (ed.) 305—316 (1963).

Hanner, O., and H. Rådström: [*1*] A generalization of a theorem of Fenchel. Proc. Amer. Math. Soc. **2,** 589—593 (1951).

Hanson, M. A.: [*1*] A duality theorem in nonlinear programming with nonlinear constraints. Austral. J. Statist. **3,** 64—72 (1961).

— [*2*] Bounds for functionally convex optimal control problems. J. Math. Anal. Appl. **8,** 84—89 (1964).

Helly, E.: [*1*] Über Mengen konvexer Körper mit gemeinschaftlichen Punkten. Jber. Deutsch. Math. Verein. **32,** 175—176 (1923).

— [*2*] Über Systeme von abgeschlossenen Mengen mit gemeinschaftlichen Punkten. Monatsh. Math. Phys. **37,** 281—302 (1930).

Hermes, H.: [*1*] Einführung in die Verbandstheorie. 2nd ed. Berlin-Heidelberg-New York: Springer 1967.

Hirschfeld, R. A.: [*1*] On a minimax theorem of K. Fan. Proc. Koninkl. Nederl. Akad. Wetensch. Amsterdam. Ser. A **61,** 470—474 (1958).

Hoffman, A. J.: [*1*] Some recent applications of the theory of linear inequalities to extremal combinatorial analysis. In Bellman-Hall [*1*] (eds.) 113—127 (1960).

—, and J. B. Kruskal: [*1*] Integral boundary points of convex polyhedra. In Kuhn-Tucker [*4*] (eds.) 223—246 (1956).

—, and M. H. McAndrew: [*1*] Linear inequalities and analysis. Amer. Math. Monthly **71,** 416—418 (1964).

Horn, A.: [*1*] Some generalizations of Helly's theorem on convex sets. Bull. Amer. Math. Soc. **55,** 923—929 (1949).

Huard, P.: [*1*] Dual programs. IBM J. of Res. and Devel. **6,** 137—139 (1962).

— [*2*] Dual programs. In Graves-Wolfe [*1*] (eds.) 55—62 (1963).

Hurewicz, W., and H. Wallman: [*1*] Dimension theory. rev. ed. Princeton: Princeton Univ. Press 1948.

Hurwicz, L.: [*1*] Programming in linear spaces. In Arrow-Hurwicz-Uzawa [*1*] (eds.) 39—102 (1958).

Hurwicz, L., and H. Uzawa: [*1*] A note on the Lagrangian saddle-points. In Arrow-Hurwicz-Uzawa [*1*] (eds.) 103—113 (1958).

Jackson, J. R.: [*1*] On the existence problem of linear programming. Pacific J. Math. **4,** 29—36 (1954).

John, F.: [*1*] Extremum problems with inequalities as subsidiary conditions. In 'Studies and essays', Courant anniversary volume. New York: Interscience 1948.

Kakutani, S.: [*1*] Ein Beweis des Satzes von M. Eidelheit über konvexe Mengen. Proc. Imp. Acad. Tokyo **13,** 93—94 (1937).

— [*2*] Two fixed point theorems concerning bicompact convex sets. Proc. Imp. Acad. Tokyo **14,** 242—245 (1938).

— [*3*] A generalization of Brouwer's fixed point theorem. Duke Math. J. **8,** 457—459 (1941).

— [*4*] Topological properties of the unit sphere of a Hilbert space. Proc. Imp. Acad. Tokyo **19,** 269—271 (1943).

Karamardian, S.: [*1*] Strictly quasi-convex (concave) functions and duality in Mathematical Programming. J. Math. Anal. Appl. **20,** 344—358 (1967).

Karlin, S.: [*1*] Operator treatment of minmax principle. In Kuhn-Tucker [*2*] (eds.) 133—154 (1950).

— [*2*] Extreme points of vector functions. Proc. Amer. Math. Soc. **4,** 603—610 (1953).

— [*3*] Mathematical methods and theory in games, programming, and economics. 2 vols. Reading, Mass.: Addison-Wesley 1959.

— [*4*] Total positivity and convexity preserving transformations. In Klee [*22*] (ed.) 329—347 (1963).

—, and L. S. Shapley: [*1*] Some applications of a theorem on convex functions. Ann. Math. **52,** 148—153 (1950).

— [*2*] Geometry of moment spaces. Mem. Amer. Math. Soc. No. 12 (1953).

Keller, O.-H.: [*1*] Die Homoiomorphie der kompakten konvexen Mengen im Hilbertschen Raum. Math. Ann. **105,** 748—758 (1931).

Kelley, J. L.: [*1*] Note on a theorem of Krein and Milman. J. Osaka Inst. Sci. Tech., part I, **3,** 1—2 (1951).

— [*2*] General topology. New York: Van Nostrand 1955.

—, and I. Namioka et al.: [*1*] Linear topological spaces. Princeton: Van Nostrand 1963.

Kirchberger, P.: [*1*] Über Tschebyschefsche Annäherungsmethoden. Math. Ann. **57,** 509—540 (1903).

Kirsch, A.: [*1*] Konvexe Figuren als Durchschnitte abzählbar vieler Halbräume. Archiv Math. **18,** 313—319 (1967).

Kirszbraun, M. D.: [*1*] Über die zusammenziehende und Lipschitzsche Transformationen. Fund. Math. **22,** 77—108 (1934).

Klee, V. L.: [*1*] The support property of a convex set in a linear normed space. Duke Math. J. **15,** 767—772 (1948).

— [*2*] A characterization of convex sets. Amer. Math. Monthly **56,** 247—249 (1949).

— [*3*] Dense convex sets. Duke Math. J. **16,** 351—354 (1949).

— [*4*] On certain intersection properties of convex sets. Canad. J. Math. **3,** 272—275 (1951).

— [*5*] Convex sets in linear spaces. I, II, III. Duke Math. J. **18,** 443—466, 875—883 (1951), **20,** 105—111 (1953).

Klee, V. L.: [*6*] Convex bodies and periodic homeomorphisms in Hilbert space. Trans. Amer. Math. Soc. **74**, 10—43 (1953).
— [*7*] The critical set of a convex body. Amer. J. Math. **75**, 178—188 (1953).
— [*8*] Common secants for plane convex sets. Proc. Amer. Math. Soc. **5,** 639—641 (1954).
— [*9*] A note on extreme points. Amer. Math. Monthly **62,** 30—32 (1955).
— [*10*] Some topological properties of convex sets. Trans. Amer. Math. Soc. **78,** 30—45 (1955).
— [*11*] Separation properties of convex cones. Proc. Amer. Math. Soc. **6,** 313—318 (1955).
— [*12*] Strict separation of convex sets. Proc. Amer. Math. Soc. **7,** 735—737 (1956).
— [*13*] The structure of semispaces. Math. Scand. **4,** 54—64 (1956).
— [*14*] Extremal structure of convex sets. Arch. Math. **8,** 234—240 (1957).
— [*15*] Extremal structure of convex sets. II. Math. Z. **69,** 90—104 (1958).
— [*16*] Some characterizations of convex polyhedra. Acta Math. **102,** 79—107 (1959).
— [*17*] Some new results on smoothness and rotundity in normed linear spaces. Math. Ann. **139,** 51—63 (1959—1960).
— [*18*] Asymptotes and projections of convex sets. Math. Scand. **8,** 356—362 (1960).
— [*19*] Polyhedral sections of convex bodies. Acta Math. **103,** 243—267 (1960).
— [*20*] Leray-Schauder theory without local convexity. Math. Ann. **141,** 286—296 (1960).
— [*21*] Convexity of Chebyshev sets. Math. Ann. **142,** 292—304 (1961).
— [*22*] (ed.) Convexity. Proc. Symp. Pure Math. **7.** Providence, R. I.: Amer. Math. Soc. 1963.
— [*23*] On a question of Bishop and Phelps. Amer. J. Math. **83,** 95—98 (1963).
— [*24*] On a theorem of Dubins. J. Math. Anal. Appl. **7,** 425—427 (1963).
— [*25*] Extreme points of convex sets without completeness of the scalar field. Mathematika **11,** 59—63 (1964).
— [*26*] Utility functions and the "lin" operation for convex sets. Israel J. Math. **2,** 191—197 (1964).
— [*27*] A combinatorial analogue of Poincaré's duality theorem. Canad. J. Math. **16,** 517—531 (1964).
— [*28*] Diameters of polyhedral graphs. Canad. J. Math. **16,** 602—614 (1964).
— [*29*] On the number of vertices of a convex polytope. Canad. J. Math. **16,** 701—720 (1964).
— [*30*] A property of d-polyhedral graphs. J. Math. Mech. **13,** 1039—1042 (1964).
— [*31*] Heights of convex polytopes. J. Math. Anal. Appl. **11,** 176—190 (1965).
— [*32*] Paths on polyhedra I. J. Soc. Indust. Appl. Math. **13,** 946—956 (1965).
— [*33*] Paths on polyhedra II. Pacific J. Math. **17,** 249—262 (1966).
— [*34*] Separation and support properties of convex sets—a survey. In Balakrishnan [*1*] (ed.) 235—303 (1969).
—, and D. W. Walkup: [*1*] The d-step conjecture for polyhedra of dimension $d<6$. Acta Math. **117,** 53—78 (1967).

Knaster, B., C. Kuratowski und S. Mazurkiewicz: [*1*] Ein Beweis des Fixpunktsatzes für n-dimensionale Simplexe. Fund. Math. **14,** 132—137 (1929).

Kneser, H.: [*1*] Sur un théoreme fondamental de la théorie des jeux. C. R. Acad. Sci. Paris **234,** 2418—2420 (1952).

Koopmans, T. C.: [*1*] (ed.) Activity analysis of production and allocation. New York: John Wiley 1951.

Köthe, G.: [*1*] Topologische lineare Räume I. Berlin-Göttingen-Heidelberg: Springer 1960.

Krabs, W.: [*1*] Lineare Optimierung in halbgeordneten Vektorräumen. Numer. Math. **11,** 220—231 (1968).

Krasnosselsky, M. A.: [*1*] Sur un critère pour qu'un domain soit étoilé (Russian, with French summary). Mat. Sb. **61** N. S. **19,** 309—310 (1946).

Krein, M.: [*1*] Sur quelques questions de la géometrie des ensembles convexes situés dans un espace linéaire normé et complet. Dokl. Akad. Nauk SSSR N.S. **14,** 5—7 (1937).

—, and Milman, D.: [*1*] On extreme points of regularly convex sets. Studia Math. **9,** 133—138 (1940).

Krelle, W., und H. P. Künzi: [*1*] Lineare Programmierung. Zürich: Verlag Indust. Organisation 1959.

Kretschmer, K. S.: [*1*] Programmes in paired spaces. Canad. J. Math. **13,** 221—238 (1961).

Kuhn, H. W.: [*1*] On a theorem of Wald. In Kuhn-Tucker [*4*] (ed.) 265—273 (1956).

— [*2*] Solvability and consistency for linear equations and inequalities. Amer. Math. Monthly **63,** 217—232 (1956).

— [*3*] Linear inequalities and the Pauli principle. In Bellman-Hall [*1*] (eds.) 141—147 (1960).

— [*4*] Some combinatorial lemmas in topology. IBM. J. of Res. and Devel. **4,** 518—524 (1960).

— [*5*] On a pair of dual linear programs. In Abadie [*2*] (ed.) 37—54 (1967).

—, and A. W. Tucker: [*1*] Nonlinear programming. In Neyman [*1*] (ed.) 481—492 (1950).

— [*2*] (eds.) Contribution to the theory of games. vol. I. Annals of Mathematics Study **24.** Princeton: Princeton University Press 1950.

— [*3*] (eds.) Contribution to the theory of games. vol. II. Annals of Mathematics Study **28.** Princeton: Princeton University Press 1953.

— [*4*] (eds.) Linear inequalities and related systems. Annals of Mathematics Study **38.** Princeton: Princeton University Press 1956.

Kumar, T. K.: [*1*] A duality theorem for continuous-time linear programming problems. Unternehmensforschung **10,** 224—236 (1966).

Künzi, H. P., und W. Krelle: [*1*] Nichtlineare Programmierung. Berlin-Göttingen-Heidelberg: Springer 1962.

Lee, E. B., and L. Markus: [*1*] Foundations of optimal control theory. New York: John Wiley 1967.

Levi, F. W.: [*1*] Eine Ergänzung zum Hellyschen Satze. Arch. Math. **4,** 222—224 (1953).

Levinson, N., and T. O. Sherman: [*1*] The sum of the intersections of a cone with a linear subspace and of the dual cone with orthogonal complementary subspace. J. Combinatorial Theory **1,** 338—349 (1966).

Loomis, L. H.: [*1*] On a theorem of von Neumann. Proc. Nat. Acad. Sci. U.S.A. **32,** 213—215 (1946).

Luenberger, D. G.: [*1*] Quasi-convex programming. J. SIAM. Appl. Math. **16,** 1090—1095 (1968).

— [*2*] Optimization by vector space methods. New York: John Wiley 1969.

Luisternik, L., and V. Sobolev: [*1*] Elements of functional analysis. New York: F. Ungar 1961.

Macbeath, A. M.: [*1*] A compactness theorem for affine equivalence-classes of convex regions. Canad. J. Math. **3,** 54—61 (1951).

Mac Duffee, C. C.: [*1*] What is a matrix? Amer. Math. Monthly **50,** 360—365 (1943).

Mangasarian, O. L.: [*1*] Duality in nonlinear programming. Quart. Appl. Math. **20,** 300—302 (1962).
— [*2*] Pseudo-convex functions. SIAM J. Control **3,** 281—289 (1965).
—, and S. Fromowitz: [*1*] The Fritz John necessary optimality condition in the presence of equality and inequality constraints. J. Math. Anal. Appl. **17,** 37—47 (1967).
—, and J. Ponstein: [*1*] Minmax and duality in nonlinear programming. J. Math. Anal. Appl. **11,** 504—518 (1965).

Marcus, M., and M. Minc: [*1*] Introduction to linear algebra. New York: MacMillan 1965.

Martos, B.: [*1*] A hiperbolikus programozás feladata. Publ. Math. Inst. Hung. Ac. Sc. **5,** ser. B, 383—406 (1960).
— [*2*] The direct power of simplicial methods in continuous programming. International Symposium on Mathematical Programming. London 1964.

Mauer, I.: [*1*] The duality principle in convex programming. Eesti NSV Tead. Akad. Toimetised Füüs.-Mat. **16,** 186—195 (1967).

Mazur, S.: [*1*] Une remarque sur l'homéomorphie des champs fonctionnels. Studia Math. **1,** 83—85 (1929).
— [*2*] Über die kleinste konvexe Menge, die eine gegebene kompakte Menge enthält. Studia Math. **2,** 7—9 (1930).
— [*3*] Über konvexe Mengen in linearen normierten Räumen. Studia Math. **4,** 70—84 (1933).

McKinney, R. L.: [*1*] Positive bases for linear spaces. Trans. Amer. Math. Soc. **103,** 131—148 (1962).

McShane, E. J.: [*1*] On multipliers for Lagrange problems. Amer. J. Math. **61,** 809—819 (1939).
—, and T. A. Botts: [*1*] Real Analysis. Princeton: Van Nostrand 1959.

Mehndiratta, S. L.: [*1*] Duality in nonlinear programming. Tata Institute of Fundamental Research. Colaba, Bombay-5. Technical Report No. 27, 1967.

Minkowski, H.: [*1*] Gesammelte Abhandlungen von Hermann Minkowski (ed. D. Hilbert). vol. I, II. Leipzig: Teubner 1911.
— [*2*] Allgemeine Lehrsätze über die konvexen Polyeder. Nachr. Ges. Wiss. Göttingen, 198—219 (1897). Also in Minkowski [*1*], vol. II.
— [*3*] Theorie der konvexen Körper, insbesondere Begründung ihres Oberflächenbegriffs. Posthum, in Minkowski [*1*], vol. II.
— [*4*] Geometrie der Zahlen. I. Lieferung. Leipzig: Teubner 1896, II. Lieferung (posthum) Teubner 1910.

Minty, G. J.: [*1*] On the simultaneous solution of a certain system of linear inequalities. Proc. Amer. Math. Soc. **13,** 11—12 (1962).
— [*2*] On the monotonicity of the gradient of a convex function. Pacific J. Math. **14,** 243—247 (1964).

Mond, G.: [*1*] A symmetric dual theorem for non-linear programs. Quart. Appl. Math. **23,** 265—269 (1965).

Moore, R. L.: [*1*] Foundations of point set theory. Rev. ed. Amer. Math. Soc. Colloq. Publ. **13,** Providence, R. I.: Amer. Math. Soc. 1962.

Moreau, J.-J.: [*1*] Fonctionelles sous-différentiables. C. R. Acad. Sci. Paris **257,** 4117—4119 (1963).
— [*2*] Théorèmes 'inf-sup'. C. R. Acad. Sci. Paris **258,** 2720—2722 (1964).
— [*3*] Proximité et dualité dans un espace hilbertien. Bull. Soc. Math. France **93,** 273—299 (1965).

Motzkin, T.: [*1*] Sur quelques propriétés caractéristiques des ensembles bornés non convexes. Rend. reale Acad. Lincei, Classe Sci. Fis., Mat. Nat. **21,** 773—779 (1935).
— [*2*] Beiträge zur Theorie der linearen Ungleichungen. Azriel: Jerusalem 1936.
— [*3*] Two consequences of the transposition theorem on linear inequalities. Econometrica **19,** 184—185 (1951).
— [*4*] New techniques for linear inequalities and optimization. In Orden-Goldstein [*1*] (eds.) 15—27 (1952).
— [*5*] The probability of solvability of linear inequalities. In Antosiewicz [*1*] (ed.) 607—611 (1955).
— [*6*] Comonotone curves and polyhedra. Bull. Amer. Math. Soc. **63,** 35 (1957).
— [*7*] Endovectors. In Klee [*22*] (ed.) 361—387 (1963).
— [*8*] A combinatorial result on maximally convex sets. Notices Amer. Math. Soc. **12,** 603 (1965).
— [*9*] Extension of the Minkowski-Carathéodory theorem on convex hulls. Notices Amer. Math. Soc. **12,** 705 (1965).
—, H. Raiffa, G. L. Thompson, and R. M. Thrall: [*1*] The double description method. In Kuhn-Tucker [*3*] (eds.) 51—73 (1953).

v. Neumann, J.: [*1*] Zur Theorie der Gesellschaftsspiele. Math. Ann. **100,** 295—320 (1928).
—, and O. Morgenstern: [*1*] Theory of games and economic behavior. 3rd edition. Princeton: Princeton Univ. Press 1953.

Neyman, J.: [*1*] (ed.) Proceedings of the Second Berkeley Symposium on Mathematical Statistics and Probability. Berkeley: Univ. of Calif. Press 1950.

Nikaidô, H.: [*1*] On a minimax theorem and its applications to functional analysis. J. Math. Soc. Japan **5,** 86—94 (1953).
— [*2*] On von Neumann's minimax theorem. Pacific J. Math. **4,** 65—72 (1954).

Norris, D. O.: [*1*] Lagrangian saddle points and optimal control. SIAM J. Control **5,** 594—599 (1967).

Orden, A., and L. Goldstein: [*1*] (eds.) Symposium on Linear Inequalities and Programming. (Project SCOOP, No. 10) Directorate of Management Analysis, DCS/Comptroller, Headquarters U.S. Air Force, Washington, D. C., 1952.

Ostrowski, A.: [*1*] Über Normen von Matrizen. Math. Z. **63,** 2—18 (1955).

Öttli, K. W.: [*1*] Die Rolle der wesentlichen Restriktionen bei einem konvexen Programm. Diss. Univ. Zürich: Juris Verlag 1964.

Pettis, B. J.: [*1*] Separation theorems for convex sets. Math. Mag. **29,** 233—247 (1956).

Phelps, R. R.: [*1*] Convex sets and nearest points. (I), II. Proc. Amer. Math. Soc. **8,** 790—797 (1957), **9,** 867—873 (1958).
— [*2*] A representation theorem for bounded convex sets. Proc. Amer. Math. Soc. **11,** 976—983 (1960).
— [*3*] Support cones and their generalizations. In Klee [*22*] (ed.) 393—401 (1963).

Pickert, G.: [*1*] Einführung in die höhere Algebra. Göttingen: Vandenhoeck und Ruprecht 1951.

Ponstein, J.: [*1*] Multiplier functions in optimal control. SIAM J. Control **6,** 648—658 (1968).

Pontryagin, L. S., V. G. Boltyanskii, R. V. Gamkrelidze, and E. F. Mishenko: [*1*] The Mathematical Theory of Optimal Processes (Russian). Moscow: Gosudarstv. Izdat. Fiz.-Mat. Lit. 1961. (English translation by K. N. Trigoroff, ed. by L. W. Neustadt. New York: Wiley 1962.)

Poritsky, H.: [*1*] Convex spaces associated with a family of linear inequalities. In Klee [*22*] (ed.) 403—436 (1963).

Price, G. B.: [*1*] On the extreme points of convex sets. Duke Math. J. **3,** 56—67 (1937).

Rademacher, H., and I. J. Schoenberg: [*1*] Helly's theorem on convex domains and Tchebycheff's approximation problem. Canad. J. Math. **2,** 245—256 (1950).

—, und E. Steinitz: [*1*] Vorlesungen über die Theorie der Polyeder, unter Einschluß der Elemente der Topologie. Berlin: Springer 1934.

Radon, J.: [*1*] Mengen konvexer Körper, die einen gemeinsamen Punkt enthalten. Math. Ann. **83,** 113—115 (1921).

Raffin, C.: [*1*] Programmes linéaires d'appui et conditions d'optimalité pour un programme convexe infini dans un espace vectoriel topologique sur R localement convexe séparé. C. R. Acad. Sci. Paris **266,** A766—A769 (1968).

— [*2*] Programme convexe infini et programme dual dans deux espaces vectoriels sur R en dualité. C. R. Acad. Sci. Paris **266,** A839—A842 (1968).

Reay, J. R.: [*1*] Generalizations of a theorem of Carathéodory. Mem. Amer. Math. Soc. No. **54** (1965).

— [*2*] A new proof of the Bonnice-Klee theorem. Proc. Amer. Math. Soc. **16,** 585—587 (1965).

— [*3*] Unique minimal representations with positive bases. Amer. Math. Monthly **73,** 253—261 (1966).

— [*4*] An extension of Radon's theorem. Illinois J. Math. **12,** 184—185 (1968).

Reidemeister, K.: [*1*] Über die singulären Randpunkte eines konvexen Körpers. Math. Ann. **83,** 116—118 (1921).

Rice, J. R.: [*1*] Tchebycheff approximation in several variables. Trans. Amer. Math. Soc. **109,** 444—466 (1963).

Rissanen, J.: [*1*] On duality without convexity. J. Math. Anal. Appl. **18,** 269—275 (1967).

Ritter, K.: [*1*] Duality for nonlinear programming in a Banach Space. SIAM J. Appl. Math. **15,** 294—302 (1967).

Rockafellar, R. T.: [*1*] Convex functions and dual extremum problems. Diss. Harvard Univ., Cambridge, Mass., 1963.

— [*2*] Duality theorems for convex functions. Bull. Amer. Math. Soc. **70,** 189—192 (1964).

— [*3*] Minimax theorems and conjugate saddle-functions. Math. Scand. **14,** 151—173 (1964).

— [*4*] Helly's theorem and minima of convex functions. Duke Math. J. **32,** 381—398 (1965).

— [*5*] Extension of Fenchel's duality theorem for convex functions. Duke Math. J. **33,** 81—89 (1966).

— [*6*] Conjugates and Legendre transforms of convex functions. Canad. J. Math. **19,** 200—205 (1967).

Rockafellar, R. T.: [7] Duality and stability in extremum problems involving convex functions. Pacific J. Math. **21,** 167—187 (1967).

— [8] A general correspondence between dual minimax problems and convex programs. Pacific J. Math. **25,** 597—611 (1968).

— [9] Monotone processes of concave convex type. Amer. Math. Soc. Mem. (to appear).

— [10] Duality in nonlinear programming. In Dantzig-Veinott [1] (eds.) 401—422 (1968).

— [11] Convex Analysis. Princeton: Princeton University Press 1969.

Roode, J. D.: [1] Generalized Lagrangian functions in mathematical programming. Thesis. Leiden 1968.

Rubin, H., and O. Wesler: [1] A note on convexity in euclidean *n*-space. Proc. Amer. Math. Soc. **9,** 522—523 (1958).

Rudin, W.: [1] Principles of mathematical analysis. New York: McGraw-Hill 1953.

Russell, D. L.: [1] The Kuhn-Tucker conditions in Banach space with an application to control theory. J. Math. Anal. Appl. **15,** 200—212 (1966).

Sandgren, L.: [1] On convex cones. Math. Scand. **2,** 19—28 (1954).

Santaló, L. A.: [1] Complemento a la Nota: Un teorema sôbre conjuntos de paralelepipedos de aristas paralelas. Publ. Inst. Mat. Univ. Nac. Litoral (Rosario) **3,** 202—210 (1942).

Schaefer, H. H.: [1] Topological vector spaces. New York: MacMillan 1966.

Schauder, J.: [1] Der Fixpunktsatz in Funktionalräumen. Studia Math. **2,** 171—180 (1930).

Schoenberg, I. J.: [1] Convex domains and linear combinations of continuous functions. Bull. Amer. Math. Soc. **39,** 273—280 (1933).

Shephard, G. C., and R. J. Webster: [1] Metrics for sets of convex bodies. Mathematika **12,** 73—88 (1965).

Sinden, F. W.: [1] A geometric representation for pairs of dual quadratic or linear programs. J. Math. Anal. Appl. **5,** 378—402 (1962).

— [2] Duality in convex programming and in projective space. SIAM J. Appl. Math. **11,** 535—552 (1963).

Sion, M.: [1] Sur une généralisation du théorème minimax. C. R. Acad. Sci. Paris **244,** 2120 (1957).

— [2] On general minimax theorems. Pacific J. Math. **8,** 171—176 (1958).

Slater, M. L.: [1] Lagrange multipliers revisited: A contribution to nonlinear programming. Cowles Commision Discussion Paper, Math. 403, November 1950.

— [2] A note on Motzkin's transposition theorem. Econometrica **19,** 185—187 (1951).

Sperner, E.: [1] Neuer Beweis für die Invarianz der Dimensionszahl und des Gebietes. Abh. Math. Sem. Univ. Hamburg **6,** 265—272 (1928).

— [2] Einführung in die Analytische Geometrie und Algebra. 2 Bde. Göttingen: Vandenhoeck & Ruprecht 1948.

Steinitz, E.: [1] Bedingt konvergente Reihen und konvexe Systeme. J. Reine Angew. Math. **143,** 128—175 (1913), **144,** 1—40 (1914), **146,** 1—52 (1916).

Stiemke, E.: [1] Über positive Lösungen homogener linearer Gleichungen. Math. Ann. **76,** 340—342 (1915).

Stoer, J.: [*1*] Duality in nonlinear programming and the minimax theorem. Numer. Math. **5,** 371—379 (1963).

— [*2*] Duality in convex programming. Habilitationsschrift. Technische Hochschule München 1964.

— [*3*] Über einen Dualitätssatz der nichtlinearen Programmierung. Numer. Math. **6,** 55—58 (1964).

Stoker, J. J.: [*1*] Unbounded convex point sets. Amer. J. Math. **62,** 165—179 (1940).

Stokes, R. W.: [*1*] A geometric theory of solution of linear inequalities. Trans. Amer. Math. Soc. **33,** 782—805 (1931).

Straszewicz, S.: [*1*] Über exponierte Punkte abgeschlossener Punktmengen. Fund. Math. **24,** 139—143 (1935).

Swarup, K.: [*1*] Some aspects of duality for linear fractional functionals programming. ZAMM **47,** 204—205 (1967).

Sylvester, J. J.: [*1*] Philos. Mag. **37,** 363—370 (1850).

Tucker, A. W.: [*1*] Some topological properties of disk and sphere. Proc. First Canad. Math. Congress, 285—309 (1945).

— [*2*] Extensions of theorems of Farkas and Stiemke. Bull. Amer. Math. Soc. **56,** 57 (1950).

— [*3*] Theorems of alternatives for pairs of matrices. In Orden-Goldstein [*1*] (eds.) 180—181 (1952).

— [*4*] A skew-symmetric matrix theorem. Bull. Amer. Math. Soc. **61,** 135 (1955).

— [*5*] Linear inequalities and convex polyhedral sets. In Antosiewicz [*1*] (ed.) 569—602 (1956).

— [*6*] Dual systems of homogeneous linear relations. In Kuhn-Tucker [*4*] (eds.) 3—18 (1956).

— [*7*] Abstract structure of linear programming ("Duality is orthogonality"). International Conference on Information Processing. Paris 1959.

— [*8*] Complementary slackness in dual linear subspaces. In Dantzig-Veinott [*1*] (eds.) 137—143 (1968).

Tutte, W. T.: [*1*] Matroids and graphs. Trans. Amer. Math. Soc. **90,** 527—552 (1959).

— [*2*] Lectures on matroids. J. Res. Nat. Bur. Standards, **69B,** 1—47 (1965).

Tychonoff, A.: [*1*] Ein Fixpunktsatz. Math. Ann. **111,** 767—776 (1935).

Uzawa, H.: [*1*] The Kuhn-Tucker Theorem in Concave Programming. In Arrow-Hurwicz-Uzawa [*1*] (eds.) 33—37 (1958).

Vajda, S.: [*1*] Nonlinear programming and duality. In Abadie [*2*] (ed.) 1—18 (1967).

Valentine, F. A.: [*1*] The problem of Lagrange with differential inequalities as added side conditions. In: Contributions to the calculus of variations, 1933—1937, 407—447. Chicago: Univ. of Chicago Press 1937.

— [*2*] Set properties determined by conditions on linear sections. Bull. Amer. Math. Soc. **52,** 925—931 (1946).

— [*3*] The dual cone and Helly type theorems. In Klee [*22*] (ed.) 473—493 (1963).

— [*4*] Convex sets. New York: McGraw-Hill 1964.

Van Slyke, R. M., and R. J.-B. Wets: [*1*] A duality theorem for abstract mathematical programs with applications to optimal control theory. J. Math. Anal. Appl. **22,** 679—706 (1968).

Varaiya, P. P.: [*1*] Nonlinear programming in Banach space. SIAM J. Appl. Math. **15,** 284—293 (1967).

Ville, J.: [*1*] Sur la théorie générale des jeux où intervient l'habilité des joueurs. In: Traité du Calcul des Probabilités et de ses Applications IV. Fasc. II. by E. Borel and J. Ville. Paris 1938.

Wald, A.: [*1*] Generalization of a theorem by v. Neumann concerning zero sum two person games. Ann. of Math. **46**, 281—286 (1945).

— [*2*] Foundations of a general theory of sequential decision functions. Econometrica **15**, 279—313 (1947).

Walkup, D. W., and R. J.-B. Wets: [*1*] Continuity of some convex-cone-valued mappings. Proc. Amer. Math. Soc. **18**, 229—235 (1967).

— [*2*] A Lipschitzian characterization of convex polyhedra. Boeing Scient. Res. Labs. Doc. D1-82-0728.

— [*3*] Some regularity conditions for nonlinear programs. Boeing Scient. Res. Labs. Doc. D1-82-0792.

Weil, A.: [*1*] Sur les espaces à structure uniforme et sur la topologie générale. Act. Sci. et Ind. Nr. **551.** Paris: Hermann 1937.

Wets, R. J.-B., and C. Witzgall: [*1*] Algorithms for frames and lineality spaces of cones. J. Res. Nat. Bur. Standards, **71 B,** 1—7 (1967).

— [*2*] Towards an algebraic characterization of convex polyhedral cones. Numer. Math. **12,** 134—138 (1968).

Weyl, H.: [*1*] Elementare Theorie der konvexen Polyeder. Comment. Math. Helv. **7,** 290—306 (1935). Translated in Kuhn and Tucker [*2*] (eds.) 3—18 (1950).

Whinston, A.: [*1*] Some applications of the conjugate function theory to duality. In Abadie [*2*] (ed.) 75—96 (1967).

Whitney, H.: [*1*] Congruent graphs and the connectivity of graphs. Amer. J. Math. **54,** 150—168 (1932).

— [*2*] On the abstract properties of linear dependence. Amer. J. Math. **57,** 509—533 (1935).

Wilde, D. J.: [*1*] Differential calculus in nonlinear programming. Operations Res. **10,** 764—773 (1962).

Wolfe, P.: [*1*] Determinateness of polyhedral games. In Kuhn-Tucker [*4*] (eds.) 195—198 (1956).

— [*2*] A duality theorem for non-linear programming. Quart. Appl. Math. **19,** 239—244 (1961).

Author and Subject Index

Page numbers in *italics* refer to the Bibliography

Satz und Druck: Zechnersche Buchdruckerei, Speyer

Die Grundlehren der mathematischen Wissenschaften in Einzeldarstellungen mit besonderer Berücksichtigung der Anwendungsgebiete

---

2. Knopp: Theorie und Anwendung der unendlichen Reihen. DM 48,—; US $ 13.20
3. Hurwitz: Vorlesungen über allgemeine Funktionentheorie und elliptische Funktionen. DM 49,—; US $ 13.50
4. Madelung: Die mathematischen Hilfsmittel des Physikers. DM 49,70; US $ 13.70
10. Schouten: Ricci-Calculus. DM 58,60; US $ 16.20
14. Klein: Elementarmathematik vom höheren Standpunkt aus. 1. Band: Arithmetik, Algebra, Analysis. DM 24,—; US $ 6.60
15. Klein: Elementarmathematik vom höheren Standpunkt aus. 2. Band: Geometrie. DM 24,—; US $ 6.60
16. Klein: Elementarmathematik vom höheren Standpunkt aus. 3. Band: Präzisions- und Approximationsmathematik. DM 19,80; US $ 5.50
20. Pólya/Szegö: Aufgaben und Lehrsätze aus der Analysis II: Funktionentheorie, Nullstellen, Polynome, Determinanten, Zahlentheorie. DM 38,—; US $ 10.50
22. Klein: Vorlesungen über höhere Geometrie. DM 28,—; US $ 7.70
26. Klein: Vorlesungen über nicht-euklidische Geometrie. DM 24,—; US $ 6.60
27. Hilbert/Ackermann: Grundzüge der theoretischen Logik. DM 38,—; US $ 10.50
30. Lichtenstein: Grundlagen der Hydromechanik. DM 38,—; US $ 10.50
31. Kellogg: Foundations of Potential Theory. DM 32,—; US $ 8.80
32. Reidemeister: Vorlesungen über Grundlagen der Geometrie. DM 18,—; US $ 5.00
38. Neumann: Mathematische Grundlagen der Quantenmechanik. DM 28,—; US $ 7.70
40. Hilbert/Bernays: Grundlagen der Mathematik I. DM 68,—; US $ 18.70
43. Neugebauer: Vorlesungen über Geschichte der antiken mathematischen Wissenschaften. 1. Band: Vorgriechische Mathematik. DM 48,—; US $ 13.20
50. Hilbert/Bernays: Grundlagen der Mathematik II. DM 68,—; US $ 18.70
52. Magnus/Oberhettinger/Soni: Formulas and Theorems for the Special Functions of Mathematical Physics. DM 66,—; US $ 16.50
57. Hamel: Theoretische Mechanik. DM 84,—; US $ 23.10
58. Blaschke/Reichardt: Einführung in die Differentialgeometrie. DM 24,—; US $ 6.60
59. Hasse: Vorlesungen über Zahlentheorie. DM 69,—; US $ 19.00
60. Collatz: The Numerical Treatment of Differential Equations. DM 78,—; US $ 19.50
61. Maak: Fastperiodische Funktionen. DM 38,—; US $ 10.50
62. Sauer: Anfangswertprobleme bei partiellen Differentialgleichungen. DM 41,—; US $ 11.30
64. Nevanlinna: Uniformisierung. DM 49,50; US $ 13.70
66. Bieberbach: Theorie der gewöhnlichen Differentialgleichungen. DM 58,50; US $ 16.20
68. Aumann: Reelle Funktionen. DM 68,—; US $ 18.70
69. Schmidt: Mathematische Gesetze der Logik I. DM 79,—; US $ 21.80
71. Meixner/Schäfke: Mathieusche Funktionen und Sphäroidfunktionen mit Anwendungen auf physikalische und technische Probleme. DM 52,60; US $ 14.50
73. Hermes: Einführung in die Verbandstheorie. DM 46,—; US $ 12.70

74. Boerner: Darstellung von Gruppen. DM 58,—; US $ 16.00
75. Rado/Reichelderfer: Continuous Transformations in Analysis, with an Introduction to Algebraic Topology. DM 59,60; US $ 16.40
76. Tricomi: Vorlesungen über Orthogonalreihen. DM 68,—; US $ 18.70
77. Behnke/Sommer: Theorie der analytischen Funktionen einer komplexen Veränderlichen. DM 79,—; US $ 21.80
78. Lorenzen: Einführung in die operative Logik und Mathematik. DM 54,—; US $ 14.90
80. Pickert: Projektive Ebenen. DM 48,60; US $ 13.40
81. Schneider: Einführung in die transzendenten Zahlen. DM 24,80; US $ 6.90
82. Specht: Gruppentheorie. DM 69,60; US $ 19.20
84. Conforto: Abelsche Funktionen und algebraische Geometrie. DM 41,80; US $ 11.50
86. Richter: Wahrscheinlichkeitstheorie. DM 68,—; US $ 18.70
88. Müller: Grundprobleme der mathematischen Theorie elektromagnetischer Schwingungen. DM 52,80; US $ 14.60
89. Pfluger: Theorie der Riemannschen Flächen. DM 39,20; US $ 10.80
90. Oberhettinger: Tabellen zur Fourier-Transformation. DM 39,50; US $ 10.90
91. Prachar: Primzahlenverteilung. DM 58.—; US $ 16.00
93. Hadwiger: Vorlesungen über Inhalt, Oberfläche und Isoperimetrie. DM 49,80; US $ 13.70
94. Funk: Variationsrechnung und ihre Anwendung in Physik und Technik. DM 120,—; US $ 33.00
95. Maeda: Kontinuierliche Geometrien. DM 39,—; US $ 10.80
97. Greub: Linear Algebra. DM 39,20; US $ 9.80
98. Saxer: Versicherungsmathematik. 2. Teil. DM 48,60; US $ 13.40
99. Cassels: An Introduction to the Geometry of Numbers. DM 69,—; US $ 19.00
100. Koppenfels/Stallmann: Praxis der konformen Abbildung. DM 69,—; US $ 19.00
101. Rund: The Differential Geometry of Finsler Spaces. DM 59,60; US $ 16.40
103. Schütte: Beweistheorie. DM 48,—; US $ 13.20
104. Chung: Markov Chains with Stationary Transition Probabilities. DM 56,—; US $ 14.00
105. Rinow: Die innere Geometrie der metrischen Räume. DM 83,—; US $ 22.90
106. Scholz/Hasenjaeger: Grundzüge der mathematischen Logik. DM 98,—; US $ 27.00
107. Köthe: Topologische lineare Räume I. DM 78,—; US $ 21.50
108. Dynkin: Die Grundlagen der Theorie der Markoffschen Prozesse. DM 33,80; US $ 9.30
110. Dinghas: Vorlesungen über Funktionentheorie. DM 69,—; US $ 19.00
111. Lions: Equations différentielles opérationnelles et problèmes aux limites. DM 64,—; US $ 17.60
112. Morgenstern/Szabó: Vorlesungen über theoretische Mechanik. DM 69,—; US $ 19.00
113. Meschkowski: Hilbertsche Räume mit Kernfunktion. DM 58,—; US $ 16.00
114. MacLane: Homology. DM 62,—; US $ 15.50
115. Hewitt/Ross: Abstract Harmonic Analysis. Vol. 1: Structure of Topological Groups, Integration Theory, Group Representations. DM 76,—; US $ 20.90
116. Hörmander: Linear Partial Differential Operators. DM 42,—; US $ 10.50
117. O'Meara: Introduction to Quadratic Forms. DM 48,—; US $ 13.20
118. Schäfke: Einführung in die Theorie der speziellen Funktionen der mathematischen Physik. DM 49,40; US $ 13.60
119. Harris: The Theory of Branching Processes. DM 36,—; US $ 9.90

120. Collatz: Funktionalanalysis und numerische Mathematik. DM 58,—; US $ 16.00
121.
122. Dynkin: Markov Processes. DM 96,—; US $ 26.40
123. Yosida: Functional Analysis. DM 66,—; US $ 16.50
124. Morgenstern: Einführung in die Wahrscheinlichkeitsrechnung und mathematische Statistik. DM 38,—; US $ 10.50
125. Itô/McKean: Diffusion Processes and Their Sample Paths. DM 58,—; US $ 16.00
126. Lehto/Virtanen: Quasikonforme Abbildungen. DM 38,—; US $ 10.50
127. Hermes: Enumerability, Decidability, Computability. DM 39,—; US $ 10.80
128. Braun/Koecher: Jordan-Algebren. DM 48,—; US $ 13.20
129. Nikodým: The Mathematical Apparatus for Quantum-Theories. DM 144,—; US $ 36.00
130. Morrey: Multiple Integrals in the Calculus of Variations. DM 78,—; US $ 19.50
131. Hirzebruch: Topological Methods in Algebraic Geometry. DM 38,—; US $ 9.50
132. Kato: Perturbation Theory for Linear Operators. DM 79,20; US $ 19.80
133. Haupt/Künneth: Geometrische Ordnungen. DM 68,—; US $ 18.70
134. Huppert: Endliche Gruppen I. DM 156,—; US $ 42.90
135. Handbook for Automatic Computation. Vol. 1/Part a: Rutishauser: Description of ALGOL 60. DM 58,—; US $ 14.50
136. Greub: Multilinear Algebra. DM 32,—; US $ 8.00
137. Handbook for Automatic Computation. Vol. 1/Part b: Grau/Hill/Langmaack: Translation of ALGOL 60. DM 64,—; US $ 16.00
138. Hahn: Stability of Motion. DM 72,—; US $ 19.80
139. Mathematische Hilfsmittel des Ingenieurs. Herausgeber: Sauer/Szabó. 1. Teil. DM 88,—; US $ 24.20
140. Mathematische Hilfsmittel des Ingenieurs. Herausgeber: Sauer/Szabó. 2. Teil. DM 136,—; US $ 37.40
141. Mathematische Hilfsmittel des Ingenieurs. Herausgeber: Sauer/Szabó. 3. Teil. DM 98,—; US $ 27.00
142. Mathematische Hilfsmittel des Ingenieurs. Herausgeber: Sauer/Szabó. 4. Teil. DM 124.—; US $ 34.10
143. Schur/Grunsky: Vorlesungen über Invariantentheorie. DM 32,—; US $ 8.80
144. Weil: Basic Number Theory. DM 48,—; US $ 12.00
145. Butzer/Berens: Semi-Groups of Operators and Approximation. DM 56,—; US $ 14.00
146. Treves: Locally Convex Spaces and Linear Partial Differential Equations. DM 36,—; US $ 9.90
147. Lamotke: Semisimpliziale algebraische Topologie. DM 48,—; US $ 13.20
148. Chandrasekharan: Introduction to Analytic Number Theory. DM 28,—; US $ 7.00
149. Sario/Oikawa: Capacity Functions. DM 96,—; US $ 24.00
150. Iosifescu/Theodorescu: Random Processes and Learning. DM 68,—; US $ 18.70
151. Mandl: Analytical Treatment of One-dimensional Markov Processes. DM 36,—; US $ 9.80
152. Hewitt/Ross: Abstract Harmonic Analysis. Vol. II. DM 140,—; US $ 38.50
153. Federer: Geometric Measure Theory. DM 118,—; US $ 29.50
154. Singer: Bases in Banach Spaces I. DM 112,—; US $ 30.80
155. Müller: Foundations of the Mathematical Theory of Electromagnetic Waves. DM 58,—; US $ 16.00
156. van der Waerden: Mathematical Statistics. DM 68,—; US $ 18.70
157. Prohorov/Rozanov: Probability Theory. DM 68,—; US $ 18.70

159. Köthe: Topological Vector Spaces I. DM 78,—; US $ 21.50
160. Agrest/Maksimov: Theory of Incomplete Cylindrical Functions and their Applications. In preparation
161. Bhatia/Szegö: Stability Theory of Dynamical Systems. In preparation
162. Nevanlinna: Analytic Functions. DM 76,—; US $ 20.90
163. Stoer/Witzgall: Convexity and Optimization in Finite Dimensions I. DM 54,—; US $ 14.90
164. Sario/Nakai: Classification Theory of Riemann Surfaces. DM 98,—; US $ 27.00
165. Mitrinovič: Analytic Inequalities. DM 88,—; US $ 24.20
166. Grothendieck/Dieudonné: Eléments de Géometrie Algébrique. En préparation
167. Chandrasekharan: Arithmetical Functions. DM 58,—; US $ 16.00
168. Palamadov: Linear Differential Operators with Constant Coefficients. DM 98,—; US $ 27.00
169. Rademacher: Topics in Analytic Number Theory. In preparation
170. Lions: Optimal Control Systems Governed by Partial Differential Equations. In preparation
171. Singer: Best Approximation in Normed Linear Spaces by Elements of Linear Subspaces. DM 60,—; US $ 16.50